Architectural, Civil, and Infrastructure Engineering in View of Sustainability

Architectural, Civil, and Infrastructure Engineering in View of Sustainability

Editors

Oleg Kapliński
Lili Dong
Agata Bonenberg
Wojciech Bonenberg

MDPI

Basel • Beijing • Wuhan • Barcelona • Belgrade • Novi Sad • Cluj • Manchester

Editors
Oleg Kapliński
Poznań University of Technology
Poznań, Poland

Lili Dong
Chongqing Jiaotong University
Chongqing, China

Agata Bonenberg
Poznan University of Technology
Poznań, Poland

Wojciech Bonenberg
Poznań University of Technology
Poznań, Poland

Editorial Office
MDPI
St. Alban-Anlage 66
4052 Basel, Switzerland

This is a reprint of articles from the Special Issue published online in the open access journal *Sustainability* (ISSN 2071-1050) (available at: www.mdpi.com/journal/sustainability/special_issues/View_of_Sustainability).

For citation purposes, cite each article independently as indicated on the article page online and as indicated below:

Lastname, A.A.; Lastname, B.B. Article Title. *Journal Name* **Year**, *Volume Number*, Page Range.

ISBN 978-3-0365-8553-6 (Hbk)
ISBN 978-3-0365-8552-9 (PDF)
doi.org/10.3390/books978-3-0365-8552-9

Contents

About the Editors

Oleg Kapliński

Oleg Kapliński is currently Full Professor at the Faculty of Architecture, Poznan University of Technology, in Poznan, Poland. He received his Ph.D. and DSc. in Civil Engineering and has authored or co-authored 280 publications (articles, reports), including 12 books (academic scripts and monographs) on related topics. His academic achievements cover the theory of decision making, including multicriteria decision aiding; construction processes organization and modeling, including an analysis of the phenomena of waiting, the phenomena of equilibrium, balancing of the construction processes in conditions of uncertainty, risk in management, network planning, and reliability of production systems; research on resentment and predilection to risk in the light of utility theory; work ethos; integrated design and management; and sustainable development. Currently, his research at the Faculty of Architecture includes interactions between architects and engineers. Kapliński is a member of the Civil Engineering Committee of the Polish Academy of Sciences and Doctor honoris causa of VGTU, Lithuania. He is an honorary member of the EURO Working Group: Operations Research in Sustainable Development and Civil Engineering (EWG-ORSDCE).

Lili Dong

Lili Dong is a Professor, the Director of Architecture and the Deputy Dean of the School of Architecture and Urbanism, Chongqing Jiaotong University, Chongqing 400074, China. She is a senior lecturer and a registered first-class architect. Dong is a registered expert of China Landscape Architecture Network Database and an expert of Chongqing Municipality Comprehensive Bid Appraisal Database. Her main interests are green building design; urban planning; landscape architecture planning and design. The designed works have won the Optimal Design Scheme of China Human Living Paradigm Gold Award and more than 10 national and provincial/ministerial awards, including the Zhan Tianyou Award. Dong has published dozens of scientific articles, has been the editor-in-chief of 6 textbooks and monographs, is a member of the editorial board of "Landscape Design", and has received 2 national patents.

Agata Bonenberg

Agata Bonenberg is a Full Professor, Head of the Institute of Interior and Industrial Design at the Faculty of Architecture at the Poznan University of Technology and a partner in EcoArch Studio Bonenberg. She graduated from the Faculty of Architecture at the Cracow University of Technology. From 2014 to 2021, she was a lecturer at the Scuola di Architettura e Società i Architettura Urbanistica Ingegneria delle Costruzioni at the Politecnico di Milano. Her professional interests are flexible interior design; design of inclusive and universal environments; colour in architecture. Her recent research interests have focused on universal interior design, and her design of a modular cabinet with mobile interior units won a gold medal in recognition of excellence in innovation at the INPEX innovation exhibition in the USA. Her main research achievements include leading the research theme 'Media–Space–Architecture'. She has published dozens of scientific articles and holds a national patent. She is also interested in painting, drawing and photography.

Wojciech Bonenberg

Wojciech Bonenberg is Full Professor at the Faculty of Architecture of the Poznan University of Technology, Poland. He is the promoter of 26 doctoral dissertations and 150 diploma theses. He is the author or co-author of over 100 scientific publications, including 9 books. His research interests are related to sustainable design, architectural revitalization and requirements engineering in architecture. He has been invited for seminars and a series of lectures by universities in Germany, Netherlands, Italy, Belarus, Ukraine, and China. He is the author of more than 200 architectural projects, many of them awarded. These include the projects of the largest trans-European road terminals: Terminal Swiecko–Frankfurt/O, Terminal Koroszczyn–Brest (route Moscow–Paris), Terminal Olszyna–Forst (route Kiev–Strasbourg). Bonenberg is a member of the Architecture and Urban Planning Committee of the Polish Academy of Sciences and of the Council of Scientific Excellence in Poland.

Preface

Contemporary challenges are deeply significant. The general public expects specialists to answer various questions, especially about the future in the context of sustainability. What hopes lie in progressive solutions in architectural, civil, and infrastructure engineering? The set of articles presented in this book is an attempt to answer these questions. The borders and topics at the interface of design–research–sustainability are multithreaded, and multidimensional. The most interesting achievements are the result of cooperation of specialists from various disciplines. The work of a designer and solving sustainable topics became a team game. The reader is presented an opportunity to learn about new developments.

Oleg Kapliński, Lili Dong, Agata Bonenberg, and Wojciech Bonenberg
Editors

sustainability

MDPI

Editorial

Architectural, Civil, and Infrastructure Engineering in View of Sustainability: Editor's Comment

Oleg Kapliński

Faculty of Architecture, Poznań University of Technology, 60-965 Poznań, Poland; oleg.kaplinski@put.poznan.pl

1. Introduction

Sustainability in engineering has been one of the most often discussed topics in recent years and is one of the key factors in the engineering and economics of a sustainable environment. The Sustainable Development Goals (SDGs) are expected to be achieved through new solutions in architecture and engineering which are going to bring economic, social, and environmental benefits. The general public expects professionals to answer various questions, above all: How will we live? Are the existing design paradigms sufficient? What innovations in architectural, civil, and infrastructure engineering are needed? Contemporary challenges are significant: they force specialists to react to climate change or compel a reaction to energy constraints (already widespread), or the depletion of raw materials. The set of articles in the current Special Issue is an attempt to answer these questions. At the same time, we see a justification for the statement that the most interesting achievements are the result of cooperation between all entities of the investment process, of all disciplines, and the concept of research should be present at all stages of this process, as well as in the advanced design. The current Special Issue is devoted to the infiltration of research, and its importance in civil engineering, technical infrastructure, and architecture. The reader has the opportunity to learn about new developments. It is worth noting that this is a continuation of the theme of the two previous Special Issues on the interpenetration of architecture and engineering. Both editions were crowned in the form of books available online [1,2].

2. Contributions

The current Special Issue aroused the interest of the scientific and academic community and received many submissions. After a rigorous scientific review by editors and reviewers, fourteen papers were accepted and published. The authors dealt with various substantive problems and proposed various models of solutions. 50 authors or co-authors took part in total, The authors originate from 10 countries, with participants from Poland and China predominating. Figure 1 presents detailed information about the distribution of authors by country.

The reader is offered a decent dose of selected references. Their total number is 724 quoted items.

The review of published articles indicates that they are all original research papers, their issues are multithreaded, and they overlap to a great extent in content. Nevertheless, several issue groups can be distinguished. These are the dominant threads, and they are: (A) energy, (B) economy, (C) pro-health activities, (D) designing. Figure 2 presents them with characteristic topics. Each group has been assigned 3 characteristic topics.

Table 1 is designed to make it easier for the reader to find answers to specific questions, to find interesting topics and instruments. It is a synthesis of the content of the works, taking into account three premises:

(a) Subject of the research,
(b) Research problem,
(c) Research techniques. Instrumentality.

Citation: Kapliński, O. Architectural, Civil, and Infrastructure Engineering in View of Sustainability: Editor's Comment. *Sustainability* **2023**, *15*, 5967. https://doi.org/10.3390/su15075967

Received: 18 March 2023
Accepted: 23 March 2023
Published: 30 March 2023

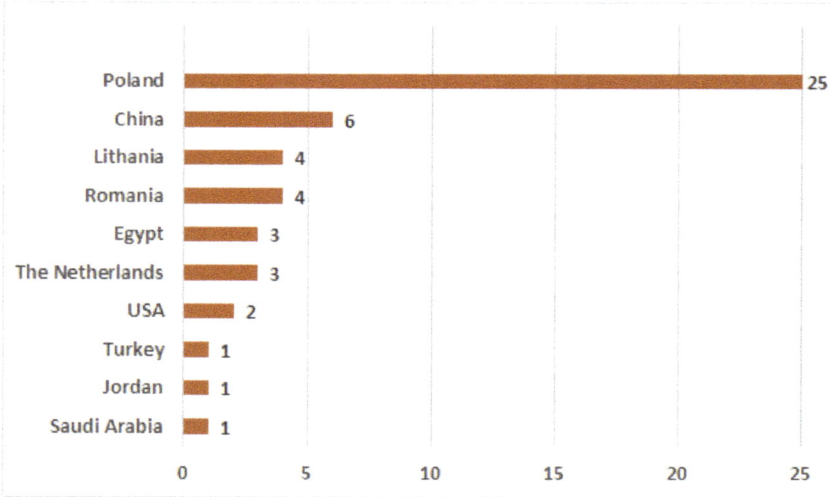

Figure 1. Distribution of authors by country.

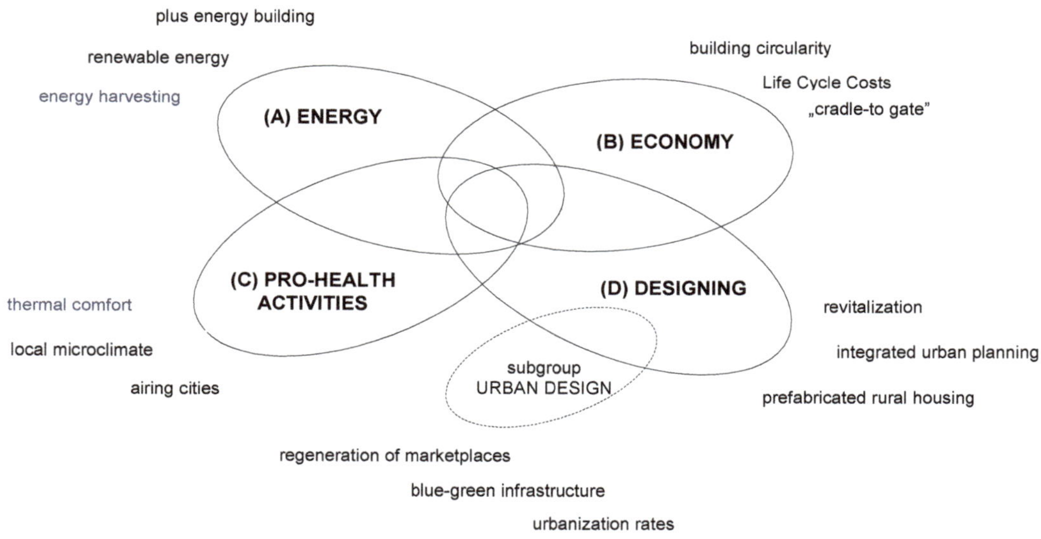

Figure 2. Dominating research threads in the current SI.

In the 'Subject of the research' field, physical objects or phenomena were indicated, in the 'Research problem' field, sought relationships and scientific values were presented. The 'Research techniques. Instrumentality' group collects characteristic research methods and tools. As you can see, this group is the most diverse. Here are just some of them: the new building circularity calculation method, the space syntax method, the automated greenery design, the multi-criteria approach, and the green BIM.

Table 1. The content of a special issue "Architectural, Civil, and Infrastructure Engineering in View of Sustainability". Synthetic collation.

Author	Subject of the Research	Research Problem	Research Techniques Instrumentality
Group (A) Energy			
Kasperski et al.	The innovative photovoltaic powered seasonal thermal storage system Plus energy building in Poland	The required size of a storage stack	TMY data (Typical meteorological year) The innovative HVAC installation Testing the numerical model
Al-Refaie & Lepkova	Renewable energy policies and photovoltaic policy in Jordan	High level of clean energy security Social acceptability	The feed-in-tariffs (FiTs) Dynamic forecast model Optimal FiTs prices, subsidies
Fernandez & Wojtkiewicz	Vibrational Energy Harvesters Bridge operation Distributed garden concept implementation USA	The cable-stayed bridge model. Integrating the weight of the harvester into the bridge structure.	The equations of motion for the nominal and modified structures.
Group (B) Economy			
Plebankiewicz & Gracki	Multi-functional buildings Maintenance costs Polish law regulations	Maintenance costs in the Total Life Cycle Costs A proposal to extend the method of determining operating costs	Net Present Value (NPV) LCC—Life Cycle Costs
Zhang et al.	Circular Economy (CE) Implementing CE in the AEC industry (Architecture, Engineering, and Construction)	Building circularity calculation method (BC) New Material Passport (MP) European Material Passport BAMB (Building as Material Banks)	The equations for building circularity scoring Proposal: New Building Circularity Calculation Method
Group (C) Pro-health activities			
Bonenberg et al.	The circular economy Integrated water management Use of rainwater Multi-family settlements in Krakow, Poland	The Bio-Morpheme reference unit. Open water reservoir	Eurostat Statistics The fractal reference model unit Comparisons with the reference model Proposed Bio-Morpheme complex
Borucka et al.	Transformation of public areas Regeneration of marketplaces in the Oliwa district (Poland)	Space syntax Evaluation of the three design models	The design variants for the revitalisation of the marketplace
Maxineasa et al.	Steel as a construction material can provide a sustainable solution for the built environment The steel cubic modules	The construction sector in view of sustainable development Three categories of indicators: Global Warming, Ozone Depletion, Human Toxicity Steel Structure Impact vs. RC Slab Impact	The cradle-to gate analysis Life Cycle Assessment
Elhadary et al.	Improving the working environment and increasing productivity. Three types of mechanical ventilation systems The forced ventilation system. The space inside the factory (Saudi Arabia)	The computational fluid dynamics (CFD) simulations	The ANSYS Fluent software The ventilation effectiveness factor (VEF)
Sędzicki et al.	Digital method of selection and design of greenery	Automated greenery design (AGD)	Greenery Scenarios New parameter sheet Test model with Grasshopper for Rhinoceros. Green BIM

Table 1. *Cont.*

Author	Subject of the Research	Research Problem	Research Techniques Instrumentality
Qi et al.	Rural areas planning The village landscape pattern and the local microclimate Adaptability of inhabitants of traditional villages (northern China)	The adaptive design mechanism A way of expressing morphological parameters. Coupling calculation method of landscape pattern and microclimate	Numerical simulation The Rhino modelling platform Grasshopper software
Badach et al.	Shaping city ventilation systems Air quality management	Integrated urban planning Urban form for city ventilation	Geographic information system (GIS) Computational fluid dynamics (CFD) simulations Autodesk CFD Historical city plans
Group (D) Designing			
Shi & Sun	Prefabricated rural housing Counties in China	Prefabrication as a sustainable construction method PEST aspects	Multi-criteria approach Model TOPSIS Weight values determined by entropy Urbanization rates
Zaleckis et al.	The perception of architectural transformations Facades of cultural heritage buildings The city centre of Kaunas (Lithuania)	The monitoring of the transformation of cultural heritage objects Visual perception	Sociological survey The space syntax method The theory of Nikos Salingaros Bill Hillier's methodology (symmetry index analysis)

The interweaving of the three premises from (a) to (c) with the four groups of substantive issues (from A to D), identified in Figure 2, with their characteristic values refined, are displayed below. Of course, this classification has features of subjective classification.

Group (A) Energy—is exceptional, unique, and has considerable potential for development because the topic of energy comes up in almost each of the fourteen articles. Kasperski et al. [3] present a plus energy building, introduce The innovative HVAC installation, and experiment with a seasonal heat storage system. This is the result of cooperation between power engineers and architects. Renewable energy policies are presented by Al-Refaie & Lepkova [4]. The authors highlight energy costs and social acceptability. Fernandez & Wojtkiewicz [5] deal with energy harvesting, or more precisely—harvesting energy during a bridge structure vibration. The idea of the authors is based on the harvester mass integration with the bridge structure.

Group (B) Economy—focuses on the life cycle, circular economy, and operating costs. Plebankiewicz & Gracki [6] deal with maintenance costs in the Total Life Cycle Costs. The analysis tool here is the Net Present Value method (NPV). The authors propose to extend the current method of determining operating costs. Circular Economy (CE) is the domain of Zhang et al. [7]. Because there is ambiguity and inconsistency in the Building Circularity (BC) assessment, the authors redefine the concept of BC assessment with three circularity cycles and five indicators. The method to be used. The three subsequent articles fall into both (B) Economy and (D) Designing categories. In the background of article [8] (Bonenberg et al.) there is a circular economy and improving the efficiency of rainwater utilisation. A Bio-Morpheme complex is proposed, including an open water reservoir. The comparison with the reference model is made on the examples of multi-family settlements in Krakow. Borucka et al. [9] focus on the transformation of public spaces, including the economic analysis of revitalization of the marketplace variants. The space syntax method is used. The small town of Oliwa on the Polish coast is the object of research. In the next article (Maxineasa et al. [10]), steel as a construction material is analysed in the "cradle-to-gate" option. The proposed structural modules can be a solution to minimize the environmental load levels.

Another set of articles is to be found in **Pro-health activities group (C)**. Colleagues from Saudi Arabia and Egypt (Elhadary et al. [11]) highlight the issue of thermal comfort. Improving the working environment and increasing productivity is the subject of the research. With the help of ANSYS Fluent software, three types of mechanical ventilation systems are discussed. Article [12] is heading in the direction of Green BIM. The authors (Sędzicki et al.) present a digital method of selecting and designing greenery. The method, called Automated Greenery Design (AGD), enables taking design decisions supported by objective measures from the early stages of design concepts. The authors suggest that the AGD method can be used in coordinating the most important elements of an architectural design. The relationship between man–rural buildings–microclimate is the subject of research completed by authors from China (Qi et al. [13]). Studies of the village landscape pattern, local microclimate, and adaptability of inhabitants are supported by the analysis of morphological parameters and the use of Rhino modelling platform and Grasshopper software. The issue of air quality management in cities is the domain of the research completed by the team of Badach et al. [4]. The authors propose Integrated Urban Planning. The urban form for ventilation is determined using the computational fluid dynamics (CFD) simulations, Autodesk CFD and GIS.

The scope of **group (D) Designing** is also wide. It includes improving methods and suggesting new solutions, projects, and applications. It most often combines elements of technology with economics. This group also includes articles discussed in the previous groups. Within group (A) Energy—designing plus energy house [3] and the cable-stayed bridge model construction [5] were discussed. Within group (B) Economy—Bio-Morpheme complex design [8], marketplace variants [9], design and testing of steel cubic modules [10] are presented. Within group (C) Pro-health activities—the following topics and articles are indicated: space within the factory [11], green design [12], rural planning [13], blue-green infrastructure [14].

Group (D) is closed with two articles on the following topics. Prefabrication as a sustainable construction method is the slogan of an article by colleagues Shi & Sun [15] from China. However, the emphasis is on the potential of implementing prefabricated houses in rural agglomerations, and on how to determine this potential. Their studies identify 16 evaluation indicators in four dimensions: political, economic, social, and technological (PEST), and use the entropy-weighted TOPSIS model. The transformation of cultural heritage structures, including the design of the facades of these buildings, is a topic of Lithuanian research (Zaleckis et al. [16]). The research context is based on sociological aspects. The space syntax method and Bill Hillier's methodology (for symmetry index analysis) are used.

3. Discussion and Comments

An overview of the achievements of the current Special Issue allows us to make a few observations.

Firstly, there is a thematic diversity despite the keystone connecting the architectural, civil, and infrastructure engineering topics in the form of sustainability. The issues presented in the articles are multi-layered and multi-threaded.

Secondly, the urban design subgroup may be extracted from the four groups synthesizing the subject matter, shown in Figure 2. It includes the following topics: multi-family settlements in the context of integrated water management [8], regeneration of marketplaces [9], landscape and greenery design [12], rural areas planning and local microclimate [13], blue-green infrastructure, urban ventilation, integrated urban planning [14], rural areas, urbanization rates [15]. The richness of this subgroup shows a gradual change in the scope of research and design from individual buildings to the development of larger spaces.

Thirdly, in addition to problem indications and technological topics, the reader is provided with an overview of methods and research instruments in the area of sustainability. The range of methods and tools is presented in the right column of Table 1. They include, among others: The space syntax method, Multi-criteria approach, Integrated urban planning, Computational fluid dynamics (CFD) simulations, Dynamic forecast model, The Building Circularity Calculation Method, Net Present Value (NPV), Life Cycle Costs (LCC), The Bill Hillier's methodology, Sociological survey. This review does not conclude the existence and development of other attractive and useful tools in the area of sustainability. The development of new sophisticated methods is exceptionally dynamic.

Fourthly, the prevalence of energy topics in the submitted and published articles is visible. Emphasizing the seriousness of this issue, i.e., energy matters may lead to the elusive conclusion that energy efficiency is almost the currency of our times.

Fifth, all articles fit into the basic analysis of EST, i.e., economic, social, and technological dimensions. Nonetheless, in this SI, there is an extension to PEST, i.e., the political dimension, which is clearly visible in two articles [4,15]. The development of the design level can be reduced to the following chain: "from conventional practice, green buildings, sustainable design (degenerating—reducing impact) to restorative design, regenerative design". The submitted articles charge at the middle part of this sequence, so it is worth interesting the reader and researchers in restorative and regenerative design (see website [17]).

Sixth, there is a certain trend: plenty of teamwork. There are 50 authors and 14 articles, which means 3.57 authors per publication. Yes, modern research or design is a team game. The co-authors of the current Special Issue come from different universities, and faculties, i.e., from different specialties and disciplines, which confirms the fact that success and concrete results are at the interface of two or more disciplines. A combination of architects with power engineers is characteristic. Work [3] is a vivid example of this statement. International works have also appeared (c.f. [4,9,11,13]. This tendency should be preserved for the sake of research dedicated to sustainability.

4. Conclusions

The interface between sustainability and engineering—especially the architectural, civil, and infrastructure classes—is multi-dimensional. Hence, the issues of research and design gain special importance.

The matter is almost unequivocal: solving sustainability problems becomes a team game. The need for such an approach arose naturally, today it becomes almost a compulsion. It is an interdisciplinary and even international game. Methods and tools exist, they are being developed, and they are becoming more and more sophisticated, thus teams of specialists are needed.

The analysis of the submitted articles shows that the environment is a widely understood concept. Man becomes an evident subject. It is satisfactory that, in addition to technological and energy issues, the authors have taken into account social aspects and presented pro-health activities. The understanding of the principles of organic design in the sphere of sustainability is increasing.

Acknowledgments: The author expresses his gratitude to the Sustainability journal for offering an academic platform for researchers where they can contribute and exchange their recent findings in architectural, civil, and infrastructure engineering.

Conflicts of Interest: The authors declare no conflict of interest.

References

1. Kapliński, O.; Bonenberg, W. (Eds.) *Architecture and Engineering: The Challenges—Trends—Achievements*; MDPI: Basel, Switzerland, 2020; 360p. [CrossRef]
2. Kapliński, O.; Bonenberg, A.; Bonenberg, W.; Lucchini, M. (Eds.) *Architecture: Integration of Art and Engineering*; MDPI: Basel, Switzerland, 2022; 476p. [CrossRef]
3. Kasperski, J.; Bać, A.; Oladipo, O. A Simulation of a Sustainable Plus-Energy House in Poland Equipped with a Photovoltaic Powered Seasonal Thermal Storage System. *Sustainability* **2023**, *15*, 3810. [CrossRef]
4. Al-Refaie, A.; Lepkova, N. Impacts of Renewable Energy Policies on CO_2 Emissions Reduction and Energy Security Using System Dynamics: The Case of Small-Scale Sector in Jordan. *Sustainability* **2022**, *14*, 5058. [CrossRef]
5. Fernandez, L.; Wojtkiewicz, S.F. Multifunctional Design of Vibrational Energy Harvesters in a Bridge Structure. *Sustainability* **2022**, *14*, 16540. [CrossRef]

6. Plebankiewicz, E.; Gracki, J. Analysis of the Impact of Input Data on the Planned Costs of Building Maintenance. *Sustainability* **2021**, *13*, 12220. [CrossRef]
7. Zhang, N.; Han, Q.; de Vries, B. Building Circularity Assessment in the Architecture, Engineering, and Construction Industry: A New Framework. *Sustainability* **2021**, *13*, 12466. [CrossRef]
8. Bonenberg, W.; Rybicki, S.M.; Schneider-Skalska, G.; Stochel-Cyunel, J. Sustainable Water Management in a Krakow Housing Complex from the Nineteen-Seventies in Comparison with a Model Bio-Morpheme Unit. *Sustainability* **2022**, *14*, 5499. [CrossRef]
9. Borucka, J.; Czyż, P.; Gasco, G.; Mazurkiewicz, W.; Nałęcz, D.; Szczepański, M. Market Regeneration in Line with Sustainable Urban Development. *Sustainability* **2022**, *14*, 11690. [CrossRef]
10. Maxineasa, S.G.; Isopescu, D.N.; Baciu, I.-R.; Lupu, M.L. Environmental Performances of a Cubic Modular Steel Structure: A Solution for a Sustainable Development in the Construction Sector. *Sustainability* **2021**, *13*, 12062. [CrossRef]
11. Elhadary, M.I.; Alzahrani, A.M.Y.; Aly, R.M.H.; Elboshy, B. A Comparative Study for Forced Ventilation Systems in Industrial Buildings to Improve the Workers' Thermal Comfort. *Sustainability* **2021**, *13*, 10267. [CrossRef]
12. Sędzicki, D.; Cudzik, J.; Bonenberg, W.; Nyka, L. Computer-Aided Automated Greenery Design—Towards a Green BIM. *Sustainability* **2022**, *14*, 8927. [CrossRef]
13. Qi, L.; Liu, R.; Cui, Y.; Zhou, M.; Bonenberg, W.; Song, Z. Study of the Landscape Pattern of Shuiyu Village in Beijing, China: A Comprehensive Analysis of Adaptation to Local Microclimate. *Sustainability* **2021**, *14*, 375. [CrossRef]
14. Badach, J.; Szczepański, J.; Bonenberg, W.; Gębicki, J.; Nyka, L. Developing the Urban Blue-Green Infrastructure as a Tool for Urban Air Quality Management. *Sustainability* **2022**, *14*, 9688. [CrossRef]
15. Shi, J.; Sun, J. Prefabrication Implementation Potential Evaluation in Rural Housing Based on Entropy Weighted TOPSIS Model: A Case Study of Counties in Chongqing, China. *Sustainability* **2023**, *15*, 4906. [CrossRef]
16. Zaleckis, K.; Doğan, H.A.; Arce, N.L. Evaluation of the Interventions to Built Heritage: Analysis of Selected Façades of Kaunas by Space Syntax and Sociological Methods. *Sustainability* **2022**, *14*, 4784. [CrossRef]
17. Regenerative Design in Architecture and Construction: The Challenges–Methods–Achievements. Special Issue. 2023. Available online: https://www.mdpi.com/topics/M4ZRQINL5I (accessed on 13 February 2023).

sustainability

MDPI

Article

A Simulation of a Sustainable Plus-Energy House in Poland Equipped with a Photovoltaic Powered Seasonal Thermal Storage System

Jacek Kasperski [1,*], Anna Bać [2] and Oluwafunmilola Oladipo [1]

1 Department of Energy Conversion Engineering, Wroclaw University of Science and Technology, Wybrzeze Wyspianskiego 27, 50-370 Wroclaw, Poland
2 Department of Architecture and Visual Arts, Wroclaw University of Science and Technology, Wybrzeze Wyspianskiego 27, 50-370 Wroclaw, Poland
* Correspondence: jacek.kasperski@pwr.edu.pl; Tel.: +48-71-320-4820

Abstract: This article describes the innovative photovoltaic powered seasonal thermal storage—PVPSTS system. It was used in the design of a plus-energy detached single-family house with a usable area of 98 m^2. This area meets the requirements of the latest building regulations in Poland. The building, with the innovative HVAC installation, was subjected to energy analysis, and a numerical model was also developed. The model was tested based on TMY data for the location of Wroclaw, Poland. Analysis of the results allowed the authors to learn the specifics of the operation of the system throughout the year and to also define its efficiency. The required size of the storage stack was determined to be 1.6 × 1.6 × 0.3 m. The photovoltaic installation, which was integrated with the roof, can produce 48 GJ of electricity per year. This is five to six times more than the building's heating needs, and any excess energy can be exported to the power grid.

Keywords: plus-energy building; thermal stack; solar energy; integrated design

Citation: Kasperski, J.; Bać, A.; Oladipo, O. A Simulation of a Sustainable Plus-Energy House in Poland Equipped with a Photovoltaic Powered Seasonal Thermal Storage System. *Sustainability* **2023**, *15*, 3810. https://doi.org/10.3390/su15043810

Academic Editors: Oleg Kapliński, Wojciech Bonenberg, Agata Bonenberg and Lili Dong

Received: 5 February 2023
Revised: 13 February 2023
Accepted: 15 February 2023
Published: 20 February 2023

1. Introduction

In many countries, including Poland, single-family houses are still the most popular type of residential buildings, and the demand for them further increased after the COVID-19 pandemic. According to the Central Statistical Office in Poland, in 2021 single-family houses accounted for 97.3% of all the buildings that were commissioned for use, with their number increasing by 17.8% when compared to the previous year. In total, 109,212 buildings were completed, including 106,261 single-family houses, with the average usable area of which being 133.3 m^2 [1]. According to the International Energy Agency (IEA), buildings account for about 30% of used final energy and more than 55% of global electricity consumption. Energy use in the building sector has increased steadily since 2000 and has an annual average growth rate of around 1.1%. This is mainly driven by the fact that there has been an increase in floor area, which has grown by around 65% since 2000. Moreover, the rapidly growing demand for energy-consuming equipment and services in buildings in emerging economies has also contributed to this growth [2].

The above data clearly prove that there is an ever-growing demand for individual forms of housing. At the same time, these data confirm the scale of the problem and the great need to look for alternative solutions that will reduce the negative impact of housing on the environment, especially in the case of single-family housing. Due to sustainable development goals (SDGs), radical action in the housing sector is needed on many levels. One of them is the search for innovative solutions and processes that allow for, e.g., a change in the consumption behavior of users, the reduction of energy consumption from non-renewable sources, the reduction of the demand for built-in and usable energy, and also the reduction of greenhouse gas emissions.

In Poland, steps are being taken to implement mechanisms in order to improve this difficult situation. To meet social expectations, the Polish government has introduced a special program that aims to simplify construction procedures for people who want to build a house with a small area [3]. Therefore, from September 2021, a single-family detached building with a gross floor area of up to 70 m^2 and with no more than two stories can be built quickly—bypassing many burdensome formal procedures. This applies to houses with a usable area of up to 90–120 m^2 that are located on a plot with a minimum area of 500 m^2.

In addition, under the recast Energy Performance Building Directive (EPBD), efforts are also underway to improve the energy efficiency of buildings. In 2014, a three-stage phase of increasing the requirements for, among others, the parameters of the thermal insulation of building partitions, as well as reducing the index of the annual demand for non-renewable primary energy, was launched. This means that from 1 January 2021, single-family buildings must be designed in such a way as to achieve an EP index of no more than 70 kWh/(m^2y) for heating, ventilation, and hot water preparation [4]. These provisions aim to implement nearly zero-energy buildings (NZEB) and plus-energy buildings (PEB), which can be achieved thanks to the use of, e.g., renewable energy sources (RES).

There is a deeply justified need to search for contemporary innovative solutions for single-family houses with respect to sustainability. A lot of research is being conducted in this direction, both around the world and in Poland. Bibliometric analysis of 2592 articles conducted by Bibal Manzoor and others (with the use of the Scopus database) showed that there are 252 publications in the field of the energy efficiency of buildings with regards to sustainability [5]. An important role in this transformation is played by the change in the approach and mentality of architects regarding designing in the old formula and also their transition to an integrated design process [6–8]. Moreover, the acceptance of the necessity to use renewable energy sources (RES) and the recognition of the opportunities they offer are also crucial. It is important to study the possibility of integrating photovoltaics (PV) with NZEB with regards to architecture [9,10]—especially building integrated photovoltaics (BIPV) systems [11].

Plus energy buildings (PEBs) are the next step beyond the current EPBD requirements. However, they are not yet widely used and do not have a clear definition. They can be referred to as "buildings that produce more energy from renewable energy sources (RES) than they import over a year" [12] or as "NZEBs that produce 30% or more of the required energy using on-site renewable energy" [13]. Tuerk A. et al. reported that PEBs have the potential to be used in cities in plus energy districts (PEDs) [14]. Research is being conducted concerning the operation of PV with battery energy storage systems (BESS) in urban residential buildings in various parts of the world [15,16]. There are studies related to PV systems and their energy potential in individual countries [17]. Another category of research concerns life cycle assessment (LCA) and life cycle carbon footprint analyses, which show that PV systems in various forms have a carbon footprint similar to that of geothermal energy. This is worse than solar collectors, wind farms, and hydroelectric plants [18].

Of particular importance are studies of the general possibilities of storing thermal energy in buildings [19] and also the possibilities of using and storing solar energy in single-family buildings in the Polish climate. Bac A. et al. [20] discussed ways of integrating the storage stack with the functional and spatial plan of a detached single-family house. From the point of view of energy efficiency, the best space for the location of the storage stack is inside the house. This is due to the fact that heat losses from the storage stack can be used for passive heating of the interior. Żabnieńska-Góra A. et al. [21] analyzed photovoltaic and thermal systems. They came to the conclusion that, apart from cooperation with seasonal energy storage, such installations are not optimal in terms of energy during periods of lower insolation. Nemś et al. [22,23] studied the storage of sensible heat in a ceramic brick stack that used air as the working medium for heating a single-family house. They determined that with intelligent control of the airflow rate, the efficiency of such a storage system can be maximized. According to this analysis, heat storage in ceramic bricks was characterized by high efficiencies—ranging from 74% to 96% at an airflow rate of 0.0068 m^3/s. Several advantages of using ceramic bricks as a storage material were also highlighted, including their non-toxicity, affordability, and resistance to high temperatures.

Ocłoń et al. [24] described a system in which a set of photovoltaic thermal (PVT) hybrid solar panels and an evacuated solar collector with a water-to-water heat pump supplied heat to underground thermal storage tanks. This system provided domestic hot water and heating for a building in mountainous conditions in Poland. Yildiz et al. [25] analyzed the amounts of PV energy that are used on average for heating domestic hot water. For this purpose, they created a unique dataset from 410 households and presented a comprehensive analysis of electricity and hot water consumption. On average, the excess generation of electricity from a 4.5 kW PV system was able to provide 48% of a household's daily energy. Nordgård-Hansen E. et al. [26] focused on detached residential houses in Norway, which were equipped with PV systems connected to electric batteries and optional ground-source heat pump systems for thermal energy storage. It was revealed that in the case of a variable electricity tariff, it is optimal to use seasonal thermal storage. It was also found that ground heat pumps contribute to greater heating stability, but they may not be economically beneficial for single-family houses. Pintanel M.T. et al. [27] described a case study of a facility in Zaragoza (Spain), which had a photovoltaic and thermal hybrid solar field with a seasonal storage tank coupled to a water-to-water heat pump. The authors stated that this innovative heating system is a good solution for social housing. Its construction costs were found to be low, especially when subsidized by the state. Ghanem R.S. et al. [28] analyzed a complex energy system for residential buildings in Amman (Jordan). The system was equipped with PV panels connected to electric batteries, a solid oxide fuel cell (SOFC) system that operated in a combined heat and power (CHP) mode, a solar thermal plant, and a thermal storage tank. Excess energy produced in the summer by the PV system was stored for later use (in winter) in a fuel warehouse. In terms of construction, the cost of SOFC-based micro-CHP systems has proven to be much higher than that of traditional technologies; however, they can be seen to provide energy self-sufficiency. Thinsurat K. et al. [29] described a hybrid solar photovoltaic thermal collector that was integrated with a strontium chloride–ammonia thermo-chemical sorption storage system. The system under analysis was located in Newcastle upon Tyne in England. The energy analyses of the authors suggested that the use of hybrid solar photovoltaic thermal collectors (with an area of 26 m^2), which are integrated with a thermochemical sorption storage system, can fully meet the annual demand for hot water for an average single-person household. Moreover, they can cover at least half of the annual electricity consumption. Chwieduk B. and Chwieduk D. [30] analyzed the utilization of energy for a PV system with a heat pump. The system was evaluated in the case of a low-energy house in the Polish climate during winter. The analyses were based on meteorological data for Warsaw. The authors showed that the reduction of primary energy consumption was not significant. However, to give a definite answer regarding the reduction of energy consumption, a study of the operation of the system over a whole year is required

The authors of this article, in their previous publications, considered the use of a concentrating solar collector and a granite seasonal storage stack in order to provide heating for a single-family building [20]. The obtained technical solution met the requirements regarding energy but strongly interfered with the architecture of the building. In addition, this solution did not meet the needs of residents in terms of electricity. Moreover, it did not create an energy-plus house. In the last few years, home PV installations have become very popular in Poland. They produce very useful electric energy, but not thermal energy, which is actually relatively more simple to produce. In addition, PV technologies are subject to systematic technological evolution, and every decade the efficiency of PV energy conversion increases by several percent when compared to previous years. It can therefore be expected that the use of PV technologies will continue to increase in the future. For this reason, the authors turned to the concept of using PV systems.

The issue of cooperation between PV panels and a thermal storage stack that is placed inside a single-family house is not recognized in the subject literature. The aim of this study is to propose an innovative concept of a plus-energy single-family house that was built in accordance with the latest legal regulations in Poland. The house is equipped with seasonal heat storage located in its residential space. Such a house will use solar energy and some form of seasonal heat storage, which will be analyzed later in the article.

2. Materials and Methods

2.1. *The Concept of a Small Plus-Energy House*

In the warm temperate climate that occurs in Poland, the basic problem of using solar energy for utility purposes is its small amount at the time that heating energy is needed. Therefore, storing the energy obtained in periods of very good availability of solar radiation is a challenge. Based on the aforementioned latest legal act concerning significant facilitations in the construction of houses in Poland, an "almost tiny house" was designed. This detached single-family house meets the requirements of sustainability thanks to the use of photovoltaics and a seasonal energy storage system and is in line with the idea of downsizing living spaces, simplifying such spaces, and essentially "living with less". It combines minimal construction and operating costs and, at the same time, responds to the needs of modern residents. Thanks to the cooperation of the architect and engineers during the integrated design process (IDP), the house has been optimized in order to achieve the energy plus standard, i.e., it produces more energy than it consumes. The IDP was used at the very beginning of preparing the early sketch design, which gave an opportunity for obtaining synergy between the architecture and the energy system.

2.2. *The Shape of the House and Its Technology*

The house has external dimensions of 5.00 × 14.00 m (see Figure 1). It is a two-story building with a shed roof, which is intended for full integration with photovoltaics. The total usable area of the ground floor and first floor is 98 m^2. This is a result of the gross floor area of the house, which is reduced by the building's envelope and the surface of the internal partitions and stairs. The house is designed for comfortable living for a maximum of four residents. On the ground floor there is a common space, a living room, dining room, kitchen, and bathroom, and upstairs there is a space for individual use and a second bathroom.

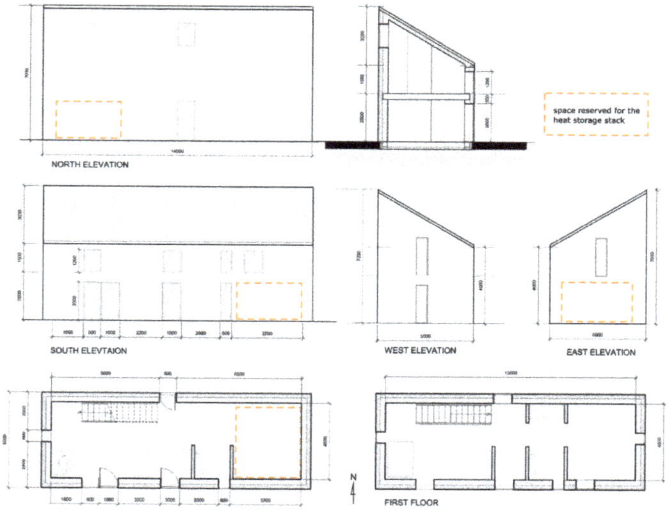

Figure 1. Technical drawing of the building (designed by Arch. Prof. Anna Bać, WUST).

The elongated shape of the building (see Figure 2) is meant to provide optimal space for both living and the assembly of PV panels on the roof. The minimalist area of the ground floor results from law restrictions. The extension of the first-floor area was abandoned in order to reduce the heated area and the volume of the building. The compact body is designed to reduce potential thermal bridges. The windows of the house face mainly south for the purpose of passive solar gains. For the same reason, there are small windows to the east and west and no windows to the north. The size of the windows has been optimized in terms of the required insolation and daylight illumination of 1/8 of the floor area.

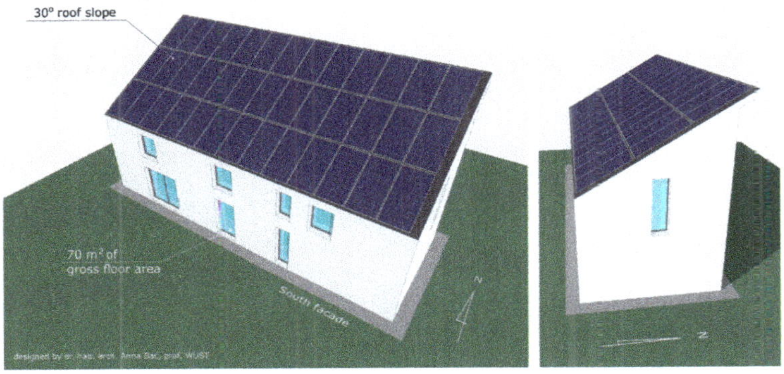

Figure 2. Views of the building.

The storage stack is located inside the building and constitutes the building's integral part on the ground floor. The storage stack's location is due to both its weight and the benefits of such an interior layout. During the heating season, heat losses through the thermal insulation of the storage stack are used to passively heat the rooms, and it can therefore be considered that waste heat is used. Access to the storage stack was provided from inside the house. Photovoltaic panels cover the entire roof area, which has southern exposure. An angle of inclination of 30° was adapted to enable the optimal use of solar radiation.

2.3. The Choice of Energy Storage Technology

There are currently many ways to store energy. Some of them are only suitable for industrial applications (hydrogen and flywheels), while others can be used for individual purposes (including single-family houses). In the case of mobile installations such as cars, planes, ships, and mobile phones, it is important to optimize energy storage technology with regard to the mass-energy stored (kJ/kg). Its value should be as high as possible. In the case of buildings, it is important to analyze the volumetric energy stored (MJ/m^3), which is due to the fact the installation should occupy as little living space as possible. Based on a review of the scientific literature and the authors' own research, Table 1 lists selected parameters of energy storage systems in the case of four technologies (sensible heat, latent heat (PCM), thermo-chemical, and an electric battery). Four basic parameters were analyzed:

- Mass energy stored (kJ/kg),
- Volumetric energy stored (MJ/m^3),
- Cost (USD/kJ),
- Lifetime (years or cycles).

Table 1. Selected features of energy storage technologies.

Storage Technology	Storage Medium	Useful Temp. Difference * (°C)	Mass Energy Stored (kJ/kg)	Volumetric Energy Stored (MJ/m^3)	Efficiency of Storage (%)	Cost (USD/kJ)	Lifetime (Years or Cycles)	Hazard
Sensible Heat	Water	0–100	335.2	335.2				
	Granite [31]	450		978		0.000038	80 cycl. [32]	
	Concrete [33,34]	200	210	504		0.00022		
	Magnesite bricks [35]	600	690	2 070		0.00014		
Latent Heat (PCM)	Paraffin wax [36]	72–76	223	191		0.047		
	Erythritol [37,38]	117	340	493		0.01		
	Palmitic acid [39,40]	61	222	219		0.28		
	Adipic acid [41]	152	275	374		0.041		
	$CH_3COONa_3H_2O$ [42]	58	265			0.025		
Termochemical	Zeolite [43]		900	1980		0.0006		
Electric battery **	Lead–acid		180	360	80	0.024	8 years	fire
	LiFePO4		393	360		0.096		fire
	NiMH		277	650		0.21		fire
	Li-ion [44]	-	720	800	95	0.092	40 years [45]	fire

* useful temperature difference or melting temperature (°C). ** related to 100 Ah, 12 V battery.

As shown in the table, the individual parameters vary greatly. In the case of sensible and latent heat storage technologies, the highest values of mass energy stored (690 kJ/kg) and volumetric energy stored (2070 MJ/m^3) are achieved in the case of magnesite bricks. Thermochemical technology with zeolite also obtains similar parameters; however, it is several times more expensive than magnesite bricks. Batteries based on Li-ion technology also achieve favorable parameters (720 kJ/kg and 800 MJ/m^3, respectively), but Li-ion technology is over 600 times(!) more expensive than magnesite bricks when it comes to construction costs. This technology also poses a significant fire hazard in the case of large installations and additionally, the production of batteries significantly burdens the environment. When considering energy (heat) storage in magnesite bricks, their additional advantage is the lack of emission of harmful or explosive gases, as well as their durability, which is estimated to be several dozen years of operation. The production of these bricks only affects the natural environment to a negligible extent. The high values of parameters calculated for magnesite bricks primarily result from the high density of this material and the very high temperature to which it can be heated. Due to it having so many advantages, it was decided to use this type of heat storage technology.

2.4. The Concept of the Energy System

Based on the experience and knowledge of the authors, and the conclusions resulting from the review of the latest publications, the elements of the energy system, which according to the authors should be used in the proposed system, were summarized. This system is proposed in order to meet the demands of space heating (without space cooling and domestic hot water usage) for the whole year. In order to properly understand the system, the main components are explained individually in the following paragraphs:

- Photovoltaic system

It was decided to use PV panels (although they are less efficient than a thermal solar collector) because they produce very useful electrical energy, and not thermal energy, which is relatively more simple to achieve. In addition, PV technology has become very popular in Poland in the last few years, and there is also a social demand for it. The system consists of a network of half-cut, p-type, monocrystalline cells of white composite foil, with anodized aluminum photovoltaic panels arranged serially on the South-end elevation of the roof of the building in order to generate electrical energy from solar radiation. The efficiency of these panels was tested to be around 20%, with the rated maximum power being 450 Wp/panel. Thirty-nine panels were used due to the fact that this number of panels provides the maximum coverage of the roof surface of the building.

- Electric heater

The heater receives energy from the photovoltaic panels and converts it to thermal energy, which is then used to heat up the thermal storage stack.

- Seasonal thermal energy storage stack

This storage stack is an enlargement of an already-existing daily thermal energy storage device (described in the next chapter). The enlargement here applies to volume and weight, as well as its charging and discharging time.

- High thermal efficiency of the building

The building was equipped with a foundation slab and thick thermal insulation placed under the slab, on the external walls, and on the roof of the building. Triple-pane windows were used, and there is a possibility of covering them with external blinds. The heat recovery system helps to recover waste heat from ventilation, as well as to maintain the quality of air. The heat recovery unit is powered by waste energy from the air leaving the building, and it is transferred in a counter-flow heat exchanger to the fresh air entering the building. Thanks to this, clean air is kept inside the living space. The counter-flow heat exchanger was chosen because of both its high efficiency and suitability for the building space.

The authors chose to call the innovative energy system described above a photovoltaic powered seasonal thermal storage system (PVPSTS). The intention of the authors is to analyze the energy possibilities of the proposed concept. For this purpose, a mathematical model was made and calculations were carried out, and the authors then attempted to evaluate the obtained results.

2.5. Thermal Storage with Air Heating

In Poland, electricity users can choose between tariffs marked as G11 and G12. The G11 tariff has a uniform cost of energy consumption around the clock. The G12 tariff has two prices: day, which is about 10% more expensive than G11, and night, which is about 50% cheaper than G12. The G12 tariff is cheaper at those times of the day when industry uses little electricity. Users can decide which tariff they want to use. The G11 tariff is intended for users who only use lighting and basic household technical equipment. In turn, the G12 tariff is intended for users who do not use gas heating, for those who heat their apartments or houses with electric dual-tariff air space heaters, and/or for those who heat (at night) domestic hot water in boilers with a large volume.

The electric dual-tariff heaters shown in Figure 3 are relatively small (approx. 20 × 50 × 80 cm) and are filled with special shaped bricks made of magnesite ceramic that has a high density. Ceramic materials of this type are resistant to high temperatures and heat up to 600 °C (and even up to 650 °C in some solutions) in electric dual-tariff heaters. The heating process is carried out by a set of electric heaters placed between the bricks. These heaters are mainly powered at night by electricity that is charged at a cheaper night tariff. The heat storage in the bricks is used to heat the rooms with hot air for 24 h a day. Heat storage bricks are surrounded by a thin (about 3 cm) layer of thermal insulation, through which a certain amount of heat flux passively penetrates for 24 h a day. In addition, when more power for heating rooms is needed, a fan is activated in the heater. It forces cold air onto the bricks from below, which is then heated up by the bricks and blown out from the bottom of the heater into a room. The air inlet and outlet are located at the bottom of the heater in order to reduce the convective escape of hot air when the fan is not operating. The small size of heaters of this type allows for a daily period of heat storage.

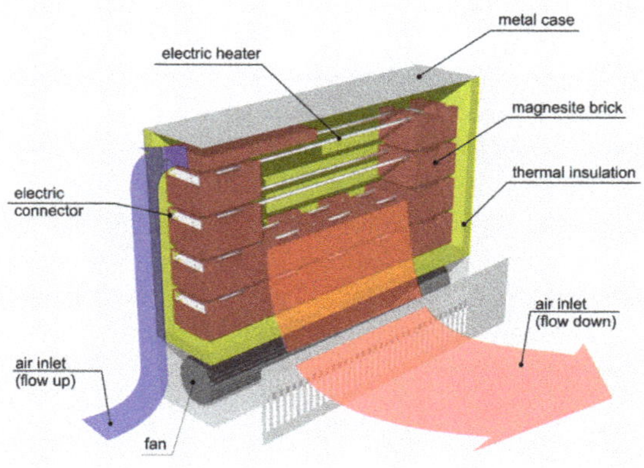

Figure 3. Cross-section of the electric dual-tariff air space heater.

The technology of electric heaters with 24 h heat storage has been known for many years and is still used today in Poland. If there is a need to use heaters of this type, but with heat storage for periods of longer than one day, they should be enlarged. Based on the preliminary calculations that were carried out by the authors of this publication, the ceramic bed storage would have to be several dozen times larger. Ceramic blocks are arranged in a lattice structure with electric heaters running parallel through the blocks. However, it is not only the number of bricks and electric heaters that would have to be increased but also the thickness of the thermal insulation. It would be necessary to use approx. 100 cm of insulation, which is due to the intended long heat retention period. However, the maximum temperature of the bricks, and the way of receiving heat from the storage stack, would not change passive heat loss through the insulation and the active heat collection by the forcing of air between the bricks using a fan. Figure 4 shows a storage stack that is suitable for the period of annual heat storage and which has a typical heater for daily heat storage.

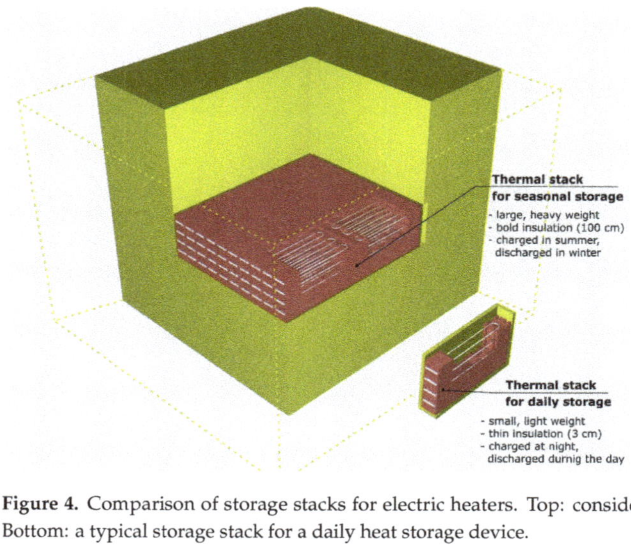

Figure 4. Comparison of storage stacks for electric heaters. Top: considered for seasonal storage. Bottom: a typical storage stack for a daily heat storage device.

2.6. The Proposed Innovative HVAC–PVPSTS System

As illustrated in Figure 5, the system receives solar energy through the south-facing photovoltaic panels installed on the roof of the building. These photovoltaic panels convert the received solar energy into electrical energy, which is then used to heat up the electric heaters running along the magnesite blocks. The electric heater heats up the blocks to a maximum of 600 °C.

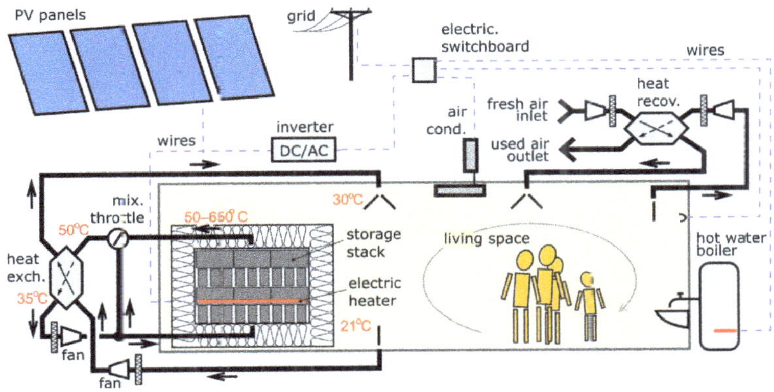

Figure 5. Scheme of the proposed HVAC–PVPSTS system.

The storage stack is located in the living space of the house in order to take advantage of the passive heat transfer (losses) through the thermal insulation of the stack. When there is a need to actively use the heat from the storage stack, the fan is used. However, the temperature of the storage stack during the year can vary from 50 to 600 °C, depending on the level it is charged to. To stabilize the operation of the heating system, hot air from the thermal storage system is sent to the mixing throttle valve (on the left in Figure 5), where it then mixes with cold air. This automatically maintains a stable air temperature of 50 °C at the outlet of the mixing valve. This hot air (at a temperature of 50 °C) is ducted into a counter-flow heat exchanger, where it crosses the air that is leaving the living space. This meeting of fluids results in a further reduction in the temperature of this hot air to about 30 °C. The heat exchanger also functions as a dust separator in order to protect the storage stack from dust contamination from the living space. Such dust could be ignited by entering the high temperature area. The purified 30 °C air is then sent into the living space, while the cold air, along with the excess hot air separated by the mix throttle valve, is directed through a fan back into the thermal storage stack to restart the cyclical movement all over again. The air returning to the thermal storage stack is once again filtered in order to prevent the system from being contaminated with dust at such high temperatures.

An important element of PV systems is their cooperation with the grid system. In the adopted solution, in order to maintain the highest efficiency of energy conversion, the power supply for the storage stack heaters is located on the DC side of the electrical installation. The generated electricity is mainly intended for heating the thermal storage. If its temperature is already too high, or further heating of the thermal storage is not needed, then the generated electricity is exported via an inverter and an electric switchboard to the grid. On the grid side, the hot water production systems and the AC unit are also supplied. The energy balancing that was conducted for the purpose of the article was focused on the needs of heating the living space.

2.7. Mathematical Modelling of the System

The "trial-and-error" method was used to find the required parameters of the energy system. According to E. G. Mbonipa et al. [46], the method involves:

- Selecting independent model parameters on the basis of previous knowledge and experience;
- Adjusting parameter values and manually changing the model input files;
- Running the model;
- Evaluating the model and comparing it with observed data trends.

Repeating the last three steps until an acceptable calibration is achieved, or until the model performance criteria are met, is what actually forms the "trial-and-error" process of the method, as can be seen in Figure 6.

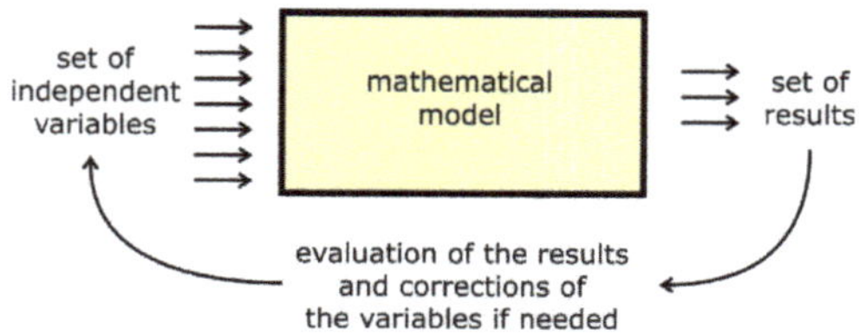

Figure 6. Trial-and-error method diagram.

In the presented case, several assumptions had to be made in order to facilitate the design of the system. One of them is the location of the building, which is related to the availability of typical meteorological year (TMY) data for selected locations in Poland. Other assumptions, including, among others, the technology and design of the building, the defining of the heating and cooling season, and the selection of the photovoltaic panel model, can be seen in Table 2. The most important of the assumed data is the number of panels that is necessary to generate the electricity required for the entire year of the building's operation, the size of the storage stack, and the choice of the day of starting the heating of the storage stack.

Table 2. Assumed parameters for the case study.

Element	Parameter	Assumed Data
House	Location	Wroclaw, Poland (Lat. 51.1°, Long. 7.0°)
	No. of people in living space	4
	Thermal insulation of walls	Styrofoam, 30 cm
	Thermal insulation of the roof	Mineral wool, 30 cm
	PVC windows	0.6 W/m^2K
	Efficiency of air heat recovery unit	80%
	Seasonal criteria	Space heating schedule: 15th of September up to 15th of May
PV	Panel module	DM450M6-72-HBW
	Number of panels	39
	Roof slope and azimuth	30°, South
Storage stack	Stack filling	Magnesite ceramic bricks
	Stack volume available range	0.5–2.5 m^3
	Min. range of temperature stack	50 °C
	Max. range of temperature stack	600 °C
	Storage insulation material	High-temperature-resistant aluminosilicate fibers

2.8. Energy Balance

The mathematical model was based on a detailed analysis of the heat exchange between the building's elements, the environment, and the thermal storage stack. The calculations took into account:

- Heat exchange through the walls with the ambient air;
- Heat exchange through the roof with the ambient air;
- Heat exchange through the floor with the heat from the ground under the building;
- Heat exchange through the windows and doors with the ambient air;
- Solar radiation coming through the windows with solar shading;

- Loss of heat and air infiltration;
- Loss of heat through intentional ventilation with heat recovery;
- Heat gains from the residents;
- Heat gains from working electrical devices;
- Loss of heat from the ceramic storage stack through its insulation to the room.

The heat exchange through the walls, floor, and roof took into account both the thickness and the material of the building structure, as well as the applied layer of thermal insulation. In the case of solar radiation entering through the windows, it was assumed that they would not be covered during the winter in order to maximize the passive gains from radiation energy. In the summer, to reduce the consumption of cold air from the air-conditioning system, up to 25% of the surface of the windows was covered (when solar radiation exceeds 500 W/m^2).

Ventilation heat recovery was assumed at a level of 80%. Heat gains from the residents and working electrical devices were considered for the purpose of calculating the work balance of the air-conditioning system in the summer. In order to carry this out, a 24 h plan of the time that people stay in the rooms of the building, including the type of activity that they perform, was created. A similar schedule was prepared for typical building equipment, i.e., high-power electrical devices (kitchen stove, oven, microwave oven, dishwasher, washing machine, fridge, electric kettle, vacuum cleaner, computer, lighting, etc.). On the basis of the obtained HEB results (see Figure 7), it was decided if the HVAC system would take energy from the storage stack and actively heat up the living space in the winter season and also whether it would activate the air-conditioning device in the summer season.

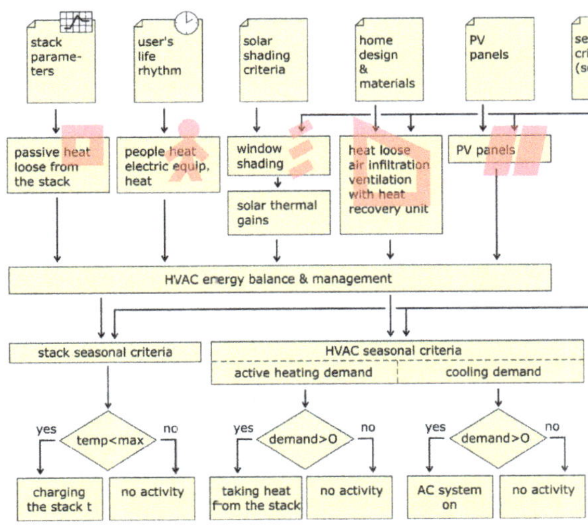

Figure 7. Scheme of hourly energy balance analysis.

The HEB calculations were carried out in one-hour steps for 24 h of the day, successively for all 365 days of the year. An example of the calculations for selected days in November is shown in Figure 8. It can be concluded from this figure that the temperature of the storage stack (red line) is variable and that it depends on the current energy balance of the HVAC–PVPSTS installation (yellow line). If the energy balance is negative (in November the building requires heating), the appropriate amount of energy is taken from the storage stack and as a result, the temperature of the storage stack decreases. If the energy balance is positive, which is the case around noon when there is solar radiation on a given day, the PV installation generates significant amounts of electricity, which in turn heats the storage stack and increases its temperature.

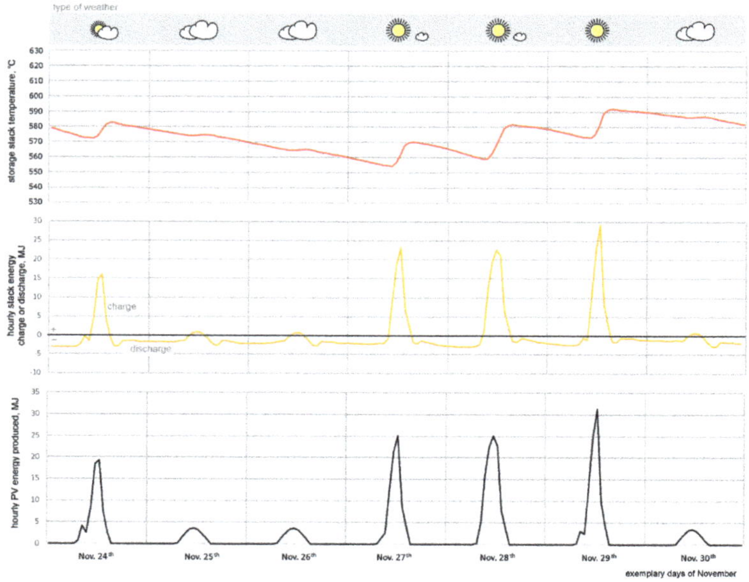

Figure 8. Changes in the selected parameters of the considered energy system using an example of a week in November.

Energy calculations for the whole year require the determination of an appropriately selected value of the storage stack's temperature (before starting the calculations). A well-selected initial value of the temperature for 1 January, 00:00, occurs when the year calculations end with the same value at 23:00 on 31 December. In order for the calculations to maintain this compliance of temperatures, two cycles of annual calculations were carried out: a preliminary and a precision one, as shown in Figure 9. In the precision calculation cycle, the value calculated in the preliminary cycle at the end of the year was assumed as the value of the initial temperature of the storage stack. After calculating the temperature for the entire year, the condition of the storage stack was assessed, and the production and export of electricity from the PV installation were summed up.

The adopted method of searching for a solution for the proposed heating system was to search for a storage stack of an appropriate size. The size of the storage stack can be correctly selected by observing the curve of changes in its temperature throughout the year. The storage stack, which is preheated in the autumn, should gradually lower its temperature until the end of the heating period (15 May). In turn, during the heating period, the temperature may fluctuate, depending on the solar radiation that is available in winter. A storage stack that is too small will cool down too quickly, without providing heating at the end of winter, whereas a storage stack that is too large will remain heated until summer, unnecessarily burdening the building's air conditioning system with its heat losses. The presented calculation method does not allow for simple calculations of how big the storage stack should be; however, it is possible to determine the required size of the storage stack using the "trial-and-error" method. When using this method, it is necessary to determine the values for all the input variables, while at the same time changing only one of them (the size of the storage stack) and evaluating each time whether the obtained results meet the needs of the correct operation of the system.

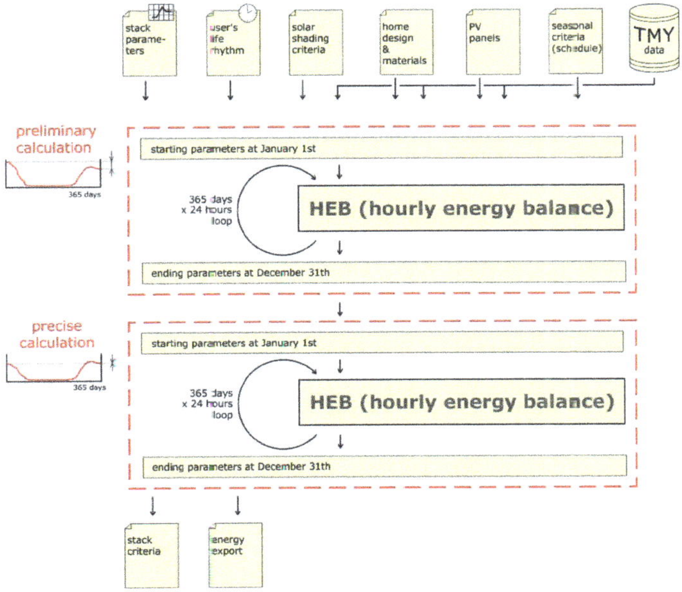

Figure 9. Two calculation cycles: preliminary and precise.

3. Results and Discussion

After conducting a series of calculations, the correct results were obtained for the ceramic storage stack, the parameters of which are shown in Table 3.

Table 3. Achieved parameters of the storage stack.

Parameter of the Storage Stack	Value
Dimensions of the ceramic filling, m	1.6 × 1.6 × 0.3
Volume of the ceramic filling, m^3	0.77
Mass of the ceramic filling, kg	2264
Thickness of the thermal insulation, m	1.00
Outer dimensions of the insulation, m	3.6 × 3.6 × 2.3
Volume of the insulated storage stack, m^3	29.8
Share of the insulated storage stack in the floor area of the living space, %	12.5
Share of the insulated storage stack in the volume of the living space, %	11.5
Maximum passive heat loss, W	414

The entire heat storage (with the obtained external dimensions) fits within the space intended for the storage stack, which is shown by the orange dashed line in Figure [illegible]. The ratio of the size of the storage stack to the living space of the building is shown in the cross-section in Figure 10.

Figure 11 presents the results of the calculations of the installation's operation over the whole year. The course of changes in the temperature of the storage stack is marked with a red line. It starts with a temperature of 436 °C in the first hour of 1 January and ends with the same temperature in the last hour of 31 December. This means that the calculations were balanced for the whole year and can be considered to be precise. The description of the operation of the installation should start from the last days of September when the storage stack is discharged to the minimum value (at the level corresponding to its temperature of 50 °C).

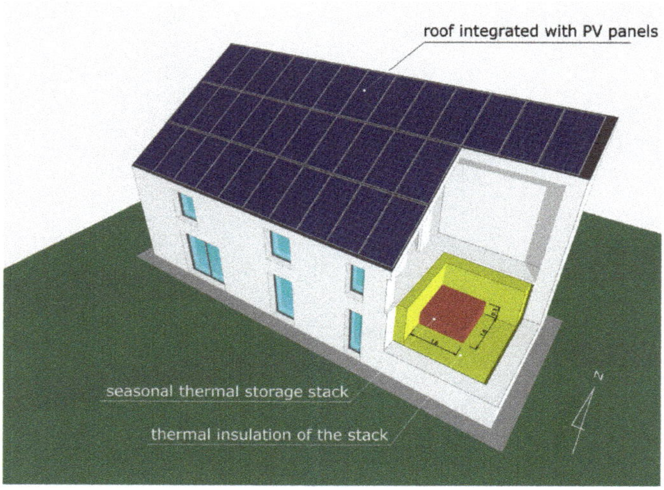

Figure 10. Ratio of the size of the storage stack to the living space of the building.

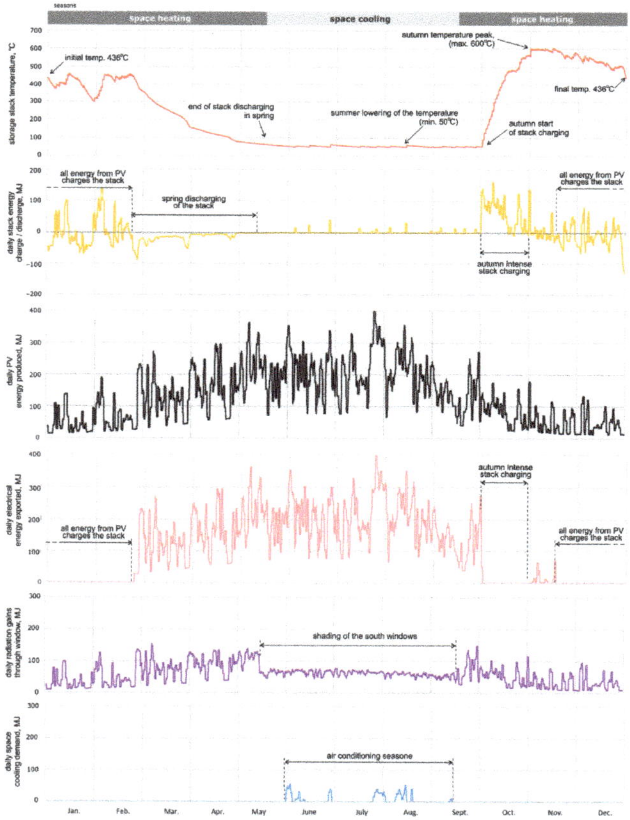

Figure 11. Results of the yearly balance calculation.

From the first day of October, the autumn process of charging the storage stack begins. It is an intensive process due to the fact that the storage stack must be fully charged quickly before the winter period. The process of intensive charging of the storage stack lasts for about one month, after which (from the beginning of November to the end of February) the availability of solar radiation significantly decreases, and the building's heating needs are at their greatest. At this time, the temperature of the storage stack drops, but there are also sunny days when the generated electricity allows for the storage stack to be recharged. As can be seen, the storage stack maintains the necessary state of charging throughout the heating period due to the fact that it cools down only at the beginning of the summer period.

In the diagram (Figure 11), the yellow line shows the energy that charges the storage stack (positive values) and also the energy that is taken from the storage stack through the ventilation system to heat the living space (negative values). This diagram clearly shows both the moment of the intensive charging of the storage stack in October and the period of the gradual discharge of the storage stack (lasting from March to May). After this time, for the summer months from May to September, the storage stack is kept at its minimum temperature, which aims to facilitate the operation of the air-conditioning system.

The amount of energy produced by the PV installation on the roof of the building is marked in Figure 11 with a black line. Understandably, with a 30° pitched roof, the largest amounts of energy are generated in the summer and the smallest are generated in the winter. As already mentioned, the electricity generated by the PV installation is only used to heat the storage stack in certain months. These are the months from the beginning of October to the end of February, which can be seen in the diagram (pink line) that shows the electricity exported from the installation to the power grid. This is surplus energy, which is not necessary for the correct operation of the building's heating system, but instead can be used for the needs of its users, e.g., for the production of hot utility water, the operation of installed electrical devices, and the operation of the air conditioning in the summer.

The amount of solar radiation energy that is acquired passively through the windows of the building is marked in purple in Figure 11. As can be seen, this energy is relatively level throughout the year, which is due to the vertical position of the windows in the walls, and the lack of windows in the roof's slope. In addition, in order to aid the work of the air-conditioning system in the summer, 75% of the area of the windows on the south side was shaded when the radiation density exceeded 500 W/m^2. In this case, the graph shows a more than two-fold decrease in passive solar gains in the period from mid-May to mid-September.

The amount of energy needed to cool the building using the air-conditioning system is marked in blue in Figure 11. It is apparent that such air-conditioning needs occur from June to mid-September, but they are not significant, which is partly due to the window shading system described above. The calculation of the annual energy balance of the installation for the above calculations is presented in Table 4.

Table 4. Energy from the annual operation of the installation.

The Amount of Energy	MJ	kWh	kWh/m^{2}a
Produced by the PV installation	48,020	13,339	
Heating needs of the building (without people's heat)	10,717	2977	33.4
Heating needs of the building (with people's heat)	8446	2346	26.4
Exported to the power grid	39,575	10,993	
Used for the needs of electric devices	8568	2380	
Used to produce hot water	19,318	5366	
Exported to the power grid but which is lowered by the building's own consumption	11,689	3247	

The selection of the day of when to start the intensive charging of the storage stack requires some discussion. Graphs showing the changes in the temperature of the storage stack with regard to the day of starting the charging process are shown in Figure 12. The conducted analysis covered five different days of when the intensive charging period started:

- 1 September;

- 15 September;
- 1 October;
- 15 October;
- 1 November.

When looking at the obtained lines that correspond with the changes in the temperature of the storage stack, it can be seen that the first three moments of the intensive autumn charging of the storage stack have no effect on its temperature in the winter and spring. However, starting the charging of the storage stack from mid-October or later results in its incomplete charging and therefore in an earlier lack of energy at the end of winter. The obtained results concern the calculations made using the TMY database, and it is thus not known what day should be recommended to the user of the considered system (for any real climatic year) for starting the charging of the storage stack. Earlier start times guarantee a greater certainty of its stable operation; however, air-conditioning of the rooms in September and October may still be required. This is due to the passive escape of heat to the living space from the prematurely heated storage stack.

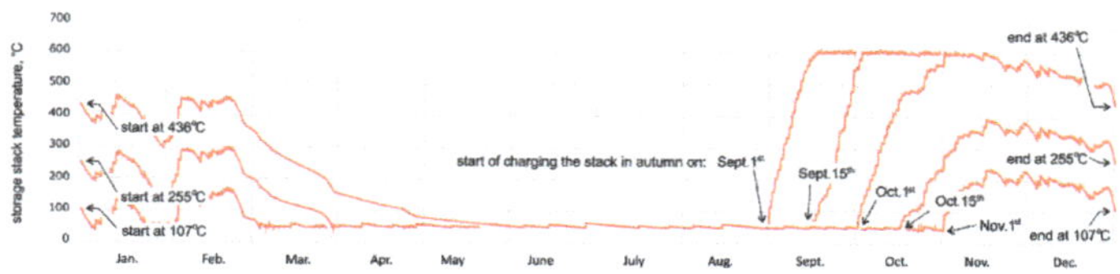

Figure 12. Influence of the day of when starting the charging of the stack.

From the obtained results, it can be seen that it is possible to build an energy installation that ensures stable heating of the building throughout the year. This type of installation is based on an HVAC–PVPSTS system that is powered by a photovoltaic installation located entirely on the roof of the building and also a heat storage stack cooperating with it. The designated size of the heat storage stack fits in the space of the building and does not dominate over the living space. On the basis of the obtained calculations, it was determined that the installation generates approx. 48 GJ (13.3 MWh) of energy, which exceeds (by several times) the heating needs of the building in the winter season (amounting to approx. 10.7 GJ (3.0 MWh)). However, the amount of electricity that can be produced in the winter season is too small, and it is necessary to store some of the required energy in the stack. There is also a need to charge the selected storage stack early enough. Taking into account that some of the exported electricity will be used for the needs of the building's inhabitants (used electric devices and the production of hot utility water), the energy balance of the installation is still positive and amounts to approx. 11.7 GJ (3.2 MWh), which in turn accounts for 24% of the energy produced by the PV installation per year. It can therefore be concluded that the building in the proposed architectural form, which is equipped with the considered energy installation, is an energy-plus building.

In future research, the authors would like to deal with energy simulations for other locations, as well as the relationship between the upper temperature of the storage stack and its volume. The developed model is also meant to be the basis for the implementation of an experimental facility.

4. Conclusions

The work considered a hypothetical innovative building, which is adapted to the new regulations that aim to simplify construction procedures. An HVAC—PVPSTS installation was selected for the building, which is in line with the latest trends in low-energy construction. The building consists of an installation of photovoltaic panels, which are integrated with its body and combined with large seasonal thermal storage that allows for heating of the living space. The power installation was mathematically modeled using TMY data, which is standard in this type of analysis. From the obtained results, it can be stated that:

- It is possible to build an energy installation that ensures year-round, stable heating of the building;
- The photovoltaic installation is located entirely on the roof of the building;
- To heat the building, it is necessary to use a seasonal heat storage stack, which is necessary to provide some of the energy that is required in the winter when solar radiation is too weak;
- The selected ceramic storage stack has dimensions of 1.6 × 1.6 × 0.3 m, and with thermal insulation, its dimensions are 3.6 × 3.6 × 2.3 m; it is located in a planned space on the ground floor of the building and does not dominate over the remaining usable space;
- The stack, with its thermal insulation, occupies only 13% of the area and 12% of the volume of the living space;
- The photovoltaic installation would produce 48 GJ (13.3 MWh) of electricity per year and fully meet the building's heating needs of 10.7 GJ (3.0 MWh) for an empty building without residents or 8.4 GJ (2.4 MWh) for an occupied building;
- The building with the installation meets the criterion for a plus-energy building;
- The amount of electricity produced exceeds the heating needs of the building by five to six times, and any excess energy can be exported to the power grid;
- If the electrical devices used in the building were powered by a photovoltaic installation, it would still generate excess energy; it would be able to export 11.7 GJ (3.2 MWh) per year to the power grid, which accounts for 24% of the energy produced by the PV installation;
- Such an installation, for correct operation, requires the storage system to be charged early enough—it was shown that it should not start later than 1 October.

Author Contributions: Conceptualization, J.K. and O.O.; methodology, J.K.; architectural design, A.B.; software, J.K. and O.O.; formal analysis, J.K.; investigation, J.K., A.B. and O.O.; resources, A.B. and O.O.; data curation, J.K. and O.O.; writing—original draft preparation, J.K., A.B. and O.O.; Writing—review and editing, J.K. and A.B.; visualization, J.K. and A.B.; supervision, J.K.; project administration, J.K.; funding acquisition, J.K. All authors have read and agreed to the published version of the manuscript.

Funding: This research was funded by Wroclaw University of Science and Technology (Project No. 8211104160).

Institutional Review Board Statement: Not applicable.

Informed Consent Statement: Not applicable.

Data Availability Statement: The relevant data underlying this study are fully available under the details provided in the references. For further questions, please contact the corresponding author.

Acknowledgments: The authors would like to thank Arch. Pelumi Olajide (M.Sc.) for the drawing in Figure 1.

Conflicts of Interest: The authors declare no conflict of interest.

References

1. Główny Urząd Statystyczny—GUS. Budownictwo w 2021; p. 3. Available online: https://stat.gov.pl/files/gfx/portalinformacyjny/pl/defaultaktualnosci/5478/13/13/1/budownictwo_w_2021_roku.pdf (accessed on 13 December 2022).
2. International Energy Agency—IEA. *The Critical Role of Buildings, Perspectives for the Clean Energy Transition*; IEA: Paris, France, 2019. Available online: https://www.iea.org/reports/the-critical-role-of-buildings (accessed on 13 December 2022).
3. Ustawa z dnia 17 Września 2021 r. o Zmianie Ustawy—Prawo Budowlane oraz Ustawy o Planowaniu i Zagospodarowaniu Przestrzennym (Dz.U. 2021 poz. 1986). Available online: https://www.nist.gov.pl/prawo/ustawa-z-dnia-17-wrzesnia-2021-r-o-zmianie-ustawy-prawo-budowlane-oraz-ustawy-o-planowaniu-i-zagospodarowaniu-przestrzennym-dz-u-2021-poz-1986,3513.html (accessed on 13 December 2022).
4. Obwieszczenie Ministra Rozwoju I Technologii z dnia 15 Kwietnia 2022 r. w Sprawie Ogłoszenia Jednolitego Tekstu Rozporządzenia Ministra Infrastruktury w Sprawie Warunków Technicznych, Jakim Powinny Odpowiadać Budynki i ich Usytuowanie, Dz. U. Poz. 1225. pp. 86–87. Available online: https://isap.sejm.gov.pl/isap.nsf/download.xsp/WDU20220001225/O/D20221225.pdf (accessed on 13 December 2022).

5. Manzoor, B.; Othman, I.; Sadowska, B.; Sarosiek, W. Zero-Energy Buildings and Energy Efficiency towards Sustainability: A Bibliometric Review and a Case Study. *Appl. Sci.* **2022**, *12*, 2136. [CrossRef]
6. Bonenberg, W.; Kapliński, O. The Architect and the Paradigms of Sustainable Development: A Review of Dilemmas. *Sustainability* **2018**, *10*, 100. [CrossRef]
7. Feria, M.; Amado, M. Architectural Design: Sustainability in the Decision-Making Process. *Buildings* **2019**, *9*, 135. [CrossRef]
8. Attmann, O. *Green Architecture. Advanced Technologies and Materials*; McGraw-Hill: New York, NY, USA, 2009.
9. Celadyn, W.; Filipek, P. Investigation of the Effective Use of Photovoltaic Modules in Architecture. *Buildings* **2020**, *10*, 145. [CrossRef]
10. Guide to BIPV. In *Building Integrated Photovoltaics*; Polysolar Limited: Cambridge, UK, 2015.
11. Heinstein, P.; Ballif, C.; Perret-Aebi, L. Building Integrated Photovoltaics (BIPV): Review, Potentials, Barriers and Myths. *Green* **2013**, *3*, 125–156. [CrossRef]
12. Salvalai, G.; Sesana, M.; Lollini, R. (Eds.) *Zero to Plus Energy Buildings: Innovation in Technologies and Methodologies for New Buildings and Deep Renovation*; Special Issue of Sustainability; MDPI: Basel, Switzerland, 2022; Volume 14. Available online: https://www.mdpi.com/journal/sustainability/special_issues/zero_to_plus_energy_buildings (accessed on 19 December 2022).
13. Attia, S. *Net Zero Energy Buildings (NZEB): Concepts, Frameworks and Roadmaps for Project Analysis and Implementation*; Elsevier: Amsterdam, The Netherland, 2018.
14. Tuerk, A.; Frieden, D.; Neumann, C.; Latanis, K.; Tsitsanis, A.; Kousouris, S.; Llorente, J.; Heimonen, I.; Reda, F.; Ala-Juusela, M.; et al. Integrating Plus Energy Buildings and Districts with the EU Energy Community Framework: Regulatory Opportunities, Barriers and Technological Solutions. *Buildings* **2021**, *11*, 468. [CrossRef]
15. Kichou, S.; Skandalos, N.; Wolf, P. Evaluation of Photovoltaic and Battery Storage Effects on the Load Matching Indicators Based on Real Monitored Data. *Energies* **2020**, *13*, 2727. [CrossRef]
16. Abdelhafez, M.; Touahmia, M.; Noaime, E.; Albaqawy, G.; Elkhayat, K.; Achour, B.; Boukendakdji, M. Integrating Solar Photovoltaics in Residential Buildings: Towards Zero Energy Buildings in Hail City, KSA. *Sustainability* **2021**, *13*, 1845. [CrossRef]
17. Kosorić, V.; Lau, S.-K.; Tablada, A.; Bieri, M.; Nobre, A.M. A Holistic Strategy for Successful Photovoltaic (PV) Implementation into Singapore's Built Environment. *Sustainability* **2021**, *13*, 6452. [CrossRef]
18. Milousi, M.; Souliotis, M.; Arampatzis, G.; Papaefthimiou, S. Evaluating the Environmental Performance of Solar Energy Systems Through a Combined Life Cycle Assessment and Cost Analysis. *Sustainability* **2019**, *11*, 2539. [CrossRef]
19. Sarbu, I.; Sebarchievici, C. A Comprehensive Review of Thermal Energy Storage. *Sustainability* **2018**, *10*, 191. [CrossRef]
20. Bać, A.; Nemś, M.; Nemś, A.; Kasperski, J. Sustainable Integration of a Solar Heating System into a Single-Family House in the Climate of Central Europe—A Case Study. *Sustainability* **2019**, *11*, 4167. [CrossRef]
21. Żabnieńska-Góra, A.; Khordehgah, N.; Jouhara, H. Annual performance analysis of the PV/T system for the heat demand of a low-energy single-family building. *Renew. Energy* **2021**, *163*, 1923–1931. [CrossRef]
22. Nemś, A.; Nemś, M. Analysis and selection criteria of photovoltaic panels for DHW. In *E3S Web of Conferences*; EDP Sciences: Les Ulis, France, 2017; Volume 13.
23. Nemś, M.; Nemś, A.; Kasperski, J.; Pomorski, M. Thermo-Hydraulic Analysis of Heat Storage Filled with the Ceramic Bricks Dedicated to the Solar Air Heating System. *Materials* **2017**, *10*, 940. [CrossRef]
24. Ocłoń, P.; Ławryńczuk, M.; Czamara, M. A New Solar Assisted Heat Pump System with Underground Energy Storage: Modelling and Optimisation. *Energies* **2021**, *14*, 5137. [CrossRef]
25. Yildiz, B.; Bilbao, J.I.; Roberts, M.; Heslop, S.; Dore, J.; Bruce, A.; MacGill, I.; Egan, R.J.; Sproul, A.B. Analysis of electricity consumption and thermal storage of domestic electric water heating systems to utilize excess PV generation. *Energy* **2021**, *235*, 121325. [CrossRef]
26. Nordgård-Hansen, E.; Kishor, N.; Midttømme, K.; Risinggård, V.K.; Kocbach, J. Case study on optimal design and operation of detached house energy system: Solar, battery, and ground source heat pump. *Appl. Energy* **2022**, *308*, 118370. [CrossRef]
27. Pintanel, M.T.; Martínez-Gracia, A.; Uche, J.; del Amo, A.; Bayod-Rújula, A.; Usón, S.; Arauzo, I. Energy and environmental benefits of an integrated solar photovoltaic and thermal hybrid, seasonal storage and heat pump system for social housing. *Appl. Therm. Eng.* **2022**, *213*, 118662. [CrossRef]
28. Ghanem, R.S.; Nousch, L.; Richter, M. Modeling of a Grid-Independent Set-Up of a PV/SOFC Micro-CHP System Combined with a Seasonal Energy Storage for Residential Applications. *Energies* **2022**, *15*, 1388. [CrossRef]
29. Thinsurat, K.; Bao, H.; Ma, Z.; Roskilly, A.P. Performance study of solar photovoltaic-thermal collector for domestic hot water use and thermochemical sorption seasonal storage. *Energy Convers. Manag.* **2019**, *180*, 1068–1084. [CrossRef]
30. Chwieduk, B.; Chwieduk, D. Analysis of operation and energy performance of a heat pump driven by a PV system for space heating of a single family house in polish conditions. *Renew. Energy* **2021**, *165*, 117–126. [CrossRef]
31. Available online: https://material-properties.org/granite-density-heat-capacity-thermal-conductivity (accessed on 20 January 2023).
32. Li, B.; Ju, F.; Xiao, M.; Ning, P. Mechanical stability of granite as thermal energy storage material: An experimental investigation. *Eng. Fract. Mech.* **2019**, *211*, 61–69. [CrossRef]
33. Available online: https://material-properties.org/concrete-density-heat-capacity-thermal-conductivity/ (accessed on 20 January 2023).
34. Cabeza, L.F.; Vérez, D.; Zsembinszki, G.; Borri, E.; Prieto, C. Key Challenges for High Temperature Thermal Energy Storage in Concrete—First Steps towards a Novel Storage Design. *Energies* **2022**, *15*, 4544. [CrossRef]
35. Bespalko, S.; Miranda, A.M.; Halychyi, O. Overview of the Existing Heat Storage Technologies: Sensible Heat. *Acta Innov.* **2018**, *28*, 82–113. [CrossRef]

36. Phase Change Material Selection, Advanced Cooling Technologies. Available online: https://www.1-act.com/products/pcm-heat-sinks/pcmselection/#:~:text=PCM%20Types%20Include%20Paraffin%20Waxes,%2C%20Hydrated%20Salts%2C%20and%20Metallics.&text=Med.&text=Paraffins%20are%20most%20common%20PCM,chemically%20compatible%20with%20most%20metals (accessed on 3 January 2023).
37. Tong, B.; Tan, Z.C.; Zhang, J.N.; Wang, S.X. Thermodynamic investigation of several natural polyols: Part III. Heat capacities and thermodynamic properties of erythritol. *J. Therm. Anal. Calorim.* **2009**, *95*, 469–475. [CrossRef]
38. Haillot, D.; Bauer, T.; Kröner, U.; Tamme, R. Thermal analysis of phase change materials in the temperature range 120–150 °C. *Thermochim. Acta* **2011**, *513*, 49–59. [CrossRef]
39. J7ankowski, N.R.; McCluskey, F.P. A review of phase change materials for vehicle component thermal buffering. *Appl. Energy* **2014**, *113*, 1525–1561. [CrossRef]
40. Zhang, Y.; Wang, L.; Tang, B.; Lu, R.; Zhang, S. Form-stable phase change materials with high phase change enthalpy from the composite of paraffin and cross-linking phase change structure. *Appl. Energy* **2016**, *184*, 241–246. [CrossRef]
41. Hasl, T.; Jiricek, I. The Prediction of Heat Storage Properties by the Study of Structural Effect on Organic Phase Change Materials. *Energy Procedia* **2014**, *46*, 301–309. [CrossRef]
42. Pathak, S.K.; Tyagi, V.V.; Chopra, K.; Kalidasan, B.; Pandey, A.K.; Goel, V.; Saxena, A.; Ma, Z. Energy, exergy, economic and environmental analyses of solar air heating systems with and without thermal energy storage for sustainable development: A systematic review. *J. Energy Storage* **2023**, *59*, 106521 [CrossRef]
43. Yue, X.; Xu, Y.; Zhou, X.; Xu, D.; Chen, H. Study on the Performance of a Solar Heating System with Seasonal and Cascade Thermal-Energy Storage. *Energies* **2022**, *15*, 7733. [CrossRef]
44. Dragonfly Energy. Why Does Energy Density Matter in Batteries? Available online: https://dragonflyenergy.com/why-does-energy-density-matter-in-batteries/ (accessed on 10 January 2023).
45. Hu, X.; Zou, C.; Zhang, C.; Li, Y. Technological Developments in Batteries: A Survey of Principal Roles, Types, and Management Needs. *IEEE Power Energy Mag.* **2017**, *15*, 20–31. [CrossRef]
46. Mbonimpa, E.; Gautam, S.; Lai, L.; Kumar, S.; Bonta, J.; Wang, X.; Rafique, R. Combined PEST and Trial–Error approach to improve APEX calibration. *Comput. Electron. Agric.* **2015**, *114*, 296–303. [CrossRef]

sustainability

Article

Impacts of Renewable Energy Policies on CO_2 Emissions Reduction and Energy Security Using System Dynamics: The Case of Small-Scale Sector in Jordan

Abbas Al-Refaie [1] and Natalija Lepkova [2,*]

1 Department of Industrial Engineering, University of Jordan, Amman 11942, Jordan; abbas.alrefai@ju.edu.jo
2 Department of Construction Management and Real Estate, Faculty of Civil Engineering, Vilnius Gediminas Technical University, Sauletekio av. 11, 10223 Vilnius, Lithuania
* Correspondence: natalija.lepkova@vilniustech.lt

Abstract: Renewable energy policies, such as feed-in-tariffs (FiTs) and subsidy policies, have been reported effective in enhancing the social acceptability to install solar photovoltaic (PV) systems. Nevertheless, a quantitative assessment approach is still needed to measure the extent to which these policies can achieve the clean energy goals and support the decision-making process. This study, therefore, develops system dynamics models to assess the impacts of PV policies on the social acceptability to install PV systems, energy security, and CO_2 emission reduction in the small-scale sector in Jordan. Simulation was then conducted from the period 2016 to 2050. The results for the FiTs (subsidy) policies showed that the predicted accumulated PV installations, power generated, and CO_2 emission reductions will reach 67.125 (88.38) Gigawatt (GW), 115.853 (152.588) Terra Wh (TWh), and 74.49 (98.114) million tons CO_2, respectively. To achieve these goals, the required cumulative FiTs and subsidy policy costs are 2.2 and 7.59 billion USD, respectively. Sensitivity analyses followed to determine the optimal FiTs price and subsidy proportion that optimize PV goals under uncertainty. In conclusion, the developed models are found valuable tools for measuring the impacts of energy policies on PV goals and thereby provide great input information to the decision-making processes when selecting the appropriate energy policies and actions. In the end, adopting FiTs and/or subsidy policies, Jordan is expected to achieve a high level of clean energy security by 2050, which enhances energy capabilities and mitigates global warming. Future research will examine the factors that affect social acceptability for PV systems.

Keywords: PV system; social acceptability; system dynamics; CO_2 emissions; feed-in-tariffs; subsidy policy

Citation: Al-Refaie, A.; Lepkova, N. Impacts of Renewable Energy Policies on CO_2 Emissions Reduction and Energy Security Using System Dynamics: The Case of Small-Scale Sector in Jordan. *Sustainability* **2022**, *14*, 5058. https://doi.org/10.3390/su14095058

Academic Editors: Oleg Kapliński, Lili Dong, Agata Bonenberg and Wojciech Bonenberg

Received: 20 March 2022
Accepted: 20 April 2022
Published: 22 April 2022

1. Introduction

Rapid economic development and growth of the world have led to continued rise in energy demand and thereby resulted in excessive depletion of natural resources, which now threaten global energy security. Therefore, saving energy, environmental awareness, and cleaner energy production are among the highest priorities and can be viewed as the critical challenges facing the world [1]. As a result, all countries have become convinced of the importance of addressing environmental problems by adopting renewable energy technologies that reduce pressure on the environment and use rates of fossil energy. The electricity generation sector has been considered as one of the largest contributors to global greenhouse gas emissions due to its great reliance on fossil fuels. The shift towards solar energy; or so-called photovoltaic (PV) system, continues and has been strengthened with technological innovations that have lower production costs, improved storage options, and reduced harmful emissions to the environment.

In Jordan, the energy sector is among the highest in the world in dependency on foreign energy sources, with 94% of the country's energy needs in 2018 coming from imported oil and natural gas, 10% of the country's gross domestic product [2,3]. Specifically, Jordan faces challenges in the energy sector because of the increasing energy demand that is from the result of population growth, increased per capita consumption and electricity tariffs, and cross-subsidies due to countries with limited resources. Studies [4,5] showed that 6% of Jordan's primary energy sources in 2025 will be generated from renewable energy. The forecasted electrical energy demand is 27.285 Terra Wh (TWh) in 2025, while it is 33,832 Giga Wh (GWh) in 2030. The growth percentage in the needed electrical energy is about 3.1%. Finally, the household consumed about 46% of the electrical energy consumption. As a result, Jordan's top priority is achieving energy supply security, eliminating its dependence on imports while meeting the growing demand for primary energy [6]. Further, the high consumption of fossil fuels increases carbon dioxide emissions and consequently causes global warming and climate change. Fortunately, the energy sector in Jordan is endowed with an abundance of solar energy, which is evident from the annual daily average solar irradiance on a horizontal surface range of between 5–7 kWh/m^2, one of the highest figures in the world. Jordan is on pace to exceed 20% of generated electricity from renewable sources by 2020 [7,8]. Recently, solar energy applications have gained more attention to substitute depletion of fossil fuel, reduce dependency on imported energy, and improve social and environmental benefits. For example, Jabera and Abul Hawa [9] conducted feasibility and reliability analyses of a PV grid-connected system and life cycle cost and payback period criterion to determine the optimal design of a PV system in a passive residential building in a Mediterranean climate. Hussein [10] examined the greenhouse gas emissions reduction potential of Jordan's utility scale wind and solar projects. The study revealed that the solar and wind project will result in a significant reduction of the country's projected greenhouse gas emissions of 1.93–3.21 mega tons of CO_2 emissions annually. Alawneh et al. [11] overviewed the off-grid PV installations and presented the present status of PV market in Jordan after activating the Renewable Energy and Efficiency Law in Jordan in year 2014. The results showed that the total PV installed capacity in Jordan exceeded 300 MW distributed in large-, medium-, and small-scale projects, according to the kind of installation, whether it is energy net metering, power wheeling, or power purchase agreement.

Further, PV energy policies can be adopted to enhance the social acceptability of installing PV systems. Nevertheless, a quantitative assessment of the impacts of PV energy policies on energy security and CO_2 emission reduction is still needed to support the decision-making process. This paper, therefore, utilizes a system dynamics approach for assessing the impacts of FiTs and subsidy policies on PV energy goals for CO_2 emission reduction and energy security in the small-scale sector in Jordan. This research develops a valuable assessment tool that assists energy decision-makers in evaluating the impacts of the deployed energy policies and evaluating their effectiveness in promoting PV installations. This study, including the introduction, is outlined as follows. Section 2 reviews relevant previous studies. Section 3 builds a system dynamics model that relates the key model factors. Section 4 conducts simulation and optimization for the proposed model and discusses research results. Section 5 summarizes conclusions and suggests recommendations for future improvements and future research.

2. Literature Review

Recently, studies on renewable energy have received significant research attention. For example, Trappey et al. [12] developed a cost-benefit methodology based on modeling system dynamics to promote renewable energy and reduce carbon emissions in Taiwan. Their study examined the impacts of adopting solar energy policies. Hsu [13] developed an assessment system dynamics model for simulating solar PV installations and CO_2 emission reduction in Taiwan, to evaluate the ability of FiTs and capital subsidies policies to promote solar PV applications. Tziogas and Georgiadis [1] investigated the causalities of cleaner and affordable electricity production mix, proposed a system dynamic approach for examining the interrelations, and identified six key areas for policy interventions affecting the social acceptability with different scenarios. Movilla et al. [14] described a simulation model of the photovoltaic energy sector in Spain. The model analyzed dynamic behavior of the PV sector under different scenarios. Jeon and Shin [15] combined system dynamics with Monte Carlo simulation to evaluate the long-term technology method for renewable energy technologies. Aslani et al. [16] discussed the role of diversification on the dependency and security of energy supply using a system dynamics model of renewable energy recourses in Finland. They analyzed and evaluated three scenarios of renewable energy policies. Aslani and Wong [17] constructed a system dynamic model to evaluate different costs of renewable energy promotion and operation and analyzed the effectiveness of renewable energy policies on increasing the security of energy supply, minimization of the total policies cost, and maximizing the number of renewable energy recourses systems in the United States. Ahmad et al. [18] constructed a system dynamics model that investigates the role of feed-in tariff policy and examined various scenarios project solar PV technology in Malaysia till 2050. They suggested insights related to capacity, finances, and environmental savings. Radomes and Arango [19] used the classic Bass diffusion theory to analyze the diffusion of a PV system in Colombia. The model incorporated both subsidy and FiT policies and analyzed the effects of policy mix tree with the breakeven analysis and cost benefit analysis. Ye et al. [20] analyzed the impacts of FiTs policies for photovoltaics in China from 2011 to 2016. Alrwashdeh [21] designed and implemented a PV system for covering the electricity required of the school of engineering at Mutah University in Jordan. Milanés-Montero et al. [22] assessed the impacts of FiT on the profitability of European PV Companies in the period 2008–2012, using a static linear panel data model. The results showed that FITs had a significant positive influence on the economic profitability of PV companies. In addition, photovoltaic companies with the highest leverage ratios were those with the largest return on investment and encouraged the adoption of PV technology. Smit et al. [23] investigated the issues that affect energy fuel choice, energy bias, and energy switching and energy access as related to the introduction of a renewable energy solution in the Enkanini informal settlement. Lan et al. [24] provided an overview of the current situation of the resources used to generate electricity in Vietnam and the proportions accounted for by sustainable resources, as well as reviewing the existing FITs policies. Some lessons were proposed for Vietnam to encourage renewable energy resource development in the future. Kumar et al. [25] presented significant achievements, prospects, projections, generation of electricity, as well as challenges and investment and employment opportunities due to the development of renewable energy in India. They also identified the various obstacles faced by the renewable sector. Azzuni et al. [26] proposed an energy transition pathway towards a 100% RE system in Jordan by 2050 and analyzed how the energy security can be enhanced. They also investigated the relationship between the decarbonization of the energy system and energy security in a qualitative approach based on a comprehensive framework. Levenda et al. [27] summarized the documented environmental impacts associated with renewable energy technologies included in many renewable energy policies globally They provided a systematic review of the literature assessing renewable energy technologies from the perspective of distributive, procedural, recognition, and capability interpretations of environmental justice. Nair et al. [28] examined the impact of renewable energy in the energy mix for total primary energy supply and electricity generation on Malaysia's short and long-term energy security using system dynamics approach. Three key dimensions of the energy security were considered to extract the indicators and their causal relations with each other. The study conclude that the energy policy structure needs to incline more towards diversification of energy sources to ensure that RE holds a major share in the energy mix as per the current energy policy objectives. Le et al. [29] reviewed the FiTs deployed in different regions of Vietnam for grid-connected solar PV applications and investigated the costs of electricity production from PV systems. The results revealed that the gap between the levelized cost of electricity and the FiTs for solar PV electricity was relatively high, particularly in regions with a lower irradiation potential. Zahedi et al. [30] investigated the environmental effects of using a hybrid wind turbine system, electrolysis system, and solid oxide fuel cell in Qazvin hybrid wind turbine-fuel cell power plant. Zahedi et al. [31] studied the importance of addressing strategic issues in the field of general policies adopted and strategies for the development of renewable energy in Iran. They introduced the types of renewable energy and general principles.

This research proposes an extension to ongoing research by developing a system dynamics model for a quantitative assessment of the impacts of FiTs and subsidy policies on the social acceptability to install PV systems, CO_2 emission reduction, and energy security in the small-scale sector in Jordan.

3. Research Methodology and Development of Dynamic Model

3.1. Research Methodology

System dynamics is a powerful social system, modeling, and computer-aided approach to policy analysis and design [32]. It produces a comprehensive representation of complex social, managerial, economic, and ecological systems, and provides interaction of nonlinear behavior of complex systems over time, information feedback, and circular causality loops. The main elements of the system dynamics method are variables in mathematical equations, presenting stocks and flows as well as causal relations, which are represented using causal feedback loops. Computer simulations follow to examine the real impacts of the social system on a policy in the laboratory to understand the implied causal feedback in the system [13,33]. The research, therefore, utilizes a system dynamics approach to analyze the impacts of renewable energy policies on CO_2 emissions reduction and energy security. The research methodology is depicted in Figure 1.

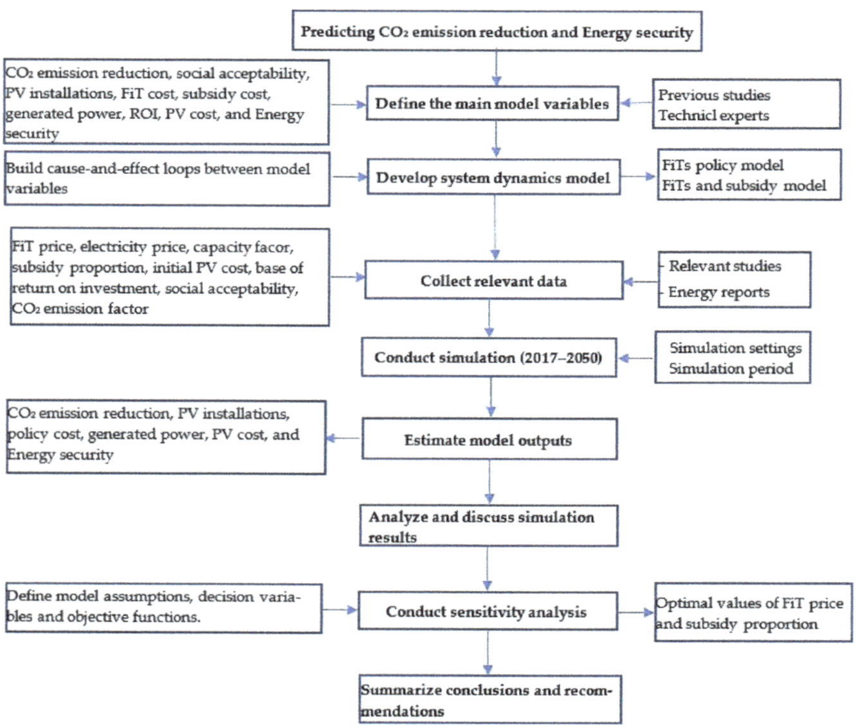

Figure 1. Framework of the research methodology.

Previous studies and technical reports are used to identify the relevant factors in the PV model, including CO_2 emission reduction, social acceptability, PV installations, FiT cost, subsidy cost, generated power, ROI, PV cost, and energy security. Figure 2 represents the overall causal feedback of the proposed PV model. In this figure, as the electricity and FiT price increase, the Return of Investment (ROI) ratio increases, which will consequently enhance the willingness of the public to install PV systems. In return, PV manufacturers will produce and install more solar PV systems and thereby increases the annual power generated and leads to significant mitigation of CO_2 emission.

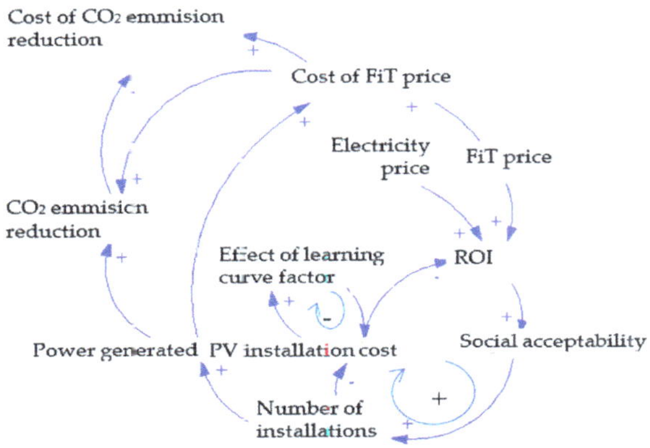

Figure 2. Causal feedback diagram for solar PV system installations.

3.2. Model Development

3.2.1. ROI, Social Acceptability of PV Installations, and FiTs Policy

Developing a strategic plan is necessary to increase local manufacture of PV technologies with the aim of creating a competitive local market, reducing the initial cost of solar power stations through appropriate tax policies, and reducing customs duties on their requirements in the future. Concurrently, it is essential to consider the social aspects that influence social acceptability to install renewable energy technology [34]. As a result, as the social acceptability (SA) to install PV systems increases, the number of PV installations will increase. This leads to manufacturing more PV systems, thereby lowering PV system cost. The annual PV installations is obtained using Equation (1).

$$\text{Annual PV installation} = \text{SA} \times \text{PV installations in the previous year} \quad (1)$$

As a step to enhancing the social acceptability move towards PV installations, governments allow consumers who wish to own renewable energy generation systems on a net measurement basis to cover their consumption by submitting applications directly to the electricity distribution company. Annual FiT costs are paid by the governments to consumers for their surplus generated electricity that was transferred into the grid. It is directly proportional to the amount of surplus from the power generated and FiT price, which is the money paid by the government to electricity supplier for every unit of electricity produced to the grid. As the FiT price increases, the ROI for the public improves, which leads to generate more profits due to increasing the annual PV system installations. Generally, the capacity factor (CF) is the annual daily average time for full loading per KW of a solar PV system. Then, the annual ROI is the multiplication of CF, a proportion of FiT price plus electricity price, and the number of days per year divided by the cost of PV system. The exact value of the base value of the ROI is unknown. Mathematically,

$$\text{ROI} = \text{CF} \times (\text{FiT price} \times 0.1 + \text{electricity price}) \times 360 \times 6.6/\text{Cost of PV system} \quad (2)$$

Figure 3 displays the system dynamics model generated using Powersim Studio 2005 (Service release 6, Powersim Software AS, Nyborg, Norway) that relates social acceptability, FiT price, and PV installations. In this figure, all processes can be characterized in terms of three elements: stocks (rectangular box, accumulated number of installation), information link (curved arrow), and flow (pipe). The difference between stock and flow is that stock is an accumulation of something and a flow is the movement. Stock and flow diagram use to show relationships among variables, which have the potential to change over time. Other symbols are also used; a variable (circle, ROI changes over time) and a constant (electricity price). It is obvious that the ROI is the driver for the growth of PV installations. However, to avoid unaffordable policy costs due to increasing PV installation and continual drop in the cost of the PV system, the ROI is assigned an upper limit.

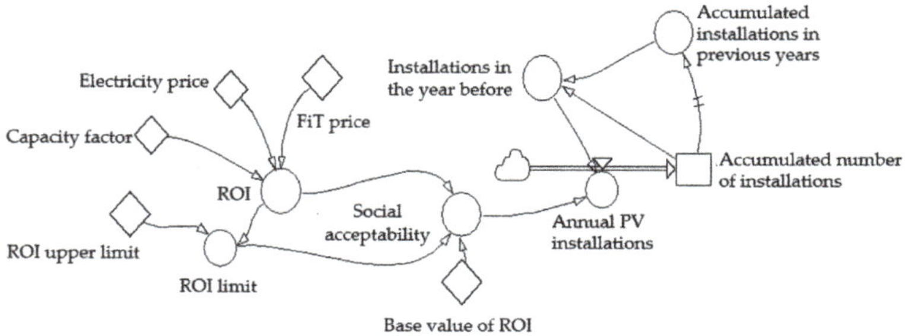

Figure 3. System dynamics of social acceptability and PV installations.

Further, net metering is a billing mechanism that credits costumers for excess self-generated electricity they feedback into the grid. Each month the consumers pay the utility the net difference, on a kWh basis, between what they consumed and what they fed back into the grid. If what they produced exceeds what they consumed, the excess is rolled over to the next month on a kWh basis. The generated electricity goes for direct consumption. If the consumption is more than the electricity generated by the PV system, the utility grid supplies the shortage; however, if the consumption is less than the amount generated, the difference is delivered back to the utility grid and an extra credit of energy will be saved for the customer for the next month until the year ends, then the customer can make a financial settlement with FiT price for each kWh generated, which is called net metering. The annual net metering (ANM) is then calculated using Equation (3).

$$\text{ANM credit (kW/year)} = 0.1 \times \text{Annual PV installations} \times \text{CF} \quad (3)$$

Finally, the implemented FiT policy for encouraging the social acceptability to install PV systems will cost the government. The annual FiT cost is the ANM credit multiplied by the FiT price (USD/kWh). The FiT cost increases directly when the FiT price increases. Therefore, the annual FiT cost is estimated as:

$$\text{Annual FIT cost (USD/year)} = \text{ANM} \times \text{FIT price} \quad (4)$$

3.2.2. Annual PV Installations, Power Generated and CO_2 Emissions

Worldwide, the generated power by PV installations provides affordable electricity and can help stabilize energy prices in the future. Renewable power generation capacity is measured as the maximum net generating capacity of power plants and other installations that use renewable energy sources to produce electricity [35]. Figure 4 depicts the relationships between the annual PV installations, power generated, and CO_2 emissions. The number of PV installations represents the total capacity of PV systems that have been installed in (KW), which increases according to the annual PV installations. The annual power generated (APG) then is the amount of power (kWh) that is generated only by the calculated number of installed PV systems through the previous year. APG can be calculated using Equation (5).

$$APG = \text{Annual PV installations} \times CF \quad (5)$$

In addition to achieving energy security, governments are promoting PV applications to reduce CO_2 emissions resulted from electricity generation. Renewable energy technology is playing a vital role in reducing the quantities of CO_2 emissions released into the atmosphere, and in doing so contributing to a reduction of the greenhouse effect. It is clean and has a much smaller impact on the environment in comparison to conventional energy technologies. The Jordanian government aims to achieve 14% CO_2 emission reduction by 2030. The annual CO_2 emission reduction is directly proportional to CO_2 emissions factor ($kgCO_2/kWh$) and power generated (kWh) by solar PV installations. The CO_2 emissions factor indicates how much CO_2 produced for every one kWh of electricity generated. Typically, when the power generated by PV installations increases, the total power generated from fossil fuel burning decreases and consequently CO_2 emission reduction increases. Mathematically,

$$\text{Annual } CO_2 \text{ emission reduction} = CO_2 \text{ emissions factor} \times APG \quad (6)$$

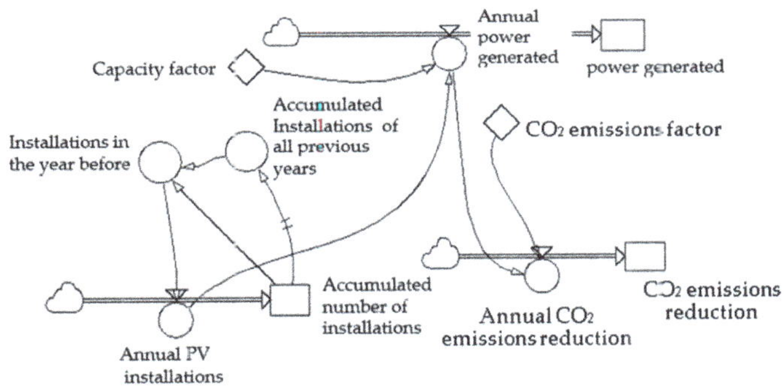

Figure 4. System dynamics of power generated and CO_2 emissions.

3.2.3. Cost of PV System

The cost of PV (CPV) system (USD/kW) includes the cost of the module, invertors, and cables. Figure 5 shows the causal model for CPV causal relations. The accumulated cost reduction is the sum of the annual cost reductions (ACR). In this research, the cost of the PV system is then estimated as the initial PV cost minus the accumulated cost reduction; assuming that the maintenance costs for the PV system are negligible. It was reported that PV cells are significantly more suitable for realizing industry-wide learning to reduce forecasted energy technology costs. Given the annual PV installation and the accumulated number of PV installations (APV), the ACR is then calculated as follows [13]:

$$ACR = CPV \times ((\text{Annual PV installations} + APV)/APV)^{\hat{}-3.32193} \quad (7)$$

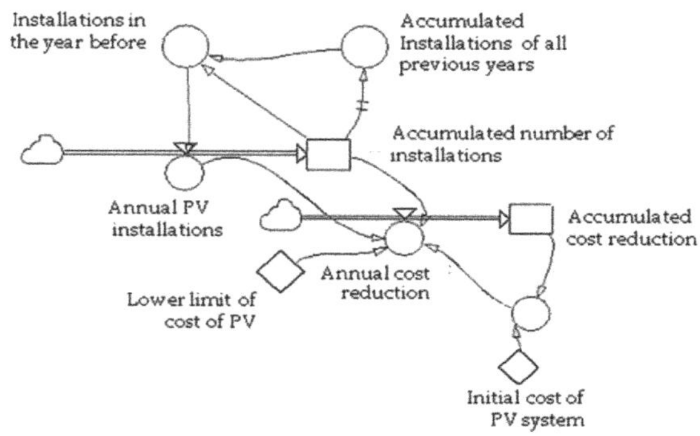

Figure 5. System dynamics model for cost of PV system.

The FiTs policy impacts on energy goals will be analyzed by integrating the presented models. The relevant formulas and parameters of this model for the PV systems in Jordan are summarized in Table 1.

Table 1. The relevant formula of the developed PV model.

Factor	Unit	Equation
Annual CO_2 emission reduction	$KgCO_2$/year	CO_2 emissions factor × Annual power generated
CO_2 emissions factor	$KgCO_2$/kWh	Constant
Annual power generated	kW/year	Annual PV installations × Capacity factor
Capacity factor	-	Constant
Annual PV installation	kW/year	Social acceptability × installation of previous year
Installation previous year	kW/year	Initial value
Social acceptability	-	IF (ROI < constant, ROI/'Base value ROI', 'ROI limit'/'Base value ROI')
Electricity price	USD/kWh	Constant
ROI	-	Capacity factor × (FiT price × 0.1 + electricity price) × 360 × 6.6/((Cost of PV system) − (subsidy cost))
Base value of ROI	-	Constant
Upper limit of ROI	-	Constant according to the upper limit of social acceptability = 0.8724
Feed in tariff price	USD/kWh	Constant
Annual net metering credit	kW/year	0.1 × Annual PV installations × Capacity factor
Annual FIT cost	USD/year	Annual net metering credit × FIT Price
Average cost of CO_2 emissions reduction	USD/kg CO_2	FiT cost/CO_2 emissions
Lower limit of PV cost	USD/kW	Assumed constant
Initial cost of PV system	USD/kW	Constant
Cost of PV system	USD/kW	Initial cost of PV system − cost reduction
Cost reduction of PV system	USD/kWh	IF (cost of PV system > lower limit PV cost, (('Cost of PV system' × ((Annual PV installations + Number of installation)/Accumulated Number of installations)$^{\wedge -3.32193}$)), 0)

4. Model Simulation and Research Results

Small-scale sectors include hospitals, residential buildings, universities, commercial and industrial enterprises, schools, and banks. The most applicable PV system configurations for small-scale sectors in Jordan is on-grid systems, which are connected to the grid via inverters; thus, they do not require batteries. Residential buildings consume energy through their life cycle, including planning, design, construction, and mostly in the building. The demand for electricity has increased in the household sector due to the increased population, high temperatures in summer, and the expansion in using air conditioning units.

In the developed PV model, it is assumed that the social acceptability to install PV systems is directly proportional to the ratio between annual ROI and the base value of ROI, where the upper limit of ROI value is calculated from the historical data and found to be 0.8724 according to the higher willingness that can be achieved. Moreover, the exact value of the base value of ROI is unknown. Hence, the suitable value of this variable is established according to annual ROI. The ROI base value (=0.73) has been decided as the average for the prices shown in Table 2.

Table 2. Electricity prices and ROI base values for small sectors.

Sector	Electricity Price ($)	ROI Base Value with FIT Price
Universities and hospitals	0.36	0.58
Residential (500–1000) kW	0.23	0.381
Residential (1000–1500) kW	0.37	0.59
Companies	0.25	0.41

Further, the capacity factor (=20%) of PV system represents the daily average time for a solar PV system with 1 kW of capacity to reach full loading in one year, it depends on location, position, and method of installation. Furthermore, there are many factors that determine electricity price, such as, the price of power generation, government subsidies, and the infrastructure for its transmission and distribution. The pricing differs not only according to the type of energy generated, but also according to the type of consumer (household, commercial, and industrial). In Jordan, the average of electricity price of 0.268 USD/kwh is considered. Finally, government spending is required to implement the FiT mechanism. The FiT price (=0.19 USD/kWh), and the FiT cost is paid by the government is the sum of the annual FiT cost. It is found that the initial cost of a PV system is 1129 (USD/kW) [2]. According to global studies, the cost of PV system installation will be reduced globally by 40–60% in 2025, so the lower limit for cost of PV systems $600 allocated to be around half of its initial cost, even if the applicable policy mechanism led to higher reduction rate. Accumulated cost reduction gives the output of the extra decrease in cost that is being calculated each simulation year, to compute the cost of a PV system in the current year. When the number of installations increases, it will cause limited reduction in cost of PV installations by the learning curve factor. The initial value of the accumulated PV installation is set to 80,000 kW.

4.1. Impact of FiT Policy on Energy Goals

Simulation of the complete model (Figure 5) was performed using Powersim Studio 5 to examine the FiTs policy on energy goals in the period ranging from 1 January 2016 to 1 January 2050. The obtained simulation results at the FiT price of (=0.19 USD/kWh) showed that:

- The expected accumulated number of installations (Figure 6) increases exponentially through the period until reaching 874 Megawatt (MW) and 5.484 Gigawatt (GW) by the end of years 2025 and 2035, respectively. However, it will reach 67.125 GW at the beginning of year 2050.

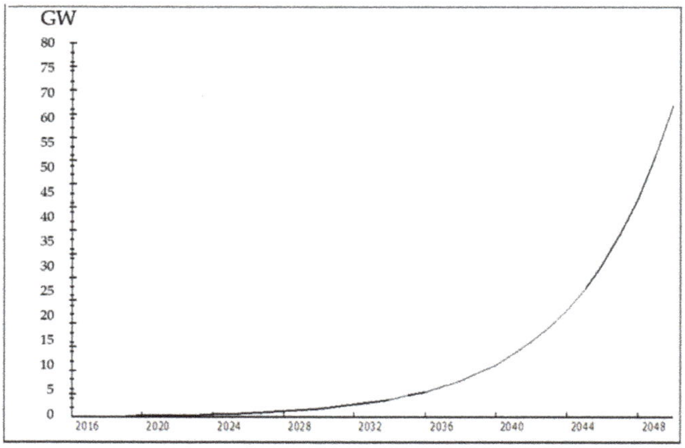

Figure 6. Accumulated number of installations versus time (FiTs policy).

- The expected accumulated power generated as depicted in Figure 7 increases as the accumulated number of PV installations increases. It is assumed that the accumulated power generated initially is zero. The accumulated power generated will reach 1.371 and 9.338 Terawatt hour (TWh) by the end of years 2025 and 2035, respectively. While the power generated by the beginning of year 2050 reaches 115.853 TWh.

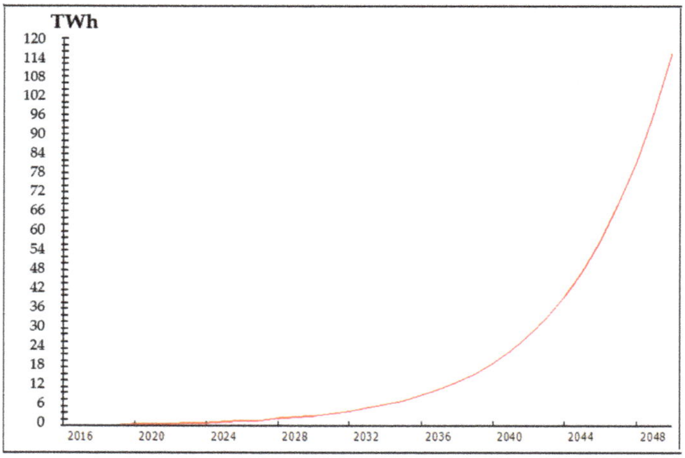

Figure 7. The cumulative power generated versus time (FiTs policy).

- In Jordan, an average value of the CO_2 emission factor is 0.643 Kg CO_2/kWh [4]. The accumulated CO_2 emissions reduction, as displayed in Figure 8, increases as the accumulated number of installations increases. Assuming that the initial CO_2 emission reduction is initially zero, the accumulated CO_2 emissions reduction will reach 0.882 and 6.0 million tons of CO_2 by the end of years 2025 and 2035, respectively. The accumulated CO_2 emission reduction will reach 74.49 million tons CO_2 at the beginning of 2050.

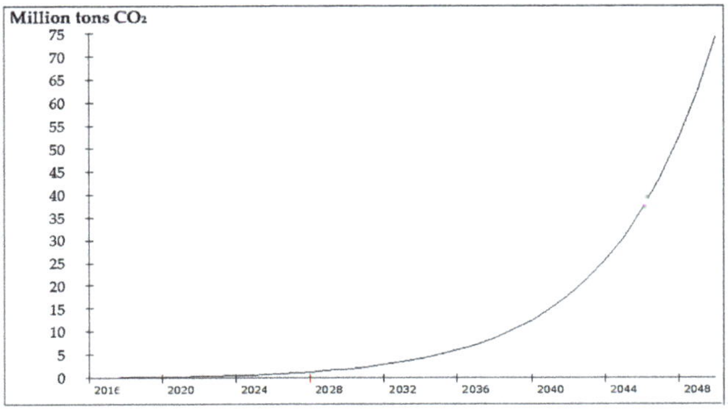

Figure 8. The CO_2 emissions reduction versus time (FiTs policy).

- The cost of a PV system decreases from its initial value of 1129.46 USD/KW till reaching the minimum cost of a PV system, which is 505.5 USD/KW. That is, the PV cost will be reduced to less than half of the initial cost, starting from the beginning of year 2019.
- The social acceptability increased from 0.6 in 2016 to 1.2 starting at the beginning of 2018 and remains constant till 2050. Similarly, the ROI increased from 0.44 in 2016 to 0.98 in 2018 and remains the same till 2050.
- The accumulated FiT cost increases as shown in Figure 9 to \$26.05 and \$177.4 million by the end of 2025 and 2035, respectively, while the accumulated FiT cost is expected to reach \$2.2 billion at the beginning of 2050.

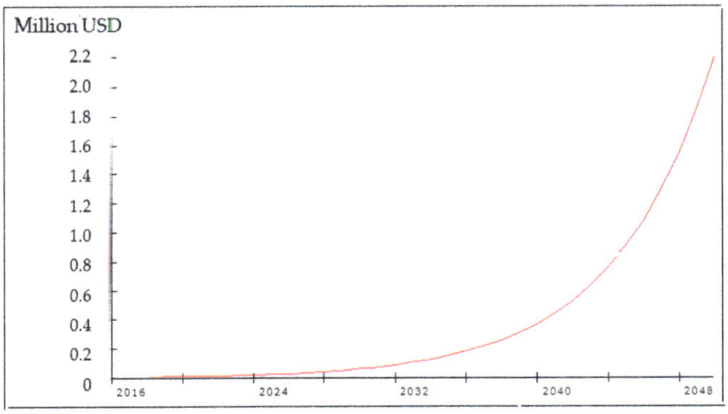

Figure 9. The FiT cost versus time.

4.2. Impacts of Ombination of FiT and Subsidy Policies on Energy Goals

Generally, the subsidy cost is grants and rebates paid by the government to cover a proportion of the investment in PV technologies to reduce the cost of PV installations. This cost is paid to consumers to promote PV installations and is implied at the beginning of each installation as a proportion of the installation cost (PV cost). Figure 10 depicts subsidy and maintenance cost relations with PV cost. The subsidy cost equals the proportion of subsidy multiplied by PV cost. In this research, a subsidy proportion is equal to 15%. The percentage of 15% was chosen after a study of the effect on different common proportions of subsidy on CO_2 emissions reduction and polices cost, which are the key factors influencing the choice of proper subsidy proportion. The lower PV price is assumed to be 550 USD. The accumulated subsidy cost is the cumulative annual subsidy cost (ASC) through the simulation period paid by the government. The ASC equals the subsidy cost multiplied to the total number of PV installations. Finally, the total of policies costs is the summation of the FiTs and subsidy policy costs. Mathematically,

$$\text{Subsidy cost} = \text{PV cost} \times \text{subsidy proportion} \times \text{annual number of PV installations} \quad (8)$$

$$\text{ASC} = \text{PV cost} \times \text{subsidy proportion} \times \text{annual number of PV installations} \quad (9)$$

$$\text{Annual policies cost} = \text{ASC} + \text{Annual FiTs cost} \quad (10)$$

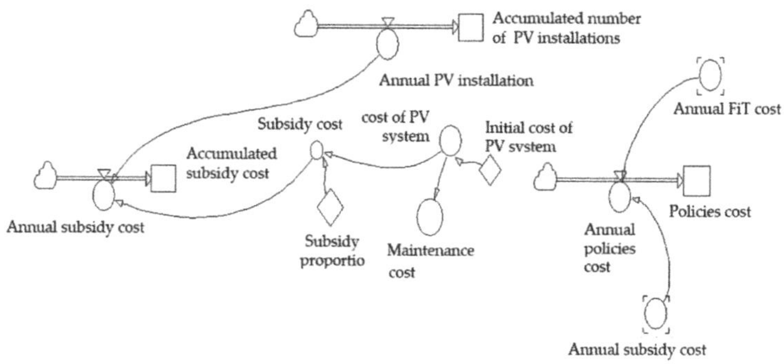

Figure 10. Subsidy and maintenance cost relations.

The main part of the PV system is the inverter, which is replaced within the 25-year life of the system. This study assumes that each solar PV system has an operational lifetime of 25 years, and the maintenance cost is 0.06 of the cost of solar PV system. When the combination of subsidy and FiTs policies are applied, the total number of solar PV installations grows and thereby the total of policy costs of promoting solar PV installation increases in a huge manner. As a result, a tremendous increase in the ROI will be generated to the public. The ROI under both policies is calculated as:

$$\text{ROI} = \text{CF} \times (\text{FiT price} \times 0.1 + \text{electricity price}) \times 360 \times 6.6/(\text{PV cost} - \text{subsidy cost}) \quad (11)$$

Further, applying the combination of subsidy policy results in larger reductions in CO_2 emissions and achieves better goals in energy dependency than FiTs policy. Similar simulation was run for the system dynamics model of the combination of subsidy and FiTs policies and the obtained results showed that:

- The accumulated number of installations versus time is shown in Figure 11, which reveals that about 1.12 and 7.19 GW are expected to be installed by the end of 2025 and 2035, respectively. While the accumulated number of installations will reach 88.38 GW at the beginning of year 2050.

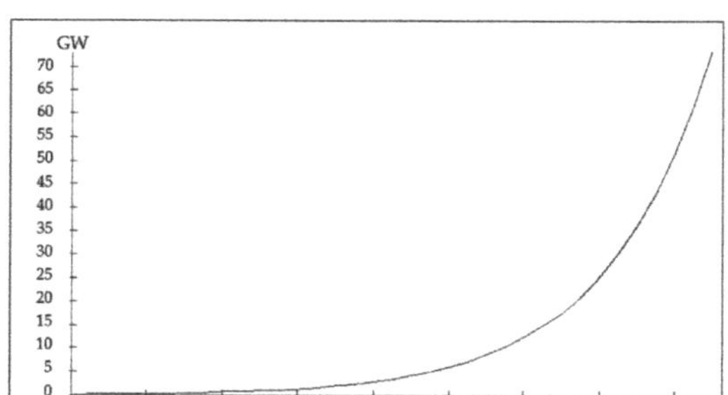

Figure 11. Accumulated number of installations versus time for combined policies.

- Figure 12 depicts the accumulated generated power versus time, where it is noted that the accumulated generated power (GW) increases nonlinearly by time passes. For example, the accumulated power generated reaches 1.794 and 12.288 TWh by the end of 2025 and 2035, respectively. While the accumulated generated power reaches 152.588 TWh at the beginning of 2050.

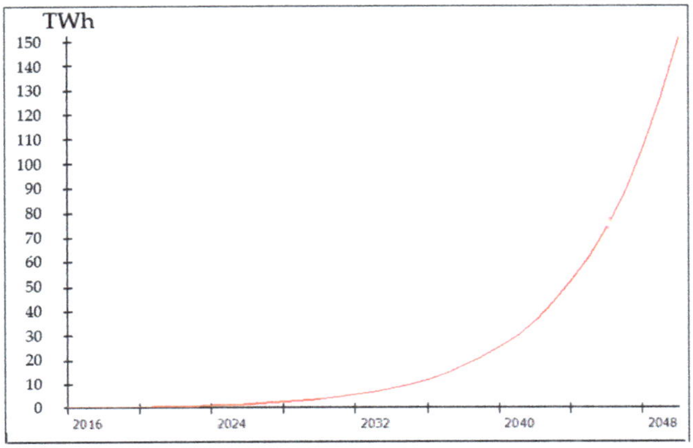

Figure 12. The cumulative power generated versus time for combined policies.

- The accumulated CO_2 emissions reduction is illustrated in Figure 13, which shows that the accumulated CO_2 emissions reduction increases by time. It is expected that the accumulated CO_2 emissions reduction by the end of 2025 and 2035 will reach 1.53 and 7.90 million tons, respectively. While the accumulated CO_2 emission reduction reaches 98.114 million tons CO_2 at the beginning of 2050.

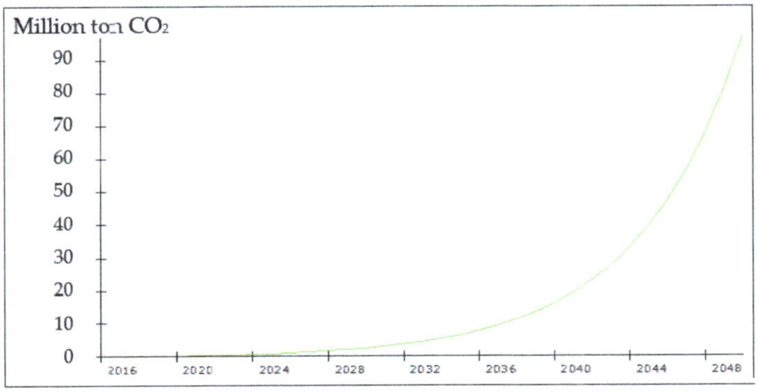

Figure 13. The CO_2 emissions reduction versus time for combined policies.

- The cost of a PV system under the combination of subsidy and FiTs policies decreases from its initial value of 1129.46 USD/KW till reaching the minimum cost of a PV system (=353.23 USD/KW) in 2019, and then it remains constant till 2050.
- The social acceptability to install PV systems increased from 0.71 in 2016 to 1.2 starting at the beginning of 2018 and then remains constant till 2050. Similarly, the ROI increased from 0.52 in 2016 to 1.65 in 2019 and then remains at a value of 1.65 till 2050.
- The accumulated total of policy costs is displayed in Figure 14, where it is noted that the total of policy costs increases by time. For example, about $102.011 and

$623.148 million $ will be incurred by the end of 2025 and 2035, respectively. While the accumulated total cost reaches $7.590 billion at the beginning of 2050.

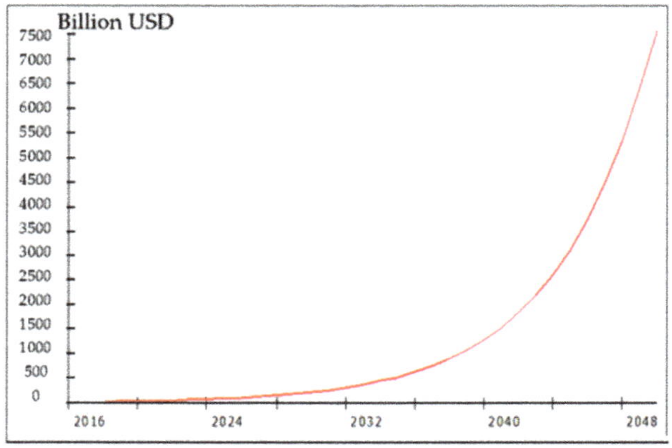

Figure 14. The cumulative policies cost versus time for combined policies.

4.3. Ecological Footprint and Energy Security

Further analyses were conducted to measure the ecological footprint and energy security policy. The obtained results are discussed as follows.

4.3.1. Ecological Footprint

Figure 15 shows the ecological footprint model. This study assumes that each installation of a PV system produces three small quantities including waste from PV energy consumption, PV installations, and PV energy operation and maintenance. Typically, the increase in the number of PV installations leads to an increase in the total volume of waste and thereby directly increases the ecological footprint. When using a clean solar energy source, however, less carbon emissions will be released into the atmosphere, which consequently decreases the ecological footprint. Simulation was then run to estimate the predicted energy waste, and the results for the energy waste and ecological footprint are depicted in Figures 16 and 17, respectively.

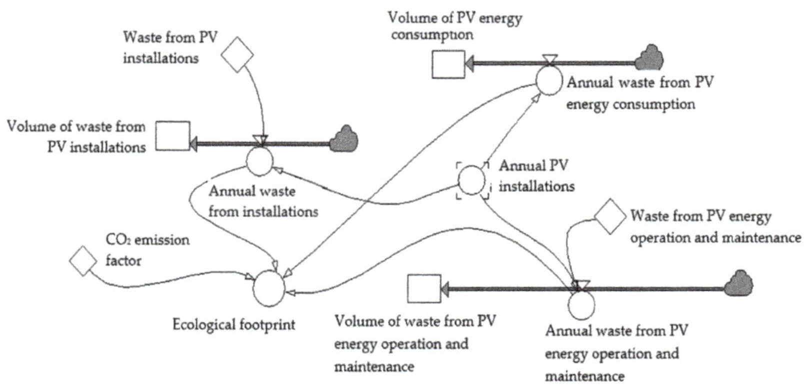

Figure 15. Ecological footprints and CO_2 emissions relations.

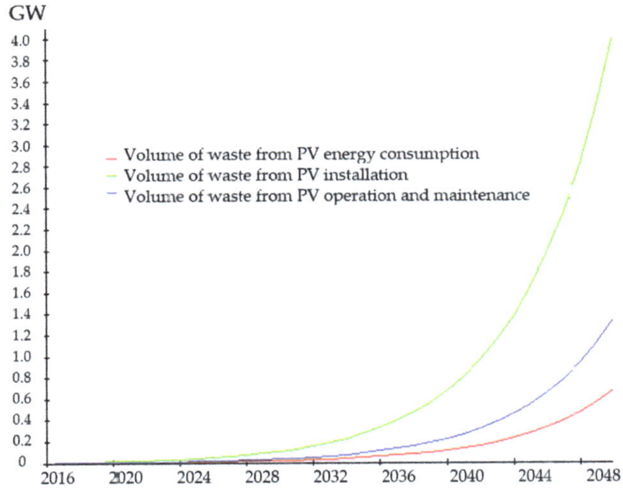

Figure 16. Energy waste versus time.

In Figure 16, the cumulative volume of PV energy consumption increases by time. The cumulative waste volumes are 10,368 and 61,603 KW by the end of 2025 and 2035, respectively. While it reaches 746,622 KW at the beginning of 2050. The cumulative waste from installation increases as time increases to 47.604 and 324.23 MW by the end of 2025 and 2035, respectively, while this cumulative waste reaches 4.023 GW at the beginning of 2050. Finally, the cumulative waste from PV energy operation and maintenance increases to 15.868 and 1.08 GW by the end of 2025 and 2035, respectively. The cumulative amount of this waste reaches 1.34 GW at the beginning of 2050.

In Figure 17, the ecological footprint increases as time increases, where it is expected to reach about 10.37 and 61.60-ton CO_2/KWh by the end of 2025 and 2035, respectively. This waste reaches 746.62 ton CO_2/KWh at the beginning of 2050.

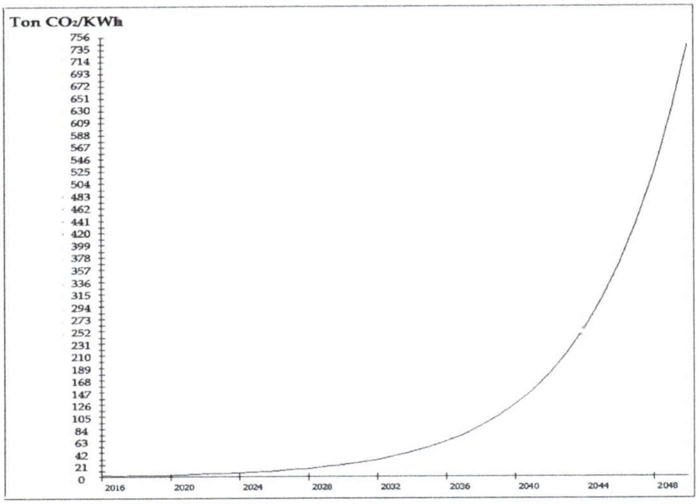

Figure 17. Ecological footprint versus time.

4.3.2. Energy Security Analysis

Energy security is defined as the availability of natural resources to consume energy throughout the year. Energy security analysis is a strong tool for policymakers. Energy reports revealed that the annual electricity demand increased from 19,390 to 23,063 GWh for the years 2016 and 2020, respectively. Jordan's demand for energy is growing at a rate of 3% annually [3–5]. The total amount of PV electricity that was generated in the year 2020 amounts to around 2250 GWh, which covered around 10% of the total electricity demand of Jordan at the said year. Small-scale sectors represent the largest percentage of sectorial consumption of electricity during 2012 to 2016 of about 60%, with an annual electricity demand growth of 7.4% [3–5]. The solar photovoltaic has a share of 92% of renewable electricity generation (when the cost of PV is US$1129.5 per kW and global solar radiation is 0.643 kg CO_2/kWh).

Figure 18 displays the relations between the level of energy security, power generated, and electricity demand. The annual electricity demand refers to the maximum amount of electrical energy that has been consumed at a given time. In this study, the level of energy security is calculated by subtracting the electricity demand from the power generated capacity at each year. When the power generated by the PV system increases each year, the level of energy security increases directly and the energy reserve become larger; unlike the electricity demand, which reduces the level of energy security as the demand increases. Table 3 lists the level of energy security under both energy policies.

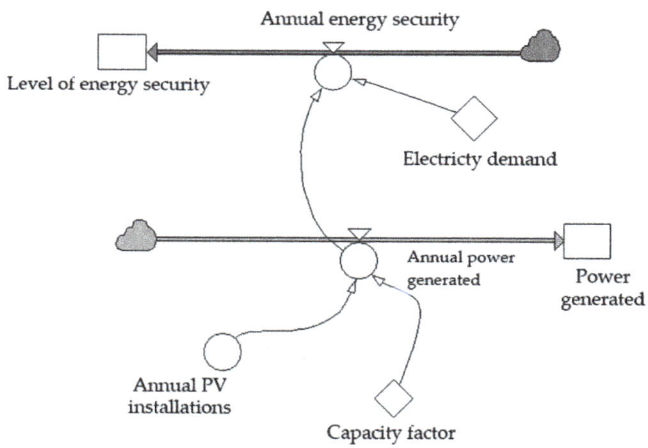

Figure 18. Developed model for level of energy security, power generated, and electricity demand.

Table 3. The estimated energy security levels.

End of Year	Electricity Demand (TWh)	Power Generated (TWh)		Energy Security Level (TWh)	
		FiTs Policy	FiT and Subsidy	FiTs Policy	FiT and Subsidy
2020	23.50	0.42	0.54	−23.08	−22.96
2025	30.00	1.37	1.79	−28.63	−28.21
2030	38.30	3.69	4.85	−34.61	−33.45
2035	48.90	9.34	12.29	−39.56	−36.61
2040	62.40	23.11	30.43	−39.29	−31.97
2045	79.60	56.68	74.64	−22.92	−4.96
2050	101.60	115.85	152.59	14.25	50.99

In Table 3, the expected energy surpluses in 2050 are 14.25 and 50.99 KWh under FiTs and subsidy policies, respectively. This surplus can be utilized to fulfill the demand for renewable energy in other sectors, such as transport. This leads to the conclusion that Jordan can achieve a 100% solar PV system in 2050 by implementing the FiTs and/or subsidy policies to increase the public acceptability to install PV system, and such a transition will positively enhance the level of energy security and significantly prevent the negative impacts of CO_2 emissions on the environment.

4.4. Sensitivity Analysis

Sensitivity analyses were conducted to analyze the possible uncertainties in developed model parameters for FiTs and subsidy polices. For the FiTs model, it is assumed that the capacity factor (mean = 0.2 and standard deviation = 0.02), electricity price (mean = $0.27 and standard deviation = 0.03), and lower limit of PV system cost (mean = $600/KWh and standard deviation = 60) are normally distributed. The decision variables are the FiT price (value ranges between 0.17 to 0.21) and subsidy proportion. Similar sensitivity analysis was conducted to determine the optimal subsidy proportion. It is assumed that the capacity factor (mean = 0.2 and standard deviation = 0.02), electricity price (mean = $0.27 and standard deviation = 0.03), and the lower limit of PV system cost (mean = $550/KWh and standard deviation = 50) and the decision variables are the FiT price (value ranges between 0.17 to 0.21) and the subsidy proportion (value ranges between 0.14 to 0.17). It is found that the optimal FiT price is $0.21, at which the FiTs policy, the accumulated PV installations, CO_2 emissions reduction, and power generated are 71.8 GW, 82.5 million ton CO_2, and 128.3 TWh, respectively. On the other hand, for the combined policies, FiTs and subsidy policies, the results showed that the optimal FiT price subsidy proportions are $0.21 and 17%, respectively. At these optimal values, the accumulated PV installations, CO_2 emissions reduction, and power generated are 97.8 GW, 112.4 million tons CO_2, and 174.8 TWh, respectively. It is clear that applying the combined FiT and subsidy policies results in more PV installations, energy generated, and CO_2 emissions reduction. Nevertheless, the FiT policy results in securing future energy at a significantly lower policy cost (=2.68 billion USD) compared to the cost of combined policies (=10.63 billion USD). Decision-makers can utilize these results to decide the applicable policy to be adopted in promoting PV installations depending on the available budget.

5. Conclusions

This research developed a valuable quantitative assessment approach to examine the impacts of PV policies on social acceptability, number of PV installations, power generated, and CO_2 emission reduction in the small-scale sector in Jordan. A system dynamics simulation was conducted for the time span 2016 to 2050, which revealed that under the FiTs and combined policies, the accumulated number of installations (cumulative power generated) could reach 67.125 (115.853 TWh) and 88.38 GW 115.853 (152.588 TWh), respectively. Further, the cost of a PV system decreases from its initial value of 1129.46 USD/KW till reaching 505.5 and 353.23 USD/KW, respectively. The expected cumulative CO_2 emission reductions reach 74.49 and 98.114 million tons CO_2 under the FiTs and subsidy policies, respectively. Finally, the cumulative FiTs and combined policies cost 2.2 and 7.590 billion USD, respectively, while ROI increases to 0.8 and 1.65, respectively. Sensitivity analyses followed to determine the optimal FiT price and subsidy proportion, where the obtained optimal FiT price and subsidy proportion were 0.21 and 17%, respectively. In conclusion, under the stated assumptions implementing PV models, adoption of renewable energy policies in Jordan will result in a high level of energy security in 2050. Policy makers in the energy sector can utilize the proposed PV models as a valuable assessment tool for deciding the applicable energy policies and determining the amount of investment that will be required to achieve economic sustainable growth in the energy sector. Future research may consider developing an interpretive structural model to identify the factors that affect social acceptability and the barriers to use PV installations.

Author Contributions: A.A.-R. conceived the study and was responsible for data collection and analysis, system dynamics model development, simulation results under FiT and subsidy policies, and results interpretation. N.L. was responsible for review of the presented literature, data analysis. All authors have read and agreed to the published version of the manuscript.

Funding: This research was supported by the University of Jordan, Research No. 2471.

Informed Consent Statement: Not applicable.

Data Availability Statement: Not applicable.

Conflicts of Interest: The authors declare no conflict of interest.

References

1. Tziogas, C.; Georgiadis, P. Investigating the causalities for cleaner and affordable electricity production mix: A System dynamics methodological approach. *Chem. Eng. Trans.* **2013**, *35*, 461–541.
2. Abu Hamed, T.; Bressler, L. Energy security in Israel and Jordan: The role of renewable energy sources. *Renew. Energy* **2019**, *135*, 378–389. [CrossRef]
3. Ministry of Energy and Mineral Resources (MEMR). *Annual Report 2018*; MEMR: Amman, Jordan, 2018. Available online: https://www.memr.gov.jo/ebv4.0/root_storage/ar/eb_list_page/56dcb683-2146-4dfd-8a15-b0ce6904f501.pdf (accessed on 10 June 2020).
4. Ministry of Energy and Mineral Resources (MEMR). *Energy 2019—Facts & Figures*; MEMR: Amman, Jordan, 2019. Available online: https://www.memr.gov.jo/ebv4.0/root_storage/en/eb_list_page/bruchure_2019.pdf (accessed on 10 June 2020).
5. National Electric Power Company (NEPCO), Annual Reports, Amman, Jordan. 2018. Available online: https://www.nepco.com.jo/store/docs/web/2018_en.pdf (accessed on 10 June 2020).
6. Abu-Rumman, G.; Khdair, A.I.; Khdair, S.I. Current status and future investment potential in renewable energy in Jordan: An overview. *Heliyon* **2020**, *6*, e03346. [CrossRef] [PubMed]
7. International Renewable Energy Agency (IRENA). Renewable Capacity Statistics 2021. Available online: https://www.irena.org/publications/2021/March/Renewable-Capacity-Statistics-2021 (accessed on 10 June 2020).
8. *International Renewable Energy Agency Report*; International Renewable Energy Agency: Bangkok, Thailand, 2016.
9. Jabera, S.; Abul Hawa, A. Optimal Design of PV System in Passive Residential Building in Mediterranean Climate. *Jordan J. Mech. Ind. Eng.* **2016**, *10*, 39–49.
10. Hussein, N. Greenhouse Gas Emissions Reduction Potential of Jordan's Utility Scale Wind and Solar Projects. *Jordan J. Mech. Ind. Eng.* **2016**, *10*, 199–203.
11. Alawneh, F.; Albatayneh, A.; Al-Addous, M.; Al-Khasawneh, Y.; Dalalah, Z. Solar Photovoltaic (PV) Power Systems in Jordan: The Past, the Present and the Future? *Adv. Sci. Technol. Innov.* **2019**, 155–159. [CrossRef]
12. Trappey, A.J.C.; Trappey, C.V.; Lin, G.Y.P.; Chang, Y.-S. The analysis of renewable energy policies for the Taiwan Penghu island administrative region. *Renew. Sustain. Energy Rev.* **2012**, *16*, 958–965. [CrossRef]
13. Hsu, C.-W. Using a system dynamics model to assess the effects of capital subsidies and feed-in tariffs on solar PV installations. *Appl. Energy* **2012**, *100*, 205–217. [CrossRef]
14. Movilla, S.; Miguel, L.J.; Blázquez, L.F. A system dynamics approach for the photovoltaic energy market in Spain. *Energy Policy* **2013**, *60*, 142–154. [CrossRef]
15. Jeon, C.; Shin, J. Long-term renewable energy technology valuation using system dynamics and Monte Carlo simulation: Photovoltaic technology case. *Energy* **2014**, *66*, 447–457. [CrossRef]
16. Aslani, A.; Helo, P.; Naaranoja, M. Role of renewable energy policies in energy dependency in Finland: System dynamics approach. *Appl. Energy* **2014**, *113*, 758–765. [CrossRef]
17. Aslani, A.; Wong, K.-F.V. Analysis of renewable energy development to power generation in the United States. *Renew. Energy* **2014**, *63*, 153–161. [CrossRef]
18. Ahmad, S.; Tahar, R.; Muhammad-Sukki, M.F.; Munir, A.B.; Abdul Rahim, R. Role of feed-in tariff policy in promoting solar photovoltaic investments in Malaysia: A system dynamics approach. *Energy* **2015**, *84*, 808–815. [CrossRef]
19. Radomes, A.A., Jr.; Arango, S. Renewable energy technology diffusion: An analysis of photovoltaic-system support schemes in Medellín, Colombia. *J. Clean. Prod.* **2015**, *92*, 152–161. [CrossRef]
20. Ye, L.C.; Rodrigues, J.F.; Lin, H.X. Analysis of feed-in tariff policies for solar photovoltaic in China 2011–2016. *Appl. Energy* **2017**, *203*, 496–505. [CrossRef]
21. Alrwashdeh, S. Assessment of the energy production from PV racks based on using different Solar canopy form factors in Amman-Jordan. *Int. J. Eng. Res. Technol.* **2018**, *11*, 1595–1603.
22. Milanés-Montero, P.; Arroyo-Farrona, A.; Pérez-Calderón, E. Assessment of the influence of feed-in tariffs on the profitability of European photovoltaic companies. *Sustainability* **2018**, *10*, 3427. [CrossRef]
23. Smit, S.; Musango, J.K.; Brent, A.C. Understanding electricity legitimacy dynamics in an urban informal settlement in South Africa: A Community Based system dynamics approach. *Energy Sustain. Dev.* **2019**, *49*, 39–52. [CrossRef]
24. Lan, T.T.; Techato, K.; Jirakiattikul, S. The Challenge of Feed-In-Tariff (FIT) Policies Applied to the Development of Electricity from Sustainable Resources–Lessons for Vietnam. *Int. Energy J.* **2019**, *19*, 199–212.

25. Kumar, J.C.R.; Majid, M.A. Renewable energy for sustainable development in India: Current status, future prospects, challenges, employment, and investment opportunities. *Energy Sustain. Soc.* **2020**, *10*, 1–36. [CrossRef]
26. Azzuni, A.; Aghahosseini, A.; Ram, M.; Bogdanov, D.; Caldera, U.; Breyer, C. Energy Security Analysis for a 100% Renewable Energy Transition in Jordan by 2050. *Sustainability* **2020**, *12*, 4921. [CrossRef]
27. Levenda, A.M.; Behrsin, I.; Disano, F. Renewable energy for whom? A global systematic review of the environmental justice implications of renewable energy technologies. *Energy Res. Soc. Sci.* **2021**, *71*, 101837. [CrossRef]
28. Nair, K.; Shadman, S.; Chin, C.M.M.; Sakundarini, N.; Yap, E.H.; Koyande, A. Developing a system dynamics model to study the impact of renewable energy in the short- and long-term energy security. *Mater. Sci. Energy Technol.* **2021**, *4*, 391–397. [CrossRef]
29. Le, H.T.-T.; Sanseverino, E.R.; Nguyen, D.-Q.; Di Silvestre, M.L.; Favuzza, S.; Pham, M.-H. Critical Assessment of Feed-In Tariffs and Solar Photovoltaic Development in Vietnam. *Energies* **2022**, *15*, 556. [CrossRef]
30. Zahedi, R.; Ahmadi, A.; Sadeh, M. Investigation of the load management and environmental impact of the hybrid cogeneration of the wind power plant and fuel cell. *Energy Rep.* **2021**, *7*, 2930–2939. [CrossRef]
31. Zahedi, R.; Zahedi, A.; Ahmadi, A. Strategic Study for Renewable Energy Policy, Optimizations and Sustainability in Iran. *Sustainability* **2022**, *14*, 2418. [CrossRef]
32. Daneshgar, S.; Zahedi, R. Investigating the hydropower plants production and profitability using system dynamics approach. *J. Energy Storage* **2022**, *46*, 103919. [CrossRef]
33. Al-Refaie, A.; Abdelrahim, D.Y. A system dynamics model for green logistics in a supply chain of multiple suppliers, retailers and markets. *Int. J. Bus. Perform. Supply Chain Model* **2021**, *12*, 259–281. [CrossRef]
34. Moula, M.M.E.; Maula, J.; Hamdy, M.; Fang, T.; Jung, N.; Lahdelma, R. Researching social acceptability of renewable energy technologies in Finland. *Int. J. Sustain. Built Environ.* **2013**, *2*, 89–98. [CrossRef]
35. Hou, G.; Sun, H.; Jiang, Z.; Pan, Z.; Wang, Y.; Zhang, X.; Zhao, Y.; Yao, Q. Life cycle assessment of grid-connected photovoltaic power generation from crystalline silicon solar modules in China. *Appl. Energy* **2016**, *164*, 882–890. [CrossRef]

sustainability

Article

Multifunctional Design of Vibrational Energy Harvesters in a Bridge Structure

Lissette Fernandez * and Steven F. Wojtkiewicz

Department of Civil & Environmental Engineering, Clarkson University, Potsdam, NY 13699, USA
* Correspondence: lfernand@clarkson.edu; Tel.: +1-315-268-6491

Abstract: This paper aims to integrate vibrational energy harvesters into bridge structures in a holistic fashion that can lessen energy demands for safe bridge operation thus potentially increasing their sustainability. Computationally efficient methodologies, that target the locality of the connection of the harvesters, are utilized to determine optimal harvester frequencies that maximize the total power generation of installed vibrational energy harvesters. Previous findings from the authors indicate that a distributed configuration of harvesters can generate equal or more power than one traditional large harvester when attached to a building structure with total equivalent harvester mass. This paper investigates whether those findings also apply to bridge structures. Results from a cable-stayed bridge model equipped with two or more harvesters along its deck are presented and discussed. Distributed gardens are investigated as a means to integrate the harvester mass with the pre-existing bridge structure. It is found that an equivalent, slightly larger, amount of power is captured by the distributed garden design compared to a single pair of large harvesters placed near the center of the bridge. This performance is very promising as the distributed garden design would enable the enhancement of the structure's aesthetics while also potentially creating ecological and environmental benefits.

Keywords: vibration; energy harvesting; green gardens; multifunctionality

Citation: Fernandez, L.; Wojtkiewicz, S.F. Multifunctional Design of Vibrational Energy Harvesters in a Bridge Structure. *Sustainability* **2022**, *14*, 16540. https://doi.org/10.3390/su142416540

Academic Editors: Oleg Kapliński, Lei Dong, Agata Bonenberg, Wojciech Bonenberg and Marinella Silvana Giunta

Received: 30 June 2022
Accepted: 6 October 2022
Published: 9 December 2022

1. Introduction

Vibrations are present in nature and in all man-made infrastructures. Civil infrastructure, such as buildings and bridges, is constantly exposed to a variety of vibration input sources, including ambient vibrations [1–8], human and vehicular traffic [4,9,10]. These motion-based vibrations can be more significant under seismic activity or earthquakes [8,11]. The vibrations caused by these energy sources are very attractive targets for infrastructure scale vibrational energy harvesters with the dual purpose of alleviating the deleterious effects caused by the vibrations and also as a source of energy capture and reuse.

Modern and economically driven architecture has based its structural design and construction techniques on the implementation of taller and lighter structures achieved through the use of less materials and with the aid of new technologies that target greater efficiency [12]. These construction techniques, that develop lighter structures, have created a secondary, but still of great importance, problem: human discomfort. The implementation of this harvesting technology in buildings and bridges, can both improve the comfort of structural occupants, by mitigating structural vibrations, while harvesting and storing the energy in a reusable electrical form.

In general, civil infrastructure systems are exposed to environmental hazards, fatigue, material aging and earthquakes; this is particularly so for bridges, which continue to degrade with the passage of time [13]. These dynamic loadings have the potential to offer sustainable energy sources that can be harvested to power electronic devices.

Bridges have become primary targets of structural health monitoring implementation due to their constant deterioration and their particular susceptibility to damage from their exposure to intermittent dynamic loadings [14]. Recently, wireless sensor node (WSN) technology has been utilized for health monitoring applications [15]. WSNs have many practical uses in environmental sensing and monitoring such as detecting vibration levels in structures [16]. These devices are made of different components as presented in [17] and shown in Figure 1.

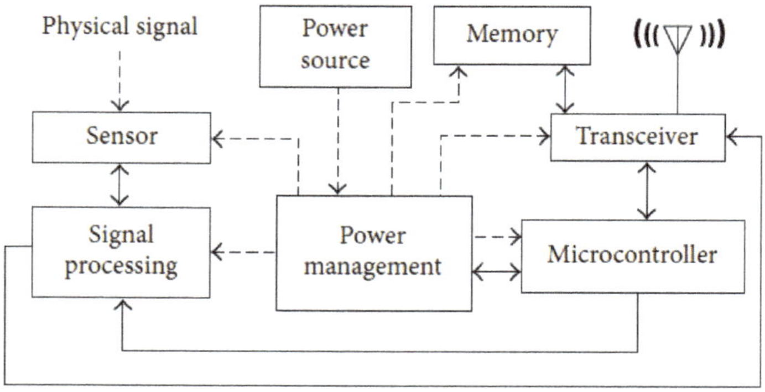

Figure 1. Architecture of wireless sensor nodes [17].

WSNs utilize battery power making their implementation challenging when they are applied at the infrastructure scale, particularly when they are spread over a large area. For this reason, their implementation in structural health monitoring applications for bridges is based on energy harvesting principles that create autonomy and self-powered capabilities for these devices [16]. Commercially available WSNs require 0.09 to 128 mW [18] of power for data transmission, however, this power requirement is on a sharp decline due to the continuing development of components. Because of this low power requirement, WSNs are very attractive for implementation in bridges, which are regularly exposed to ambient vibrations from traffic and wind loads that can, on average, generate power on the order of hundreds of microwatts with proof masses on the order of tens of grams. However, when devices require larger power, large proof masses are also required since these are proportional to the energy harvester's output power capabilities [14].

Many studies on the application of vibrational energy harvesting to bridge structures have been conducted, considering mostly vibration based electromagnetic [19–21] and piezoelectric [22] harvester types. Recently, some researchers have also investigated the effects of multifunctional composite materials combined with energy harvesting for various applications including health monitoring in bridge infrastructure [23]. However, most of these studies have been limited to very small harvesters with proof masses on the order of tens of grams. The literature reports piezoelectric energy harvesters for bridge applications with average power generation capabilities in the ranges of 30 μW to 10 mW for acceleration excitations on the order of 0.1 g to 4.4 g [1–3,24]. Bridge electromagnetic harvesters have also been reported to provide average power values in the range of 2 μW to 26 mW under accelerations from 8 mg to 3 mg [4–7].

The ratio of harvester mass to structural mass has oftentimes been found to be the most important parameter to consider when determining the efficiency of vibrational energy harvesters when augmenting a real, large-scale structure, with a larger mass ratio dictating better performance [25]. However, as civil structures often possess masses in the order of thousands of tons, a multifunctional harvester mass should be contemplated to avoid adding huge inert masses to a building or bridge.

Motivated by these considerations, this paper investigates the impact of multiple, smaller vibrational energy harvesters distributed throughout a cable-stayed bridge versus the traditional design of a single vibrational energy harvester attached to the structure.

Also, to account for the problem of the large among of mass required, a multifunctional harvester mass will be considered.

In order to accomplish that, first the locality of the harvester relative to the cable-stayed bridge example is exploited to reduce the complexity of the system in order to perform design optimization to maximize power generation. This is achieved by implementing the computationally efficient methodology described in [25] to optimally design the frequencies of the devices augmenting the bridge example shown.

The green garden concept is incorporated as the potential multifunctional design for the vehicular cable-stayed bridge. This multifunctional feature can provide both environmental and aesthetic advantages. In the particular case of vehicular bridge structures, green gardens can help decrease the environmental pollution by absorbing carbon dioxide emissions as typical passenger vehicles emit about 4.6 metric tons of carbon dioxide per year [26]. This multifunctional design solution would also be beneficial for the ecology since they can facilitate habitat provisions for birds and other species [27].

Incorporating the greenery concept within the structural design can alleviate the climatic issue created by solar gain, since plants have the ability to absorb heat and cool the environment [12]. The implementation of the garden concept into the harvester design will have the added benefit of increasing the environmental comfort of the structure.

More specifically, implementing the green garden concept as the multifunctional harvester's mass in the example presented will guarantee significant power generation under different levels of vibration-based input loads considering that the proof mass in this case would be on the order of thousands of kilograms. This potential amount of power will not only guarantee implementation of more complex structural health monitoring technologies as schematically shown in Figure 2, but will also serve as a potential source to the power grid, lighting, traffic devices and more complex wireless sensing and communication systems.

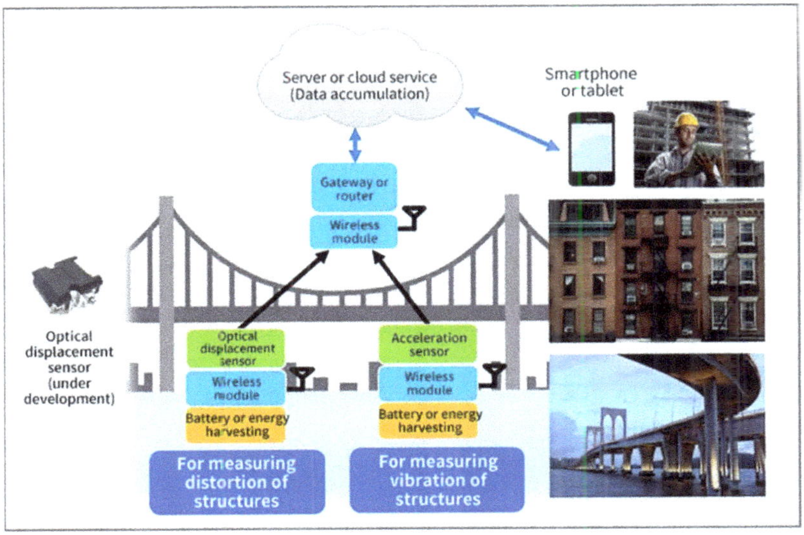

Figure 2. Schematic bridge health monitoring representation [28].

2. Problem Formulation

The efficient computational methodology [29] adapted and described in [25] to perform analysis and design optimization of electromagnetic vibrational energy harvesters at the infrastructure scale was applied to design and optimize two and more devices attached to a cable-stayed bridge example. The equations of motion for the nominal (without harvesters) and modified (with harvesters) systems are defined by (1) and (2) in [25], with vectors $\mathbf{u}_b$ and $\mathbf{U}_b$, of size $n_b \times 1$, representing the system's response in relative coordinates without and with the connection of the devices, respectively. $\mathbf{u}_h$ and $\mathbf{U}_h$, of size $n_h \times 1$, represent the response of the harvester(s), in the two cases.

The system's degrees of freedom are defined by $n_{dof} = n_b + n_h$, where n_b is the number of degrees of freedom of the superstructure and n_h is the number of harvesters attached.

The electrical properties of the harvesters are defined by the diagonal $n_h \times n_h$ matrix $\mathbf{B}_L$ containing the magnetic flux of each harvester and $\mathbf{B}_{LL}$ and $\mathbf{R}_{LL}$, which are also diagonal matrices of size $n_h \times n_h$ containing the ratio of individual magnetic flux to inductance and resistive load to inductance, of each harvester, respectively. These matrices are assembled as explained in Equations (11)–(17) in [25].

The nominal system can be represented in state space form as defined by (1)–(4), with the response $\mathbf{x}(t)$ written in terms of the nominal system's impulse response in the pattern of the matrix $\mathbf{B}$ with initial conditions $\mathbf{x}(0) = \mathbf{x}_0$ as shown in (5) with $\mathbf{H}_\mathbf{B}(t) = e^{\mathbf{A}t}\mathbf{B}$.

$$\dot{\mathbf{x}}(t) = \mathbf{A}\mathbf{x}(t) + \mathbf{B}\mathbf{f}(t), \tag{1}$$

$$\mathbf{y}(t) = \mathbf{C}\mathbf{x}(t), \tag{2}$$

$$\mathbf{A} = \begin{bmatrix} \mathbf{0}_{n_{dof} \times n_{dof}} & \mathbf{I}_{n_{dof}} & \mathbf{0}_{n_{dof} \times n_h} \\ -\mathbf{M}^{-1}\mathbf{K} & -\mathbf{M}^{-1}\mathbf{C} & \begin{bmatrix} \mathbf{0}_{(n_{dof}-n_h) \times n_h} \\ -\mathbf{M}_h{}^{-1}\mathbf{B}_L \end{bmatrix} \\ \mathbf{0}_{n_h \times (2n_{dof}-n_h)} & \mathbf{B}_{LL} & -\mathbf{R}_{LL} \end{bmatrix}, \tag{3}$$

$$\mathbf{B} = \begin{bmatrix} \mathbf{0}_{n_{dof} \times n_b} \\ \mathbf{M}_b^{-1}\mathbf{P} \\ \mathbf{0}_{2n_h \times n_b} \end{bmatrix}, \tag{4}$$

$$\mathbf{x}(t) = e^{\mathbf{A}t}\mathbf{x}_0 + \int_0^t \mathbf{H}_\mathbf{B}(t-\tau)\mathbf{f}(\tau)d\tau, \tag{5}$$

The $\mathbf{x}(t)$ state vector of size $n_{dim} \times 1$ consists of the displacements and velocities of each DOF of the superstructure and harvester along with the current generated by each harvester, with $n_{dim} = 2n_{dof} + n_h$. Desired system outputs, $\mathbf{y}(t)$, can be represented as linear combination of the states $\mathbf{x}(t)$, and $\mathbf{x}(t) = \left[\mathbf{u}_b\ \mathbf{u}_h\ \dot{\mathbf{u}}_b\ \dot{\mathbf{u}}_h\ \mathbf{i}\right]^T$.

The modified system is represented by Equations (6)–(8) in state space with the system states defined by $\mathbf{X}(t) = \left[\mathbf{U}_b\ \mathbf{U}_h\ \dot{\mathbf{U}}_b\ \dot{\mathbf{U}}_h\ \mathbf{I}\right]^T$ of size $n_{dim} \times 1$. $\Delta\mathbf{K}$ and $\Delta\mathbf{C}$ represent the stiffness and damping matrices arising from connections of the harvester(s) to the system.

$$\dot{\mathbf{X}}(t) = (\mathbf{A} + \Delta\mathbf{A})\mathbf{X}(t) + \mathbf{B}\mathbf{f}(t), \tag{6}$$

$$\mathbf{Y}(t) = \mathbf{C}\mathbf{X}(t), \tag{7}$$

$$\Delta\mathbf{A} = \begin{bmatrix} \mathbf{0}_{n_{dof} \times n_{dof}} & \mathbf{0}_{n_{dof} \times n_{dof}} & \mathbf{0}_{n_{dof} \times n_h} \\ -\mathbf{M}^{-1}\Delta\mathbf{K} & -\mathbf{M}^{-1}\Delta\mathbf{C} & \mathbf{0}_{n_{dof} \times n_h} \\ \mathbf{0}_{n_h \times n_{dof}} & \mathbf{0}_{n_h \times n_{dof}} & \mathbf{0}_{n_h \times n_h} \end{bmatrix}, \tag{8}$$

Using the superposition principle, the modified states, $\mathbf{X}(t)$, can be calculated using (9) with the nominal states, $\mathbf{x}(t)$, and the convolution of the pseudoforce vector, $\mathbf{p}(t)$ of size $n_h \times 1$, arising from the connection of the harvester(s) to the structure and the impulse response, $\mathbf{H}_\mathbf{L}(t) = e^{\mathbf{A}t}\mathbf{L}$ of size $n_{dim} \times n_h$, in the pattern of the modification, where the modification is given by the addition of the harvesters to the superstructure.

$$\mathbf{X}(t) = \mathbf{x}(t) + \int_0^t \mathbf{H}_\mathbf{L}(t-\tau)\mathbf{p}(\tau)d\tau, \tag{9}$$

The modification can be compactly written as shown in Equations (10)–(13), where matrix $\mathbf{R}$, of size $n_{dof} \times n_h$, maps which superstructure degree of freedom each harvester is connected to and where $\mathbf{I}_{n_h}$ is the identity matrix of size $n_h \times n_h$.

$$\mathbf{S}_{(i,j)} = \begin{cases} 1, & \forall j = 1, \ldots, nh \; \& \; i = \mathbf{pos}(j) \\ 0, & otherwise \end{cases}, \tag{10}$$

$$\mathbf{R} = \begin{bmatrix} \mathbf{S} \\ -\mathbf{I}_{n_h} \end{bmatrix}, \tag{11}$$

$$\overline{\mathbf{L}} = -\mathbf{M}^{-1}\mathbf{R}, \tag{12}$$

$$\mathbf{L} = \begin{bmatrix} \mathbf{0}_{n_{dof} \times n_h} \\ \overline{\mathbf{L}} \\ \mathbf{0}_{n_h \times n_h} \end{bmatrix}, \tag{13}$$

The pseudoforce can be efficiently computed [29] as given in (14), where $\delta\mathbf{K}$ and $\delta\mathbf{C}$ are diagonal matrices containing the stiffness and damping of the harvesters, respectively.

$$\mathbf{p}(t) - \tilde{\mathbf{x}}(t) - \int_0^t \tilde{\mathbf{H}}_{\mathbf{L}}(t-\tau)\mathbf{p}(\tau)d\tau = 0, \tag{14}$$

$$\tilde{\mathbf{x}}(t) = \overline{\mathbf{q}}\mathbf{G}\mathbf{x}(t), \tag{15}$$

$$\tilde{\mathbf{H}}_{\mathbf{L}}(t) = \overline{\mathbf{q}}\mathbf{G}\mathbf{H}_{\mathbf{L}}(t), \tag{16}$$

$$\overline{\mathbf{q}} = \begin{bmatrix} \delta\mathbf{K} & \delta\mathbf{C} & \mathbf{0}_{n_h \times n_h} \end{bmatrix}, \tag{17}$$

$$\mathbf{G} = \begin{bmatrix} \mathbf{R}^{\mathrm{T}} & \mathbf{0}_{n_h \times n_{dof}} & \mathbf{0}_{n_h \times n_h} \\ \mathbf{0}_{n_h \times n_{dof}} & \mathbf{R}^{\mathrm{T}} & \mathbf{0}_{n_h \times n_h} \\ \mathbf{0}_{n_h \times n_{dof}} & \mathbf{0}_{n_h \times n_{dof}} & \mathbf{I}_{n_h} \end{bmatrix}, \tag{18}$$

In the results reported here, the low dimensional pseudoforce was calculated using trapezoidal rule and fast Fourier transforms as presented in [29]. The outputs of interest for the modified system, $\overline{\mathbf{X}}(t)$ of size $n_h \times 1$, can be computed by the sum of the outputs of interest of the nominal system, $\overline{\mathbf{x}}(t)$ of size $n_h \times 1$, and the convolution of the corresponding portion of the impulse response to the outputs of interest, $\overline{\mathbf{H}}_{n_h}(t)$ of size $n_h \times n_h$, in the pattern of the modification, and the low order pseudoforce, $\mathbf{p}(t)$.

$$\overline{\mathbf{X}}(t) = \overline{\mathbf{x}}(t) + \int_0^t \overline{\mathbf{H}}_{n_h}(t-\tau)\mathbf{p}(\tau)d\tau, \tag{19}$$

$$\overline{\mathbf{x}}(t) = \mathbf{G}_{n_h}\mathbf{x}(t), \tag{20}$$

$$\overline{\mathbf{X}}(t) = \mathbf{G}_{n_h}\mathbf{X}(t), \tag{21}$$

$$\overline{\mathbf{H}}_{n_h}(t) = \mathbf{G}_{n_h}\mathbf{H}_{\mathbf{L}}(t), \tag{22}$$

$$\mathbf{G}_{n_h} = \begin{bmatrix} \mathbf{0}_{n_h \times n_{dof}} & \mathbf{0}_{n_h \times n_{dof}} & \mathbf{I}_{n_h} \end{bmatrix}, \tag{23}$$

Optimization

The efficient approach to solve for the modified system outputs of interest previously described allows one to exploit the locality of the harvester(s) and considerably reduce the cable-stayed bridge example presented in order to repeat the required analysis to perform design optimization.

The optimization was performed using the *patternsearch* algorithm in MATLAB. Figure 3 schematically represents the *patternsearch* optimization algorithm flowchart as described in [30]. The algorithm uses an adaptative mesh of design points to find a specified function's minimum value. A sequence of approximation points that approach an optimal solution is located by the algorithm, by comparing the objective function from a point in the sequence to the next as its value either decreases or remains the same.

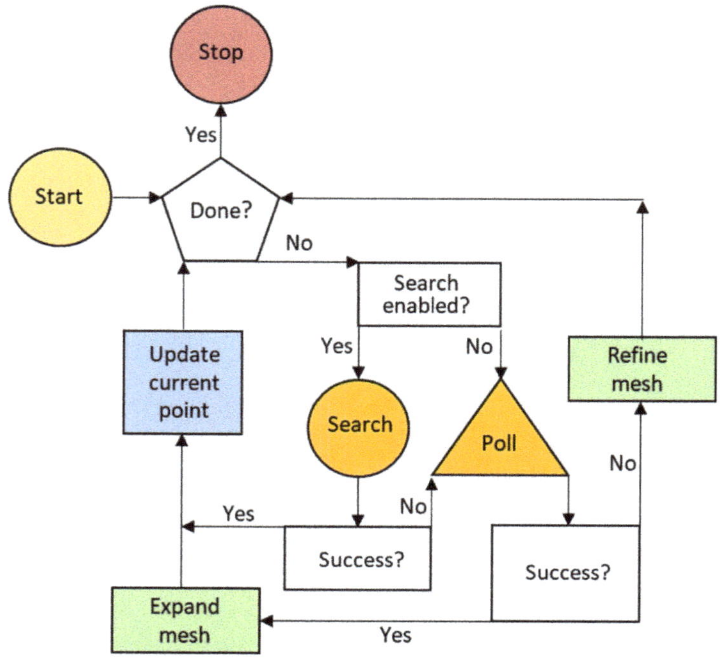

Figure 3. *Patternsearch* optimization flowchart [30].

In this paper, the objective function to minimize was chosen to be the negative of the total energy extracted from all the different harvester(s) configurations and defined in (24), where $\mathbf{I}(t)$ is a $nh \times 1$ vector of currents generated by the harvesters calculated using the efficient approach described in the problem formulation, and $\mathbf{R}_L$ is a $n_h \times n_h$ diagonal matrix containing each resistive load of the harvester(s).

$$f(\mathbf{x}) = -\int_0^t \mathbf{I}(t)^{\mathrm{T}}\mathbf{R}_L\mathbf{I}(t)\, dt, \tag{24}$$

The design variable was chosen to be a vector containing the frequencies of the harvesters, ω_h, with prescribed lower and upper bounds that were selected after an initial parameter screening. An initial value of 1 rad/s was selected for the design variables for all results reported.

$$\omega_{lower} \leq \omega_h \leq \omega_{upper},$$
$$\omega_{lower} = 0.9 \text{ rad/s}$$
$$\omega_{upper} = 25 \text{ rad/s}$$

For this particular example, the damping of the harvesters was neglected and not included as a design variable. In order to do this, maximum structure and harvester displacements were assessed after each optimization run to guarantee that they were within acceptable and safe values.

3. Example: Cable-Stayed Bridge

3.1. Model Description

The example considered herein consists of the cable-stayed Bill Emerson Memorial Bridge (2003) located between Cape Girardeau, Missouri and East Cape Girardeau, Illinois, spanning the Mississippi river. The bridge model, adapted from [31], consists of 579 nodes, 128 cable elements, 162 beam elements, 420 rigid links and 134 nodal masses. The structure has a total of 3474 degrees of freedom and a total mass of 51,987,767.58 kg. The fundamental frequency of the bridge is 1.0172 rad/s and the second modal frequency is 1.765 rad/s. Dyke et al. [32] developed a finite element model of the bridge as shown in Figure 4. In the model, the connections between the bridge deck and the tower are purely through the cables, this allows one to place energy dissipation devices between the deck and the tower. The initial model consisting of 3474 DOFs was reduced by imposing boundary conditions, removing slave DOFs and applying a static condensation to eliminate DOFs with a small contribution to the global system response. The resulting reduced model, consisting of 419 degrees of freedom as presented in [31] and further described in [32], will be used for the remainder of this section. The model was cast in the form of Equations (1) and (2) with vectors $\mathbf{u}_b$ and $\mathbf{U}_b$ representing the system's response in relative coordinates. In order to maintain the symmetry of the structure with respect to the main longitudinal axis, X, of the deck span, two different scenarios were considered: (1) two harvesters attached at two symmetric joints of the bridge and (2) sets of two harvesters attached at symmetric joints of the bridge with identical equivalent total harvester mass. Each harvester was considered to be a uniaxial device oriented along the X-axis. Table 1 summarizes the details for both scenarios.

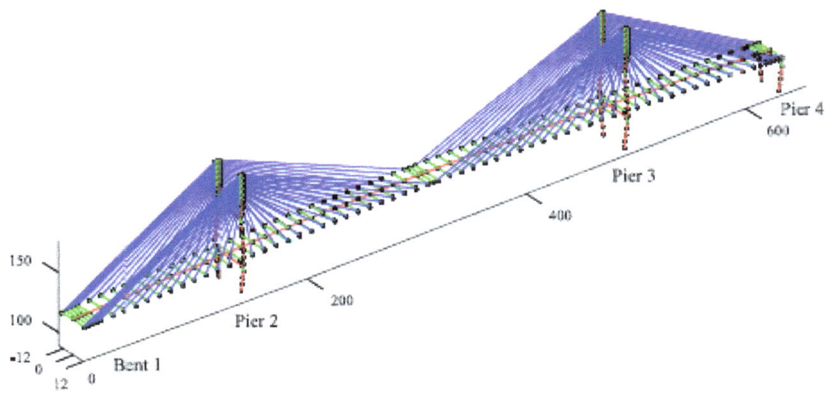

Figure 4. Representation of finite element model of the cable-stayed bridge [31].

Table 1. Harvester(s) configurations parameters for bridge model.

	Scenario 1	Scenario 2
Number of harvesters	2	n_h
Location	Symmetric joints	Symmetric joints
Mass of harvester	259,938.835 kg	519,877.67/n_h kg
Total harvester mass	519,877.67 kg	519,877.67 kg
Mass ratio	0.01	0.01
System DOFs	421	$419 + n_h$

The electrical properties of the harvester(s) were scaled using an existing electromagnetic vibrational energy harvester as explained in [25] and as shown in Table 2. The total harvester mass was considered to be 1% of the total structure mass, or more specifically

519,877.67 kg, and was kept constant for both cases in order to more accurately compare the performance of the system in all different configurations.

Table 2. Scaled harvester(s) electrical properties.

Harvester Electrical Properties	Harvester Mass m_h (kg)	Resistive Load R_L (Ω)	Magnetic Flux B_L (N/A)	Inductance L_e (H)
Base harvester	2.4	4118	452	0.04
Scaled harvester	m_h	4118 L^{-1}	452L	0.04L
	where $L = \sqrt[3]{m_h} / \sqrt[3]{2.4\text{ kg}}$			

3.2. Input Load

To more realistically assess the power generation capabilities of the harvester(s) when augmenting a realistic bridge model, the El Centro earthquake record was utilized as the ground excitation in the longitudinal direction for a duration of 40 s. This loading was recorded at the Imperial Valley Irrigation District substation of the North-South component in El Centro, California in 18 May 1940. There are some limitations associated with the earthquake record for the analyses since the loading record has a sampling time of 0.02 s. In order to address this limitation, and to more accurately compute the response of the system, a subsequent linear interpolation of the earthquake record was performed, reducing the sampling time to 0.0013 s, but still maintaining a realistic loading history without loss of accuracy. Figure 5 contains the original and interpolated sample El Centro earthquake record. The load was applied in the reduced bridge model at the kept X-direction degrees of freedom.

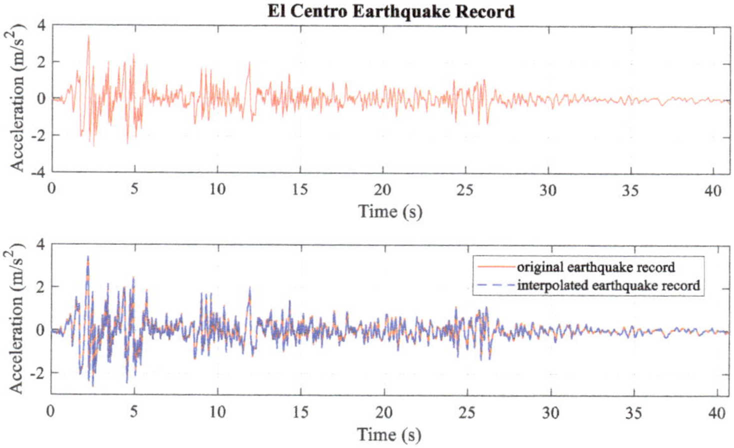

Figure 5. Representation of original and interpolated El Centro earthquake record.

Given that the Bill Emerson Memorial Bridge is located in the New Madrid seismic zone and constitutes the principal crossing of the Mississippi River, its design strongly accounts for seismic activity [32]. Under this consideration, the El Centro input load was selected as a good representation to assess the structure's performance when augmented with vibrational energy harvesters and exposed to moderate seismic loads.

A second loading was employed to better assess the power generation capabilities of the vibrational energy harvesters under more common loads. In this particular case, a loading possessing the magnitude of typical traffic loads was utilized to more realistically predict the amount of power and possible applications from the implementation of these devices. The literature reports maximum peak acceleration values of about 0.1 cm/s^2 at the Bill Emerson Bridge due to ambient/traffic induced vibrations [33,34]. Based on these considerations, the El Centro earthquake record was scaled to a peak acceleration value of 0.000214 m/s^2 and used as the input load to optimize the power generation capabilities of the harvesters under loads of magnitude of the daily, realistic loads. Figure 6 shows the scaled loading for a duration of 40 s.

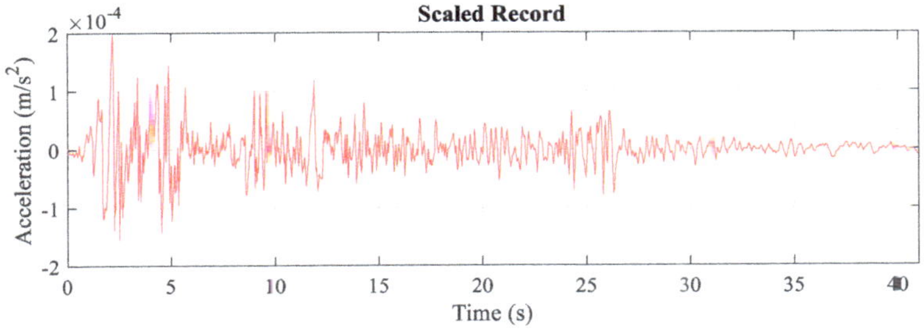

Figure 6. Scaled input loading.

4. Results and Discussion

4.1. Optimization Design

Multiple studies were performed to determine the optimal location of different harvester(s) configurations under the El Centro earthquake and the scaled input loads. These were conducted by optimizing the frequencies of the devices to maximize the total power generation from implementation of 2, 4 and 8 harvesters, with harvester frequency bounds set from 0.9 to 25 rad/s, which were chosen after an initial parameter screening. In all cases, the devices were placed symmetrically about the main bridge axis (X) to maintain the structure's symmetry. For the sake of brevity, in this paper, only the results for the 8-harvester configuration are provided. Details of all results for all configurations can be found in [35].

De et al. in [31] reported optimal locations for the implementation of pairs of passive control devices, including viscous dampers and TMDs, on the Bill Emerson Bridge. Based on the tabulated results, the optimal joint combinations reported in [31] were chosen as the initial set of potential locations for the 8-harvester configuration. Additionally, locations were also investigated taking into consideration the desired multifunctionality of the harvester masses and also keeping in mind the goal of an overall aesthetic bridge design. For this reason, different joint combinations, selected to perform design optimization, outside those reported in [31], were considered where harvesters were attached at locations corresponding to the outer deck on both sides of the bridge.

The optimal power generation for the 2, 4 and 8-harvester configurations was found to be 1489 kW, 1613.6 kW and 1623.8 kW under El Centro input and 4.914 mW, 5.325 mW and 5.359 mW under the scaled input, respectively. From the resulting optimal power generation for all three configurations, one can note that four or more harvesters can generate more power than a single harvester pair of equivalent total mass, supporting the assertion that multiple smaller devices result in better performance and ease of implementation.

Results from the optimization of the total power generation as a function of location are presented in Tables 3 and 4 for the 8-harvester configuration under both input loads, respectively. In this particular case, the optimal harvester's location consisted of a combination of joints as found in [31] with additional harvesters on the outer deck of the bridge.

Table 3. Optimized frequencies for eight harvesters at different locations under El Centro input. The resulting optimal location and frequencies that maximizes power generation are shown in bold.

		8 Harvesters					
	Location		**Frequencies [rad/s]**			**Power [kW]**	**Energy [W-h]**
1	(117 184 204 209 459 461 525 527)	9.13 9.17	9.13 9.17	13.18 9.17	12.86 9.17	1253.1	348.10
2	(117 184 205 210 318 323 459 525)	12.24 13.47	11.71 13.47	13.09 9.21	12.87 9.21	1368.6	380.16
3	(119 186 206 211 319 324 459 525)	12.26 13.39	11.70 13.39	21.47 9.20	12.79 9.20	1442	400.57
4	**(117 184 206 211 319 324 461 527)**	**11.66 13.46**	**11.67 13.46**	**12.39 9.20**	**12.91 9.20**	**1623.8**	**451.05**
5	(119 186 205 210 318 323 461 527)	11.71 13.47	12.25 13.47	12.87 9.20	13.09 9.20	1370.2	380.61
6	(83 119 150 186 461 494 527 560)	9.15 9.15	12.28 9.12	9.15 9.15	9.13 9.13	1564.2	434.50

Table 4. Optimized frequencies for eight harvesters at different locations under scaled input. The resulting optimal location and frequencies that maximizes power generation are shown in bold.

		8 Harvesters					
	Location		**Frequencies [rad/s]**			**Power [mW]**	**Energy [μW-h]**
1	(117 184 204 209 459 461 525 527)	9.13 9.17	9.13 9.17	13.18 9.17	12.86 9.17	4.135	1.149
2	(117 184 205 210 318 323 459 525)	12.24 13.47	11.71 13.47	13.09 9.21	12.87 9.21	4.516	1.255
3	(119 186 206 211 319 324 459 525)	12.26 13.39	11.70 13.39	21.47 9.20	12.79 9.20	4.759	1.322
4	**(117 184 206 211 319 324 461 527)**	**11.66 13.46**	**11.67 13.46**	**12.39 9.20**	**12.91 9.20**	**5.359**	**1.488**
5	(119 186 205 210 318 323 461 527)	11.71 13.47	12.25 13.47	12.87 9.20	13.09 9.20	4.522	1.256
6	(83 119 150 186 461 494 527 560)	9.15 9.15	12.28 9.12	9.15 9.15	9.13 9.13	5.162	1.434

The optimal configuration is shown in Figure 7. It is important to note that from all selected joints for all different configurations, only the optimal locations for the 8-harvester scenario considered connecting the devices at joints on both Piers 2 and 3, specifically the ones located at deck level. The strategic placement of these devices along the outer deck not only maximizes the potential power generation but also allows for flexibility of implementation for the multifunctional harvester masses or more specifically gardens on each side of the structure.

Since damping was not included in the optimization process for all different configurations, maximum structure displacements and accelerations were computed to guarantee they were within acceptable values as shown in Table 5.

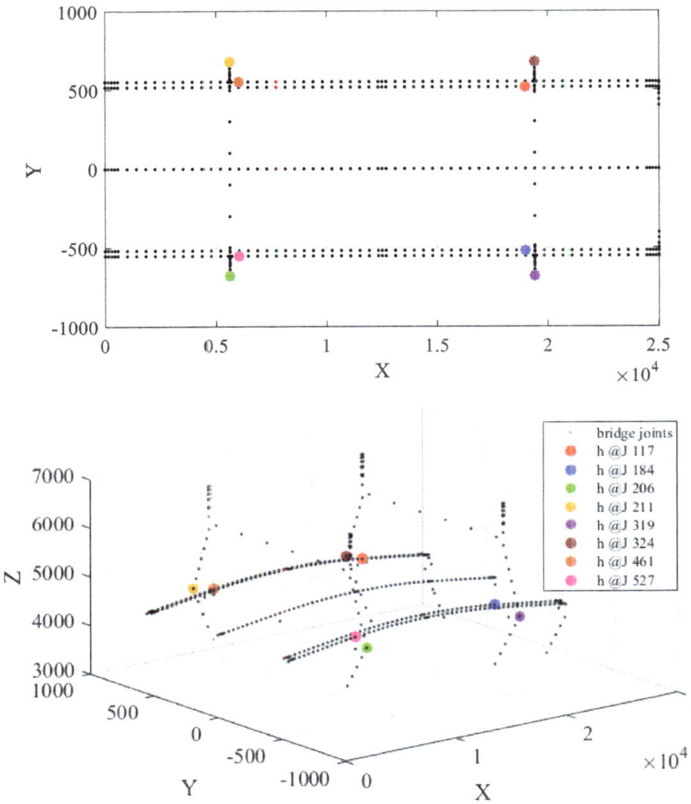

Figure 7. Optimal configuration for eight harvesters with top and 3D views.

Table 5. Maximum bridge nominal (n), without harvesters, and modified (m), with harvesters, displacement and acceleration responses for all configurations.

Cable-Stayed Bridge Model, El Centro Input						
h	x_n **[m]**	x_m **[m]**	**Joint**	**DOF**	$\ddot{x}_n$ **[m/s^2]**	$\ddot{x}_m$ **[m/s^2]**
2		0.3395				17.08
4	0.3703	0.3383	9	49	17.4	17.03
8		0.3544				16.99

Figure 8 shows the displacement time history for the 8-harvester configuration while Figure 9 shows the corresponding displacement frequency response functions. The resulting optimal frequencies display a wide distribution since multiple local frequency peaks are observed in this case, as shown in Figure 9.

To further investigate the behavior of the resulting optimized frequencies, the frequency content of the input load was determined and shown in Figure 10. We can clearly observe that the highest frequency happens around 9.19 rad/s with some high peaks in the orders of 12 to 13 rad/s and hence, the tuning of all harvesters in that range.

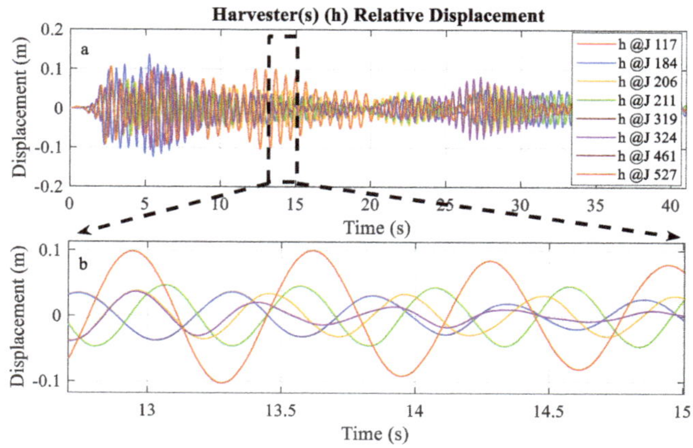

Figure 8. (**a**) Eight-harvester displacement time histories with no damping with (**b**) zoom in for 12.5–15 s of (**a**).

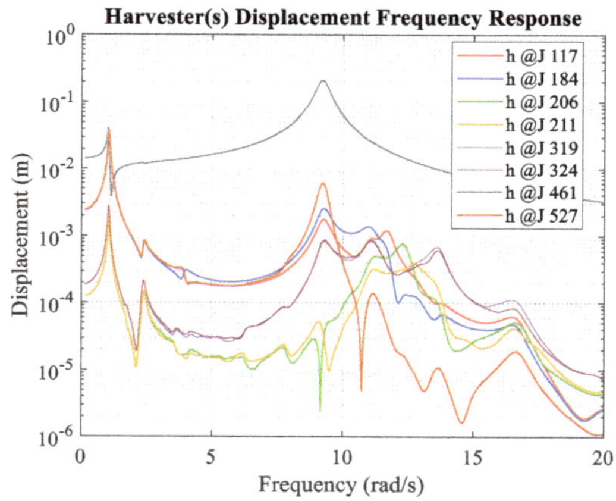

Figure 9. Eight-harvester displacement frequency response functions with no damping.

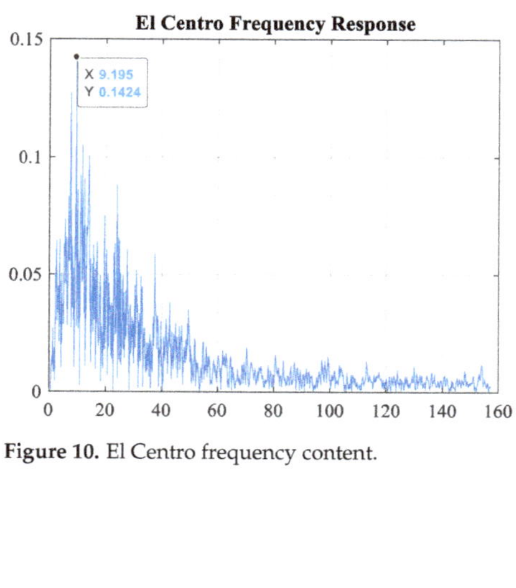

Figure 10. El Centro frequency content.

The amount of power extracted under seismic excitation and shown in Table 3 is considerable and is sufficient to potentially power wireless communication devices that can alert or prevent structural damage during catastrophic events. The applications of this technology and the potential energy extracted are significant for implementation of structural health monitoring in bridges. Although the energy obtained from the scaled load is significantly smaller in this example, it is still sufficient to power WSNs, considering that their power requirements for data transmission are on the order of 0.09 to 128 mW [18]. In this particular example under this load, the potential power generation is on the order of 5 mW in all cases.

After the optimized frequencies and the associated energy harvested were found, further analyses were conducted to compare the optimum power generation of the 8-harvester configuration to two other scenarios, when the harvester(s) was tuned to the fundamental bridge frequency of 1.0172 rad/s and also to the scenario where they were tuned to different constant values selected from the optimal results or more specifically 9.15 rad/s in this case. The El Centro earthquake record was selected as the input load to perform this comparison.

The results are shown in Figure 11. One can clearly observe that tuning the devices to the fundamental bridge frequency underestimates the power generation capabilities in this configuration. The power is increased by 76% when optimizing each individual frequency for all harvesters with respect to tuning the devices to the fundamental frequency of the structure. The difference between the results from optimizing each individual device versus fixing a close to optimal value represents a 50% increase considering that, for this particular configuration, there is a wide distribution of optimized frequencies. The same comparison for 2, 4 and 16-harvester configurations can be found in [35].

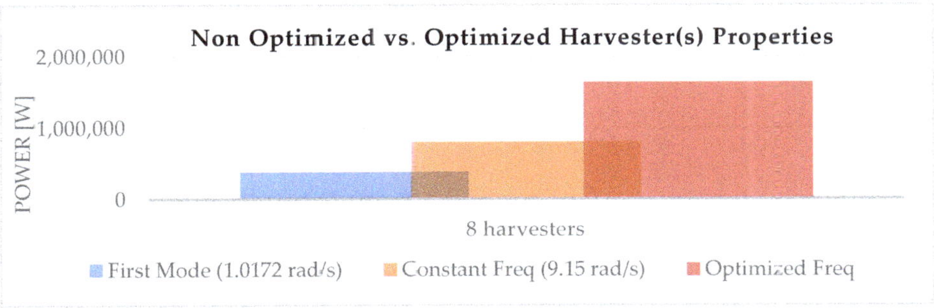

Figure 11. Influence of non-optimized versus optimized harvester(s) properties in the power generation capabilities of 8 harvesters.

Table 6 contains a comparison of required computational times including up-front costs and the required number of function evaluations for the bridge example. The computational times were calculated using the cputime command in MATLAB on a computer with two 3.20 GHz Intel(R) Xeon(R) CPU E5-2667 v4 processors, 256 GB RAM, Windows 10 and running MATLAB 2019b with parallel pool and 12 workers. The computational speed increase was compared with the projected cost of the conventional method using lsim in MATLAB for the number of function evaluations required for the proposed approach for each harvester configuration. For this particular example, the computational efficiency was increased two to three orders of magnitude. It is important to note that as the number of harvesters increases so does the required number of function evaluations which causes an increase in the computational efficiency of the proposed method. In particular, the design optimization process for the 8-harvester case was computed in approximately 11 min using the proposed method versus the approximate 3 days that it would have taken to completely replicate the process using conventional methods.

Table 6. Comparison of required computational time for the cable-stayed bridge model.

Number of Harvesters	Function Evaluations	Conventional Method (Projected)	Proposed Method (Actual)	Computational Speed Up
2	239	10,870 s	164 s	66.3
4	903	54,687 s	324 s	168.8
8	3703	231,290 s	684 s	338.1

4.2. Multifunctional Concept Design

Green gardens were considered as the multifunctional design concept to be implemented as the harvester's mass for the cable-stayed bridge model. For ease of analysis, roof gardens were adapted and implemented in this example. These types of gardens can be categorized as shallow (ultra-extensive), medium depth (extensive) and deep (intensive) systems [36]. The first refers to gardens that have an approximate growth media depth of 2.5″ to 4″ and require little maintenance; they are suitable for implementation at inaccessible areas and can accommodate sedums, herbs and grasses. Extensive gardens have a growth media depth of 5″ to 8″ and also require relatively low maintenance. They can include sedum, herbs grasses and other types of vegetation. Irrigation systems may be required for this category to support more diverse plants and when installed in semi-arid climates. Intensive gardens typically exceed a growth media depth of 8″ to accommodate planting systems that require deeper media. This type of system requires high maintenance such as watering, fertilizing, mowing/weeding, needs to incorporate an irrigation system and they also impose the greatest dead load.

Under these considerations, extensive gardens were selected for implementation in the cable-stayed bridge model, considering their low maintenance requirements and their potential to accommodate different types of vegetation. A schematic representation including different components of the green garden can be found in [36].

This concept, when applied to bridge structures, would have significant impact on the environment since it would improve and reduce the pollution created by traffic carbon dioxide emissions considering the ability of plants to absorb carbon emissions, and can reduce the urban heat island effect since sunlight is used for growth as it is absorbed by vegetation instead of becoming heat energy [36]. The concept would also improve the aesthetics of the structure as well as creating ecological habitats for birds and other species. One conceptual garden implementation for bridge infrastructure is shown in Figure 12. The Friedrich Bayer Bridge in Brazil is another example of incorporating the garden concept into bridge infrastructure [37].

Figure 12. Garden Bridge by Michael Tefft [38].

As previously mentioned, green gardens were selected as the multifunctional concept to implement for the Bill Emerson cable-stayed bridge. The solution considered connecting the garden mass to joints located towards the center of the bridge deck to maximize the power generation, maintain the symmetry of the structure and strategically improve the bridge's aesthetics in general. Recent experimental studies on piezoelectric energy harvesting from vehicle–bridge coupling vibration show that higher energy harvesting efficiency can be achieved when the harvester is installed at the center of the structure [39]. These results validate the approach of implementing the proposed multifunctional vibrational energy harvesting concept at the center of the Bill Emerson cable-stayed bridge.

Based on these considerations, the total garden mass was divided into two sections and placed on joints located between piers 2 and 3, and more specifically 466, 474, 482 and 490 on the outer right deck and 532, 540, 548 and 556 on the outer left deck as schematically shown in Figure 13. In this particular case a total of eight harvesters were considered, but different configurations can be selected without affecting the efficacy and practicality of the employed computational method. The total garden area was divided into eight equal tributary areas where each is supported by a harvester at the specified joints and schematically shown in Figure 14. A schematic representation of the green gardens connected to both sides the bridge outer deck is shown in Figure 15.

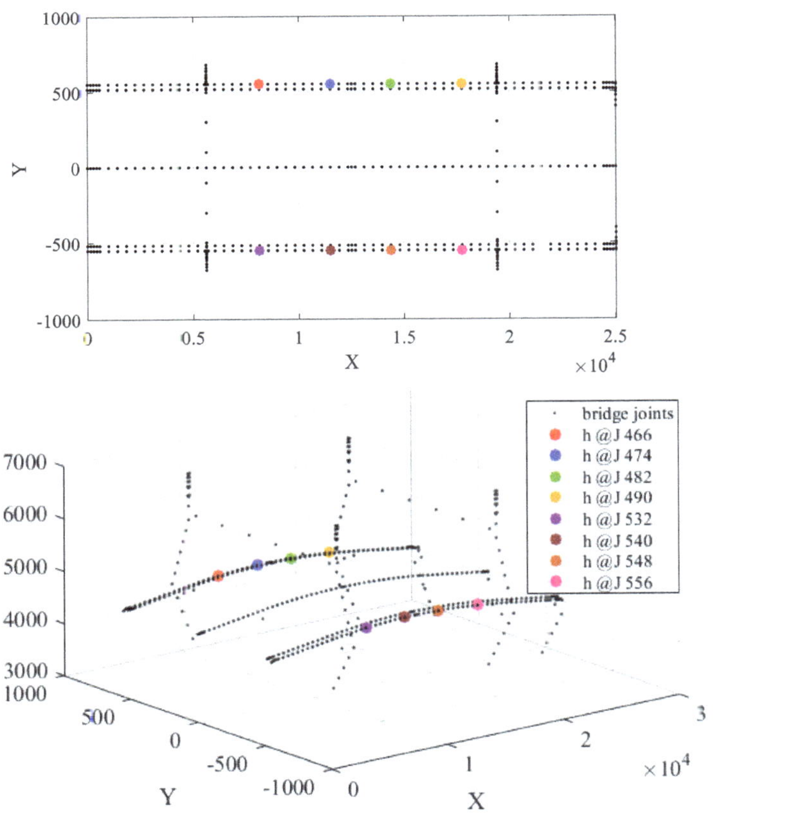

Figure 13. Optimal configuration for eight harvesters, with top and 3D views, for green garden implementation.

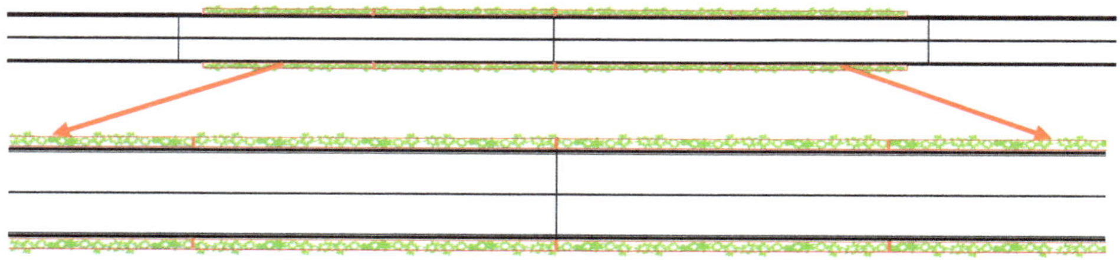

Figure 14. Schematic representation of bridge deck top view with green garden connected on both sides of the outer deck between piers 2 and 3 with zoom in portion.

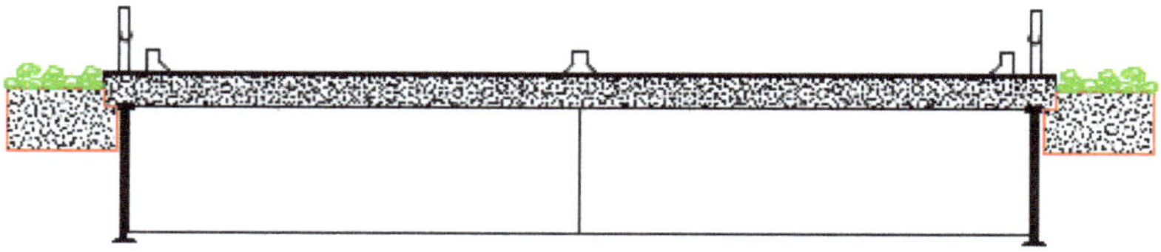

Figure 15. Schematic representation of bridge deck cross-section with green gardens attached at both sides of the outer deck.

Detailed calculations were performed to determine the total required mass for the green gardens, with each harvester tributary area defined by a length of 82.29 m, considering the spacing between joints, and a width of 3 m, selected to achieve proportionality within the structure. Table 7 shows the different components and general specifications for the design and implementation of the green gardens as presented in [27]. Based on the calculations performed, each harvester's mass was found to be approximately 67,093.7 kg resulting in a total harvester mass of 536,749.6 kg.

Table 7. Components and specifications for the implementation of green garden at the cable-stayed bridge model.

System Components	Specifications	Weight	Total
Number of harvesters	8		
Area of each h	246.89 m^2		1975.1 m^2
Growth media depth	0.2 m		
Volume of each h	246.89 $m^2 \times 0.2$ m		395.02 m^3
Drainage Composite	1.22×15.24 m roll	31.75 kg each	1406.7 kg (44 rolls)
Protection Fabric	3.81×60.69 m each	0.5425 kg/m^2	6428.95 kg (6)
Growth Media	Saturated Weight	1204.6 kg/m^3	475,841.1 kg
Plants, Sedum Mats	2.32 m^2 each	26.85 kg/m^2	53,072.78 kg (852)
Total harvester mass			**536,749.6 kg**

The same two input loads were considered in the design process as described in Section 3.2, including both El Centro earthquake record and a scaled version of it to simulate typical vibration levels from traffic loads at the Bill Emerson cable-stayed bridge. The optimization again consisted of maximizing the power generated under both input loads by optimizing each harvester frequency. Results from the optimization process are shown in Table 8 for both input cases.

Table 8. Optimized eight-harvester frequencies with no damping for distributed garden concept implementation.

Optimized Frequencies in rad/s for the Cable-Stayed Bridge Garden Application							
Load	**Location**	**Frequencies [rad/s]**				**Power [W]**	**Energy [W-h]**
El Centro	(466 474 482 490 532 540 548 556)	9.14 9.14	9.14 9.14	9.14 9.14	12.25 9.14	1,574,902.46	437.47
Scaled	(466 474 482 490 532 540 548 556)	9.14 9.14	9.14 9.14	9.14 9.14	12.25 9.14	0.0052	1.44×10^{-5}

In this particular example the resulting frequencies follow a similar pattern, with most of them tuned around 9.14 rad/s under both input loads, consistent with their highest peaks located around 9.14 rad/s in Figure 16. There is an exception to this behavior for the harvester connected at joint 490 which is tuned at 12.25 rad/s, also consistent with its highest peak located between 12.2 and 12.4 rad/s in Figure 16. Although the total harvester mass was considered to be similar to that described in Section 4.1 or more specifically 1.03% of the total structure mass, the harvester's placement was chosen taking into consideration aesthetic factors, with a difference of about only 3% in total power generation with respect to the optimal results for the 8-harvester case shown in Table 5.

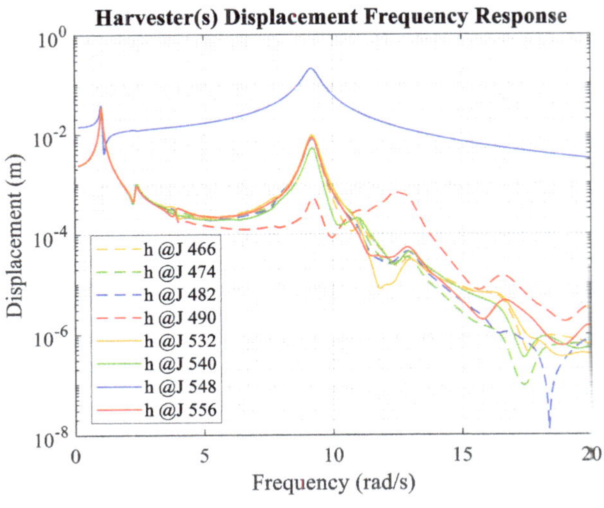

Figure 16. Eight-harvester displacement frequency response functions for El Centro input and no damping for the distributed garden concept implementation.

Figure 17 shows the displacement time histories for all devices subjected to the El Centro earthquake record input load for the implementation of the green garden concept using a total of eight harvesters and no damping. In all cases, maximum harvester displacements are within acceptable and safe values. For this particular example, only displacements under El Centro earthquake record load were investigated considering that the peak acceleration from the traffic induced scaled input load is considerably smaller than the earthquake load.

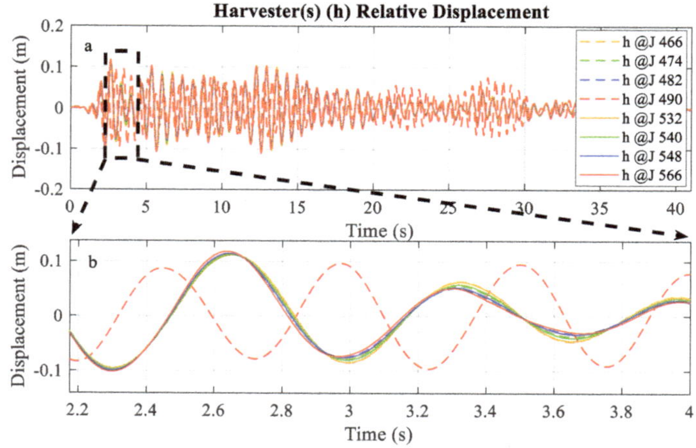

Figure 17. (**a**) Eight-harvester displacement time histories with (**b**) zoom in for 2.2–4 s of (**a**) for El Centro input and no damping for the distributed garden concept implementation.

5. Conclusions

Earthquake vibrations were employed to assess the power generation capabilities of the different harvester configurations applied to a realistic bridge structure. Even though civil infrastructure is not constantly exposed to earthquake loads, the results obtained during these moderate to extreme vibrations validate the benefits of incorporating this technology into the overall structural system design and their significant power generation capabilities (the order of thousands of kilowatts over a 40 s period). Although the amount of electrical power obtained from traffic induced level vibrations in the particular bridge example considered are only on the order of milliwatts, the scavenged energy is still sufficient to power WSNs that can monitor the structural health of bridge structures.

To further investigate the effects of multiple input loads in the performance and power generation capabilities of vibrational energy harvesters in bridges, wind and traffic loads can be included in the analyses.

Pre-defined symmetric locations were considered in this paper to optimize the total power generation of all harvesters for the purpose of implementing the multifunctional green garden concept. However, a topology optimization can be performed without alterations to the proposed methodology and is one future direction of the authors' research.

The green garden concept was investigated and presented as potential multifunctional solution to holistically integrate the mass of the harvesters into the overall structural system design for the cable-stayed bridge example. The proposed design solution maximized the harvester's power generation capabilities while utilizing the harvester's proof mass as gardens to enhance the structure's aesthetics while benefitting both the ecology and the environment.

The efficient design, analysis, and validation of vibrational energy harvesters is critical to their successful implementation in bridges and buildings. This research, when implemented as multifunctional devices, will enable energy efficient civil structural systems capable of dissipating vibrations while integrating the mass of the energy harvester to improve the functionality and aesthetics of structures. Since vibrational energy harvesters will not only dissipate structural vibrations but will also provide a localized energy source, this multifunctional concept can also potentially contribute to the sustainability of varied infrastructure systems.

Author Contributions: Conceptualization, L.F. and S.F.W.; methodology, L.F. and S.F.W.; software, L.F. and S.F.W.; validation, L.F. and S.F.W.; formal analysis, L.F. and S.F.W.; investigation, L.F. and S.F.W.; resources, L.F. and S.F.W.; data curation, L.F. and S.F.W.; writing—original draft preparation, L.F.; writing—review and editing, S.F.W.; visualization, L.F. and S.F.W.; supervision, S.F.W.; project administration, L.F. and S.F.W.; funding acquisition, L.F. and S.F.W. All authors have read and agreed to the published version of the manuscript.

Funding: This research was funded by the Dwight David Eisenhower Transportation Fellowship Program (DDETFP), grant number 69JJ32145071.

Institutional Review Board Statement: Not applicable.

Informed Consent Statement: Not applicable.

Data Availability Statement: Not applicable.

Acknowledgments: We would like to thank James Gibert from Purdue University for the harvester base data provided and his valuable insights on the mechanics of vibrational energy harvesters. We would also like to thank Subhayan De from Northern Arizona University for providing the cable-stayed bridge model data.

Conflicts of Interest: The authors declare no conflict of interest.

References

1. Rhimi, M.; Lajnef, N. Tunable Energy Harvesting from Ambient Vibrations in Civil Structures. *J. Energy Eng.* **2012**, *138*, 185–193. [CrossRef]
2. Baldwin, J.D.; Roswurm, S.; Nolan, J.; Holliday, L. *Energy Harvesting on Highway Bridges*; Department of Transportation: Oklahoma City, OK, USA, 2011.
3. Cahill, P.; Nuallain, N.A.N.; Jackson, N.; Mathewson, A. Energy Harvesting from Train-Induced Response in Bridges. *J. Bridge Eng.* **2014**, *19*, 1–11. [CrossRef]
4. Sazonov, E.; Li, H.; Curry, D.; Pillay, P. Self-Powered Sensors for Monitoring of Highway Bridges. *IEEE Sens. J.* **2009**, *9*, 1422–1429. [CrossRef]
5. McEvoy, T.; Dierks, E.; Weaver, J. Developing innovative energy harvesting approaches for infrastructure health monitoring systems. In Proceedings of the ASME 2011 International Design Engineering Technical Conferences and Computers and Information in Engineering Conference, Washington, DC, USA, 28–31 August 2011; pp. 325–339.
6. Jo, B.; Lee, Y.; Yun, G.; Park, C.; Kim, J. Vibration-induced energy harvesting for green technology. In Proceedings of the International Conference on Chemical, Environmental and Civil Engineering, Dubai, United Arab Emirates, 24–25 March 2012; pp. 167–170.
7. Galchev, T.; Kim, H.; Najafi, K. Micro Power Generator for Harvesting Low-Frequency and Nonperiodic Vibrations. *J. Microelectromech. Syst.* **2011**, *20*, 852–866. [CrossRef]
8. Zhang, Y.; Cai, C.S. A retrofitted energy harvester for low frequency vibrations. *Smart Mater. Struct.* **2012**, *21*, 075007. [CrossRef]
9. Ali, S.F.; Friswell, M.I.; Adhikari, S. Analysis of energy harvesters for highway bridges. *J. Intell. Mater. Syst. Struct.* **2011**, *22*, 1929–1938. [CrossRef]
10. Kwon, S.D.; Park, J.; Law, K. Electromagnetic energy harvester with repulsively stacked multilayer magnets for low frequency vibrations. *Smart Mater. Struct.* **2013**, *22*, 055007. [CrossRef]
11. Elvin, N.G.; Lajnef, N.; Elvin, A.A. Feasibility of structural monitoring with vibration powered sensors. *Smart Mater. Struct.* **2006**, *15*, 977–986. [CrossRef]
12. Ali, M.; Armstrong, P.J. Sustainability and the tall building: Recent developments and future trends. In *Sustainability and the Tall Building*; AIA Illinois Central Symposium: Springfield, IL, USA, 2010; pp. 1–12. [CrossRef]
13. Farrar, C.R.; Lieven, N.A.J. Damage prognosis: The future of structural health monitoring. *Philos. Trans. R. Soc. A* **2007**, *365*, 623–632. [CrossRef]
14. Shen, W.; Zhu, S.; Zhu, H. Experimental study on using electromagnetic devices on bridge stay cables for simultaneous energy harvesting and vibration damping. *Smart Mater. Struct.* **2016**, *25*, 1–17. [CrossRef]
15. Wang, H.; Jasim, A.; Chen, X. Energy harvesting technologies in roadway and bridge for different applications—A comprehensive review. *Appl. Energy* **2018**, *212*, 1083–1094. [CrossRef]
16. Khan, F.; Ahmad, I. Review of energy harvesters utilizing bridge vibrations. *Shock. Vib.* **2016**, *2016*, 1340402. [CrossRef]
17. Khan, F.; Iqbal, M. Electromagnetic bridge energy harvester utilizing bridge's vibrations and ambient wind for wireless sensor node application. *J. Sens.* **2018**, *2018*, 1–18. [CrossRef]
18. Khan, F.U. Review of non-resonant vibration based energy harvesters for wireless sensor nodes. *J. Renew. Sustain. Energy* **2016**, *8*, 044702. [CrossRef]
19. Beeby, S.P.; Torah, R.N.; Tudor, M.J.; Glynne-Jones, P.; O'Donnell, T.; Saha, C.R.; Roy, S. A micro electromagnetic generator for vibration energy harvesting. *J. Michromechanics Microeng.* **2007**, *17*, 1257–1265. [CrossRef]

20. Khan, F.; Sassani, F.; Stoeber, B. Copper foil-type vibration-based electromagnetic energy harvester. *J. Michromechanics Microeng.* **2010**, *20*, 125006. [CrossRef]
21. Shen, W.; Zhu, S. Harvesting energy via electromagnetic damper: Application to bridge stay cables. *J. Intell. Mater. Syst. Struct.* **2015**, *26*, 3–19. [CrossRef]
22. Saadon, S.; Sidek, O. A review of vibration-based MEMS piezoelectric energy harvesters. *Energy Convers. Manag.* **2011**, *52*, 500–504. [CrossRef]
23. Alsaadi, A.; Shi, Y.; Pan, L.; Tao, J.; Jia, Y. Vibration energy harvesting of multifunctional carbon fibre composite laminate structures. *Compos. Sci. Technol.* **2019**, *178*, 1–10. [CrossRef]
24. Peigny, M.; Siegert, D. Piezoelectric energy harvesting from traffic-induced bridge vibrations. *Smart Mater. Struct.* **2013**, *22*, 095019. [CrossRef]
25. Fernandez, L.; Wotjkiewicz, S.F. Computationally efficient analysis and design optimization of vibrational energy harvesters at the infrastructure scale. *Mech. Syst. Signal Process.* **2021**, *170*, 108780. [CrossRef]
26. Green Vehicle Guide. Available online: https://www.epa.gov/greenvehicles/greenhouse-gas-emissions-typical-passenger-vehicle (accessed on 1 November 2021).
27. SkyGarden Green Solutions. Available online: https://www.sky-garden.co.uk (accessed on 1 November 2021).
28. Bridge Health Monitoring System. Available online: https://www.signaguard.com/bridge-health-monitoring-system/ (accessed on 1 June 2021).
29. Gaurav; Wojtkiewicz, S.F.; Johnson, E.A. Efficient uncertainty quantification of dynamical systems with local nonlinearities and uncertainties. *Probabilistic Eng. Mech.* **2011**, *26*, 561–569. [CrossRef]
30. Fu, Q.; Jiang, B.; Wang, C.; Zhou, X. A novel deblocking quantization table for luminance component in baseline JPEG. *J. Commun.* **2015**, *10*, 629–637. [CrossRef]
31. De, S.; Wotjkiewicz, S.F.; Johnson, E.A. Efficient optimal design and design-under-uncertainty of passive control devices with application t a cable-stayed bridge. *Struct Control. Health Monit.* **2017**, *24*, e1846. [CrossRef]
32. Dyke, S.J.; Caicedo, J.M.; Turan, G.; Bergmn, L.A.; Hague, S. Phase I benchmark control problem for seismic response of cable-stayed bridges. *J. Struct. Eng.* **2003**, *129*, 857–872. [CrossRef]
33. Yan, D.; Wang, W.; Chen, G.; Hartnagel, B.A. Condition assessment of bill emerson memorial cable-stayed bridge under postulated design earthquake. *Transp. Res. Rec.* **2010**, *2172*, 159–167. [CrossRef]
34. Chen, G.; Yan, D.; Wang, W.; Zheng, M.; Ge, L.; Liu, F. *Assessment of the Bill Emerson Memorial Cable-Stayed Bridge Based on Seismic Instrumentation Data*; Missouri University of Science and Technology: Jefferson, MO, USA, 2007.
35. Fernandez, L. *Computationally Efficient Analysis, Optimization and Holistic Design of Vibrational Energy Harvesters at the Infrastructure Scale*; Clarkson University: Potsdam, Germany, 2021.
36. Roof Garden Systems. 2010. Available online: https://www.carlislesyntec.com/en/Roofing-Products/Specialty/Roof-Garden (accessed on 1 June 2021).
37. ODS Outdoor Design Source. Available online: https://www.outdoordesign.com.au/news-info/innovative-spinning-bridge-design/3407.htm (accessed on 1 June 2021).
38. Available online: https://www.flickr.com/photos/mtefft/27730943583 (accessed on 20 November 2021).
39. Zhang, Z.; Xiang, H.; Shi, Z.; Zhan, J. Experimental investigation on piezoelectric energy harvesting from vehicle-bridge coupling vibration. *Energy Convers. Manag.* **2018**, *163*, 169–179. [CrossRef]

sustainability MDPI

Article

Analysis of the Impact of Input Data on the Planned Costs of Building Maintenance

Edyta Plebankiewicz * and Jakub Gracki

Faculty of Civil Engineering, Cracow University of Technology, 31-155 Cracow, Poland; jakub.gracki@pk.edu.pl
* Correspondence: edyta.plebankiewicz@pk.edu.pl

Abstract: The aim of the article is to analyze the method of determining the maintenance costs of buildings based on the method proposed in Polish legal regulations. The analysis of the sensitivity of the method shows that the assumed number of product use cycles during the calculation period has the greatest impact, while the adopted warranty period has the lowest impact. A multi-functional building combining housing, office, service and commercial was analyzed in order to obtain a broader picture of the model's operation. The results of the analyses allow us to conclude that despite the higher price of materials, the most durable solutions, which are the most expensive to purchase, turn out to be the most advantageous in the entire life cycle of the building. The method proposed in Polish law regulations has certain limitations. In order to level them, it was proposed to extend the method by using NPV (Net Present Value) for calculations and extending the life cycle of the building to 80 years.

Keywords: life cycle costs; maintenance costs; warranty period; number of use cycles

Citation: Plebankiewicz, E.; Gracki, J. Analysis of the Impact of Input Data on the Planned Costs of Building Maintenance. *Sustainability* **2021**, *13*, 12220. https://doi.org/10.3390/su132112220

Academic Editors: Oleg Kapliński, Lili Dong, Agata Bonenberg and Wojciech Bonenberg

Received: 8 September 2021
Accepted: 3 November 2021
Published: 5 November 2021

1. Introduction

The costs associated with a construction object are influenced by many factors. In the broadest scope, these costs are captured as LCC—Life Cycle Costs.

Life cycle costing interests EU bodies. The EU document regulating the application of the idea of life cycle costs is the Directive 2014/24/UE of the European Parliament and the Council of 26th February 2014 on public procurement and repealing [1]. Article 67 of the Directive concerning criteria states that the most economically advantageous tender from the point of view of the contracting authority is determined based on the price or cost, using a cost-effectiveness approach such as life cycle costing in art. 68. Article 68 indicates that life cycle costing in an appropriate range shall cover some or all of the following costs during the life cycle of a product, service, or works:

(a) Costs incurred by the contracting authority or other users, such as:
- Acquisition-related costs;
- Costs of use, such as consumption of energy and other resources;
- Maintenance costs;
- End-of-life costs, such as demolition and recycling costs;

(b) Costs attributed to environmental externalities associated with a product, service or work over their life cycle, provided that their monetary value can be determined and verified; such costs may include greenhouse gas and other pollutant emission costs and other climate change mitigation costs.

According to the Directive, when contracting authorities estimate costs using a life cycle costing approach, they specify in the procurement documents the data that bidders should provide and the method that the contracting authority will use to determine the life-cycle costs based on these data.

There are various forms of implementation of the Directive in the European Union countries. In reference to the EU provisions, provisions regarding the calculation of life cycle costs have been included in the Polish Public Procurement Law (PPL). As a consequence of the provisions of the Public Procurement Law, on 13 July 2018, the Regulation of the Minister of Investment and Development of 11 July 2018, on the method of calculating the life cycle costs of buildings and the method of presenting information on these costs was published [2].

One of the components of life cycle costs, which are particularly difficult to determine, are the costs incurred during the operation of the building, in particular the costs of building maintenance. They are related to the natural process of decreasing the utility value of an object over time and the need to carry out construction works that restore the technical and functional features of construction objects [3].

The article analyzes the calculation of some chosen building maintenance costs (i.e., windows, doors, roofing and flooring) based on the method proposed in the regulation. The purpose of this paper was the attempt of finding weaknesses in the regulation method based on calculations given for some chosen products. The second authors' goal was the implementation of some novelties which possibly could be useful in broadening the models' implication possibilities. The influence of the input data on the results was analyzed, and the possibility of extending the method by NPV calculation as well as the calculation period extension up to 80 years.

2. Literature Summary

Proper maintenance of the building aims to ensure that it is maintained in a good technical and aesthetic condition throughout the entire life cycle of the building. For this purpose, the building owner/facility manager undertakes a number of activities, mainly consisting of the maintenance and repair of individual building elements.

The owner of the building, apart from the obligatory activities related to maintaining the proper technical condition of the building, resulting from the applicable regulations, has a choice of many options for repairing and improving the standard of the building. Each of them is associated with the corresponding costs. Therefore, one of the basic issues affecting the costs incurred at the operational stage of the facility is defining the building maintenance strategy.

The ISO standard 15686-5:2017 Buildings and constructed assets –Service life planning Part 5: life cycle costing [4] and standard EN 13306:2017-Maintenance-Maintenance terminology characterizes [5] synthetically two main types of maintenance. The first of the given strategies is called preventive maintenance, and the next one is called corrective maintenance. In the literature [6–9], other variants with a wider range of maintenance types can be found.

Making decisions regarding the maintenance of buildings requires taking into consideration many factors. Therefore, this problem is often solved using MCDM (Multiple Criteria Decision Making) methods [10–18]. Peach and Visser [19] analyze the problem of including human factors in maintenance. Much of the articles are devoted to the proper planning schedule repairs [20–22].

There are many attempts in the literature to develop models that would estimate the cost of building maintenance. Kim et al. [23] propose a model that determines the maintenance and repair costs by means of a statistical analysis of actual cost data. The authors conducted a study of maintenance and repair expenses in educational establishments to determine key performance indicators (KPIs) and to develop an integrated facility management cost estimation model. The study used a multiple regression analysis method to generate a model for determining maintenance and repair costs.

The authors also used the example of a university. Farahani et al. [24], to investigate the problem, presented a case study on the maintenance phase of four university buildings on the campus of the National Taiwan University. Using historical data on maintenance and repair over 42 years, a life cycle cost prediction (LCC) model was determined using three different methods: simple linear regression (SLR), multiple regression (MR), and finally, backpropagation artificial neural network (BNP). In [25], the benefits of proper monitoring of the building's condition on the example of roofing material have been presented.

The proposed models for determining the maintenance costs of building objects use both a simple and a complex mathematical apparatus. A brief overview of simple models supporting the estimation of the budget for maintenance of buildings we can find, for instance, in [26]. The slightly more complex models include the one developed by Kwon et al. [27] based on case-based reasoning and genetic algorithm. Fregonara and Ferrando [28] developed the Stochastic Annuity Method for Supporting Maintenance Costs Planning. Determining the costs of building maintenance is often based on incomplete and imprecise data, which is well represented by fuzzy sets and neural networks [29–32]. Models in this area also often capture the risk [33]. There have also been attempts to use BIM technology to solve this problem [34–37].

3. Calculation of Maintenance Costs in Accordance to Polish Regulations

One of the options for calculating the building maintenance costs is presented in the Regulation of the Minister of Investment and Development of 11 July 2018, on the method of calculating the life cycle costs of buildings and the method of presenting information on these costs. This regulation is based on the Directive 2014/24/EU. According to the ordinance, the life cycle costs of a building are calculated as the sum of the costs of purchasing, using and maintaining a building according to the formula:

$$C_g = C_n + C_{uz} + C_{ut} \tag{1}$$

where: C_g—life cycle costs during a 30-year life cycle of the building, called "calculation period", C_n—acquisition costs, defined based on the offer price, C_{uz}—costs of use, C_{ut}—maintenance costs.

The cost of maintenance included in Formula (1) is primarily the costs of repairs and ongoing maintenance, which allow the building to maintain in a proper technical and aesthetic condition.

These costs should be calculated according to the formula:

$$C_{ut} = \sum (A_i - B_i) \tag{2}$$

where: i—every further product, A_i—maintenance cost of ith product in the calculation period, B_i—producers' warranty value of the ith product.

Maintenance costs of the ith product:

$$A_i = I \times K \times N \tag{3}$$

where: I—number of product units, K—cost of replacement of the product unit, N—number of product use cycles during the calculation period.

Producers' warranty value of the ith product:

$$B_i = A_i \times O_g / 30 \tag{4}$$

where: O_g—warranty period of the ith product given in years.

According to the regulation, the calculation of the life cycle cost of the building is calculated as the sum of the costs of acquisition, use and maintenance. Thirty years of building life is adopted as a calculation period. The cost of maintenance is calculated as the sum of the unit cost of maintaining products during the calculation period reduced by the contractor's guarantee value for a given product. The unit costs of maintaining products are designated as the multiplication of the number of individual units, the costs of replacing the product unit, and the number of product use cycles during the calculation period

In accordance with the regulation, the contracting authority specifies, among others, types of products taken into account in determining the maintenance costs paying attention to their impact on the functioning of the building and the estimated cost of products including assembly. Products listed in the regulation are: windows; doors; flooring; installations (plumbing, gas, electricity, air conditioning, etc.); elevators; façade; roofing and other products recognized by the contracting authority as valid.

The product use cycle is the period after which the product should be replaced. When specifying the number of life cycles of the product, not only the physical life of the product may be taken into consideration but also its expectations related to the need for its more frequent or rare exchange due to the nature of the building. The number of use cycles for products specifies the Annex to the Regulation. For most products, a cycle varies between 1–3 but, for example, for PVC panels floors 3–5 cycles; for roofing 2–4 cycles.

The contractor determines the product warranty period. It can apply to the whole product (e.g., a whole window or its elements; for example, a warranty for the profile is 5 years, and for the whole window only 3 years).

4. Model Sensitivity Analysis

The basic data to be assumed for the calculation of maintenance costs are therefore: I—number of product units, K—cost of replacement of the product unit, N—number of product use cycles during the calculation period as well as O_g—warranty period of the ith product given in years. The authors decided to perform a simple sensitivity analysis. It shows how the adopted data influence the final results.

The following data were assumed:

$I = 2500$ m^2;
$K = 200$ EUR;
$N = 2$;
$O_g = 3$ years.

In the following variants, one of the input data (independent variable) was changed, the remaining data were taken as constants, and its influence on the obtained C_{ut} values (dependent variable) was examined. The scope of the adopted variables and the final results are presented in Table 1. The obtained results in each variant are presented in the charts shown in Figures 1–4.

Table 1. Assumed variants and results presentation.

Variant	Independent Variable	Range of the Variables	Interval	The Dependent Variable Difference (EUR)
I	I (m^2)	1800–2200	100	144,000
II	K (EUR)	180–220	10	144,000
III	N (cycle)	1–4	1	1,080,000
IV	O_g (lata)	1–4	1	8000

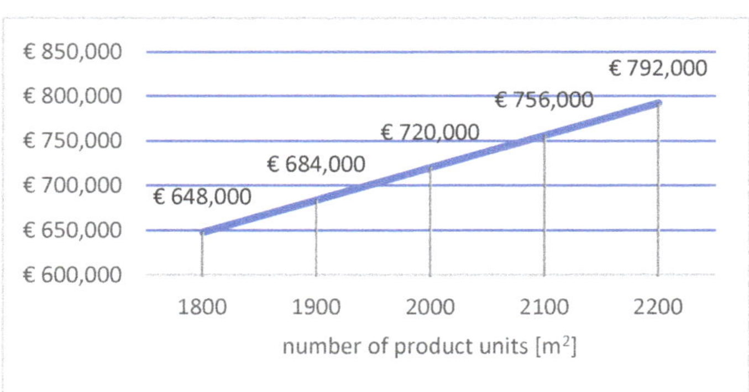

Figure 1. Obtained results for variant I.

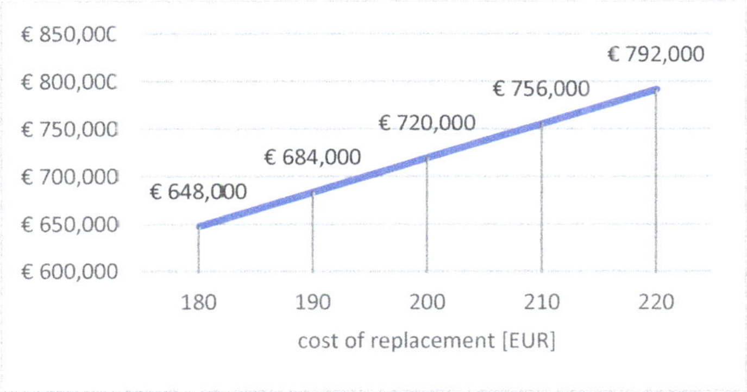

Figure 2. Obtained results for variant II.

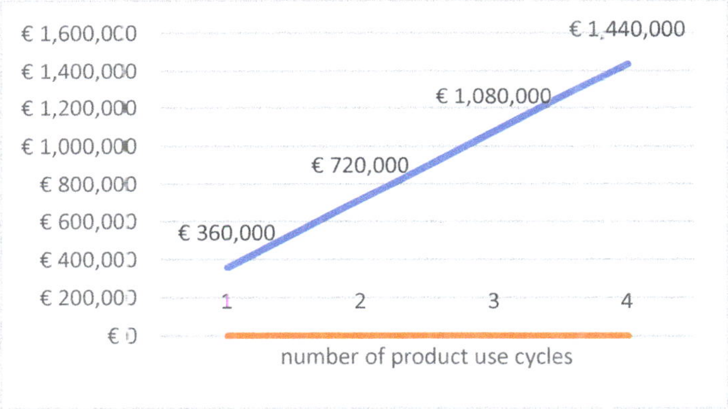

Figure 3. Obtained results for variant III.

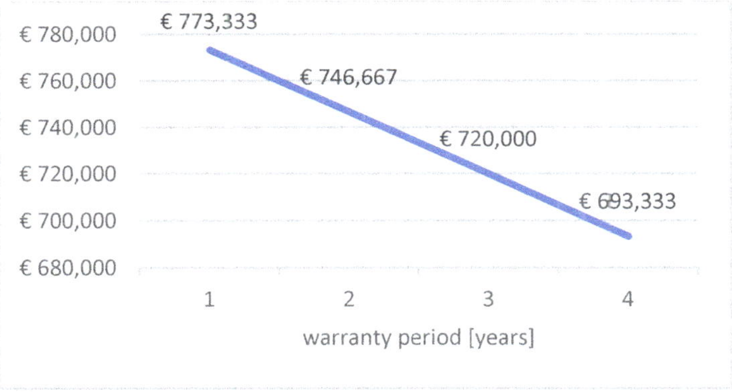

Figure 4. Obtained results for variant IV.

The presented results of the sensitivity analysis show that the influence of all variables is linear. The variable N has the greatest impact. Increasing it by one cycle causes an increase in the value of the cost of living by about 25%. A change in the number of units of a product by 100 affects the end results as much as a change in the cost of replacing a unit by 10 EUR, which equals 5%. The adopted warranty period has the least impact. Extending the warranty period by 1 year reduces costs by less than 4%.

5. Model Analysis on the Example of A Multi-Functional Building

In real circumstances, dependencies in the analyzed formula, allowing the determination of maintenance costs, are more complex. For example, higher purchase costs and, at the same time, replacing a given element generate higher durability translating into a smaller number of exchange cycles and often also a longer warranty period provided by the manufacturer. It shall also be noted that the regulation does not consider the refurbishment of some items, such as, for instance, doors or windows, which require periodic maintenance to maintain their condition and operability. The formula limitations may arise from the fact that the model given in the regulation that is analyzed was designed only for a comparison of given offers and shall help in the decision of which offer is the best during public auctions.

In order to obtain a broader image of the model action, the building presented in BCO (Bulletin of Building Prices) for the first quarter of 2021, building 1122-407 on the name "Housing, office and service apartment building", was chosen. The building has 15 stories with a 3-story underground garage. It is a multi-functional building combining housing and office functions located on the overground floors as well as service and commercial parts on the ground floor section. The property also has three underground floors where parking spaces in the garage space and technical infrastructure are located. The object is placed on a reinforced concrete foundation slab. Concrete monolithic structural walls (in the underground made in diaphragm wall technology) and all slabs are also reinforced, and monolithic concrete inverted flat roof is designed.

The analysis included four elements of the building, i.e., windows and external doors, internal doors, floors with terracotta and roofing. For each of the analyzed elements, the number of product use cycles was established during the calculation period, consistent with the compartment given in Annex 1 to the Regulation of the Minister of Investment and Development from 11 July 2018. Specified solutions were adopted for these elements, and the necessary data were established to determine the costs of living in accordance with the Regulation.

It must be noted that the analyses took into account only selected elements of the building that the contracting authority may take into account in the analysis of the maintenance costs. It is practically impossible to determine the total maintenance costs in accordance with the ordinance. As was already mentioned, the formula provided in the regulation takes into account only the cost of replacing building elements without taking into account the costs of their ongoing maintenance. For this reason, following the regulation, it is difficult to determine the maintenance costs of elements, such as pumps, elevators, HVAC equipment, escalators, lamps or water and fire devices. The assumption adopted in the regulation allows, for instance, for the inclusion of the maintenance costs of elevators, but only in the case of their replacement 1–3 times. In practice, such actions do not take place with a building's 30-year life cycle.

5.1. Windows and External Doors

The overall surface of the windows and external doors in the analyzed building equals 5 315 m^2. Four different solutions were found for windows and external doors, in which both the number of product use cycles during the calculation period, as well as the warranty period, has been changed depending on the assumed solution. For various variants, prices were found in the range of 128 EUR/m^2–628 EUR/m^2, and they were PVC windows, wooden windows as well as aluminum windows. Details of data adopted in specified variants are presented in Table 2.

Table 2. Data given for the windows and external doors.

Variant	K (EUR/m^2)	N	O_g (Years)	C_{ut} (EUR)
PVC windows (1)	128	4	2	2,549.421
PVC windows (2)	260	3	3	3,733.015
wooden windows	318	2	5	2,821.026
wooden coated aluminum windows	628	1	10	2,224.956

Figure 5 shows the costs of maintaining the presented element for four solutions. The cheapest maintenance costs were obtained for the most durable solution, and thus the most expensive; however, the least durable solution did not turn out in this case, the most expensive solution from the maintaining cost point of view an element throughout the entire computing period of the building. It turned out that medium-lasting solutions are the least cost-effective, presumably due to the fact that they are more expensive than the least-durable variant, and at the same time, the number of product use cycles during the calculation period is not sufficiently reduced to comply with higher purchase costs.

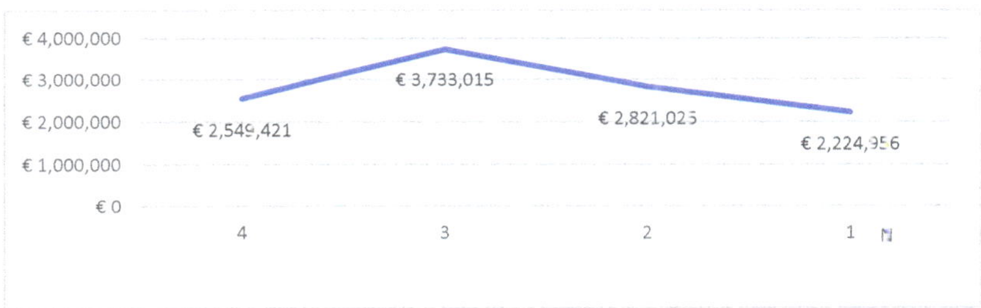

Figure 5. Summary of the assumed solutions for the widows and external doors.

5.2. Internal Doors

The overall surface of the internal doors in the analyzed building equals 2 495 m^2. Three different solutions for internal doors were found. For various variants. prices were found in the range of 192 EUR/m^2–322 EUR/m^2. In this case, the economic profitability of the proposed solutions increases with increasing the durability of the element and thus the increase in the purchase price. Details of data adopted in these variants are presented in Table 3.

Table 3. Data given for the interior doors.

Variant	K (EUR/m^2)	N	O_g (Years)	C_{ut} (EUR)
interior doors (1)	192	3	2	1,339,891
interior doors (2)	230	2	3	1,033,640
interior doors (3)	322	1	5	669,953

The internal doors maintenance costs during the whole life cycle are shown on the Figure 6.

Figure 6. Summary of the assumed solutions for the interior doors element.

5.3. Roof

The overall surface of the roof in the analyzed building equals 3823 m^2. Three different solutions for the roof were designed. For various variants, prices were found in the range of 135 EUR/m^2–174 EUR/m^2. For the solutions adopted for the roof as well as the floors, similar conclusions can be drawn as for interior doors. In the case of a roof, the most durable solution turns out to be the most cost-effective considering the maintenance costs of this structure element over the entire life cycle calculation period. Details of the data adopted in these variants are presented in Table 4.

Table 4. Data given for the roof.

Variant	K (EUR/m^2)	N	O_g (Years)	C_{ut} (EUR)
roofing	135	3	5	1,293,088
green roof (1)	162	2	7	951,713
green roof (2)	174	1	10	444,065

The roof maintenance costs during the whole life cycle are shown on the Figure 7.

Figure 7. Summary of the assumed solutions for the roof element.

5.4. Flooring

The overall surface of the flooring in the analyzed building equals 34,459 m^2. Two different solutions for the terracotta flooring were designed. The difference between these variants was the durability of the flooring. The cheaper variant assumed one charge of flooring in the building during the calculation period, whereas the more expensive one, as well as more durable one, assumed one flooring during the whole buildings' life cycle. The first option was priced at 28 EUR/m^2, while the second one was estimated at the point of 41 EUR/m^2. Taking into account the prices, the number of product life cycles in the calculation period as well as the warranty periods, the more durable option turned out to be the more cost-effective one throughout the life cycle. Details of the data adopted in the two designed variants are presented in Table 5.

Table 5. Data given for the floors.

Variant	K (EUR/m^2)	N	O_g (Years)	C_{ut} (EUR)
flooring (1)	28	2	2	1,788,560
flooring (2)	41	1	6	1,124,347

The flooring maintenance costs during the whole life cycle are shown on the Figure 8.

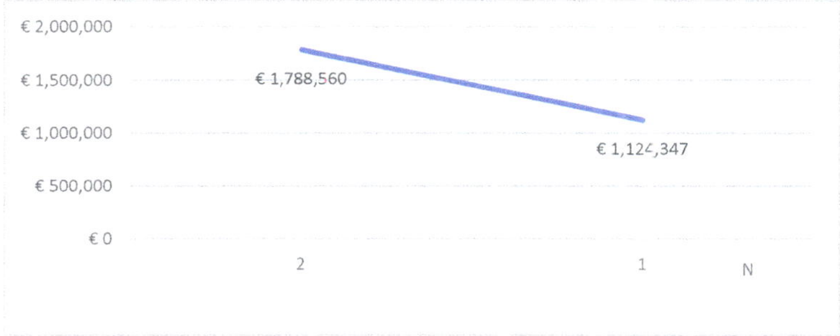

Figure 8. Summary of the assumed solutions for the terracotta flooring element.

5.5. The Solutions Analysis

The analyses carried out in the article allowed determining the impact of variables on the maintenance costs and to select the variants generating the highest and the lowest cost values. The summary of these results is presented in Figure 9.

Figure 9 allows concluding that, despite the higher price of materials at the beginning of production, the most beneficial solutions in the entire life cycle of the facility turn out to be the most durable solutions, which are the most expensive to purchase. This trend has worked well for all four cases considered.

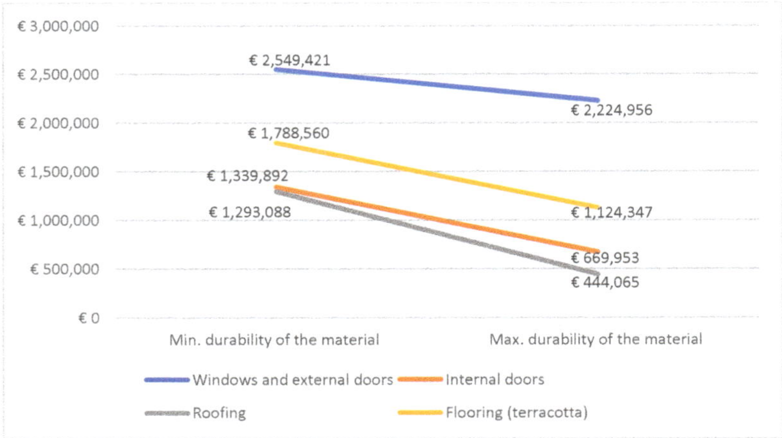

Figure 9. The most and least preferred solutions summary.

5.6. Share of Maintenance Costs in The Total Life Cycle Costs

The calculated maintenance costs of the facility during the calculation period were also analyzed in relation to the entire life cycle costs of the building over the 30-year lifetime of the building. For this purpose, acquisition costs were assumed in the amount of construction costs specified in the BCO and operating costs based on the average demand for utilities per m^2 of the facility. On the basis of the calculated values, it was found that when using solutions with a lower durability, the share of calculated building maintenance costs in the total costs incurred in 30 years of use is about 5% higher compared to higher durability solutions. The conclusions described above are presented in Figure 10.

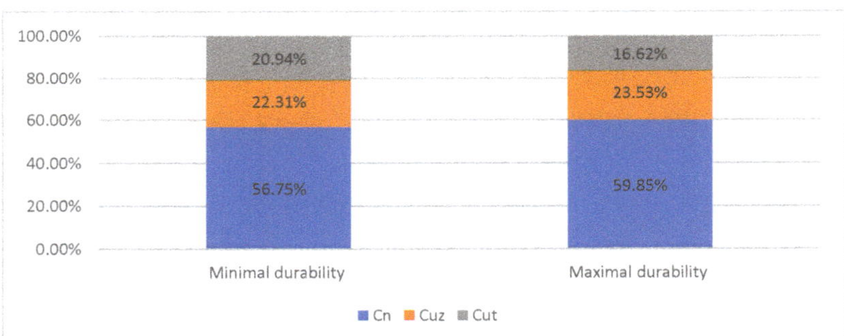

Figure 10. Percentage life cycle costs summary of the most and the least preferred solutions summary for the maintenance costs.

6. Extension of the Method Adopted in The Regulation

The method which is applied in the regulation is a simple method—does not take into account changes in the value of money over time. Another limitation is also the 30-year life cycle that has to be imposed.

6.1. NPV Addition

In the first stage, the obtained results were compared to the method determining the present value of NPV.

The basic calculation formula is as follows:

$$LCNPV = \sum_{i=0}^{ESL} \frac{CF_i}{(1+r)^i} \tag{5}$$

where: *CFi*—cash flow in *i*th year, *ESL*—estimated service life in years, *i*—subsequent year, *r*—discount rate.

Figures 11 and 12 show the results for the maintenance costs of windows and external doors determined by the simple and NPV methods, assuming different interest rates. The same cost assumptions as for the simple method and the 5-year warranty period for all variants were adopted.

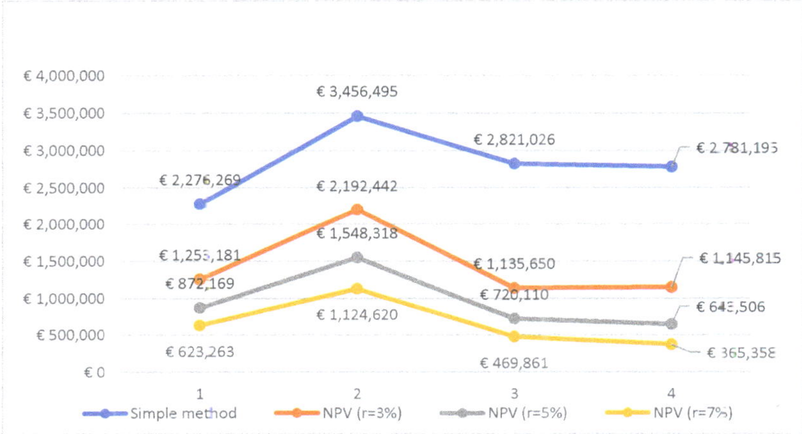

Figure 11. Cost breakdown for windows and external doors element for each variant calculated using simple and NPV methods.

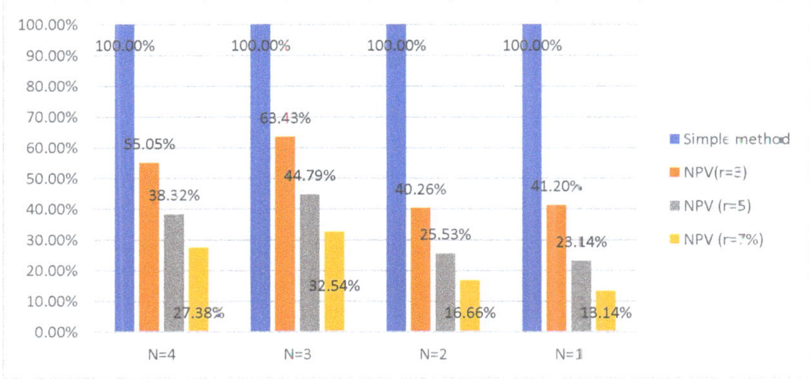

Figure 12. Percentage life cycle costs summary of each variant of windows and external doors element calculated using simple and NPV methods.

The results presented above show that the most durable variant is the most profitable, assuming that the NPV discount rate is equal to 7%. Besides the calculation method choice, the most durable option turned out to be the most profitable option each time, while the least profitable one is the one assuming 3 product life cycles during the assumed calculation period.

Figure 13 presents the average percentage results for all materials included in the study. The presented results also show that the most durable materials turn out to be the cheapest to maintain, taking into account the entire life cycle of the facility.

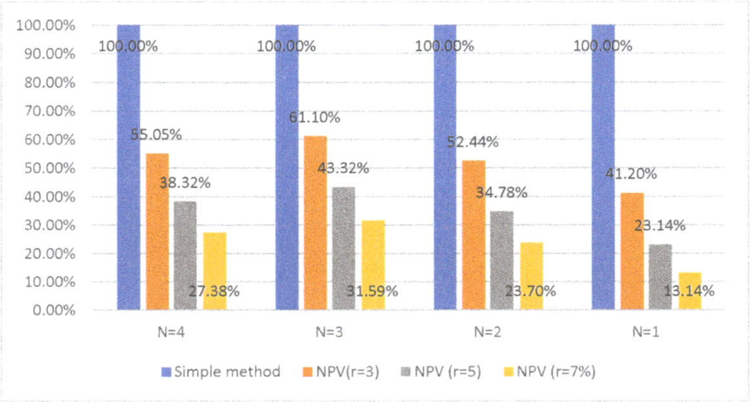

Figure 13. Percentage average results summary of all materials calculated using simple and NPV methods.

Figure 14 shows the results of the analysis for the total maintenance costs of the facility in the calculation period in relation to the entire life cycle costs of the building over the 30-year lifetime of the building. The list was prepared for the NPV variants with the assumption of a 3%, 5% and 7% discount rate.

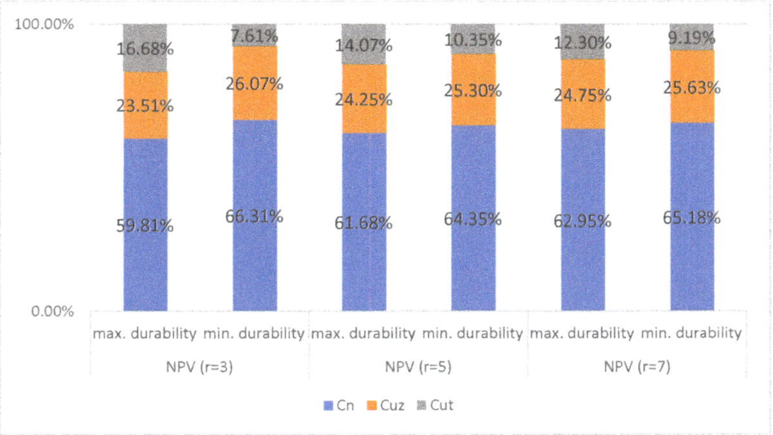

Figure 14. Percentage life cycle costs summary of the most and the least preferred solutions summary for the maintenance costs. (NPV).

On the basis of the calculated values, it was found that when using solutions with a lower durability, the share of calculated building maintenance costs in the total 30 years life cycle is always higher than when using solutions with higher durability. However, this difference is smaller than in the case of calculations performed in accordance with the regulation and decreases with the increase in the discount rate. Assuming a discount rate of 3%, the difference equals 2.56%, assuming a 5% discount rate, the difference equals 1.05%, while assuming a 7% discount rate, the difference equals only 0,88%.

6.2. The Life Cycle Equals 80 Years

An attempt to extend the life cycle was also made. The 30-year life cycle of the building (imposed by the regulation) was extended to 80 years.

Figure 15 shows the results for the maintenance costs of windows and external doors. The same cost assumptions were adopted as for the simple method, and the number of product life cycles in the calculation period was assumed proportionally greater than the designed life cycle period.

Figure 15. The cost breakdown for each variant of windows and external doors element assuming 80-year life cycle period.

The presented data show that in the case of windows and external doors, assuming an 80-year life cycle of the building, the least durable solution is the cheapest solution (11 life cycles of the product were assumed). Intermediate solutions are the most expensive, and the most durable option is the second most economically profitable option.

Other elements, i.e., internal doors, roof and floors, remain the cheapest in the case of choosing the most durable solutions and the most expensive in the case of choosing the least durable. Figure 16 shows the results described above.

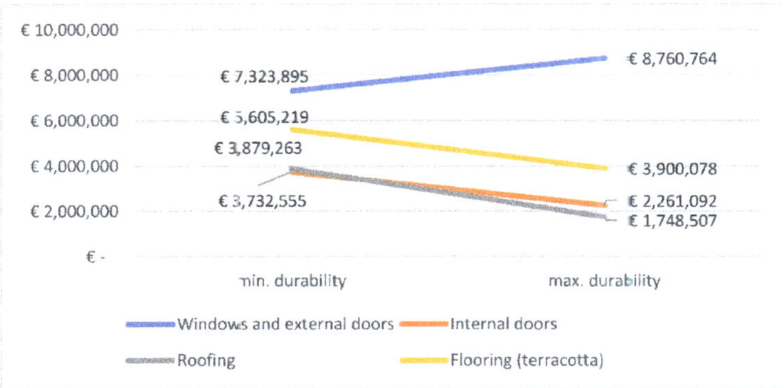

Figure 16. Cost breakdown for each element assuming 80-year life cycle period.

7. Discussion

The article analyzes the determination of building maintenance costs based on the method proposed in Polish legal regulations. It should be noted here that the method proposed in Polish regulation has certain limitations. The formula is fixed, and some variables are not easy to implement in a realistic way. One of the main limitations is taking into account only the costs of replacing individual elements without the possibility of calculating refurbishment costs, which, in many cases, constitute a significant piece of the maintenance budget. The assumption made in the regulation regarding the calculation of maintenance of some elements is directly contrary to real conditions. For instance, the inclusion of the maintenance costs of elevators requires their replacement 1–3 times, and such actions do not take place in real conditions with a building's 30-year life cycle. The method used for calculations does not take into account changes in the value of money over time. Another limitation is also the life cycle has to be designated as 30 years.

Another limitation is, for instance, that the warranty period is fixed to one period of time for the whole product calculation, which does not take into account the possibility that, in the whole calculated period, products with different warranty periods will be implemented.

The formula limitations may arise from the fact that the model analyzed, which is given in the regulation, was designed only for the comparison of given offers and shall help in making a decision of which offer is the best during the public procurement procedure. The restrictions imposition is aimed at adopting the same assumptions by all tenderers, which facilitates the evaluation of tenders by the procurer, giving a picture of how the higher initial cost translates into savings during the operation of the building. However, all the above-mentioned limitations make it practically impossible to determine the total maintenance costs in accordance with the regulation. For this reason, the authors decided to carry out a maintenance cost calculation, limited to the cost of replacement, of the selected building elements.

8. Conclusions

One of the components of the life cycle costs of a building, which is particularly difficult to determine, is its maintenance costs incurred during its operation. The article analyzes the determination of building maintenance costs based on the method proposed in Polish legal regulations.

The method's sensitivity analysis shows that the assumed number of life cycles has the greatest impact. An increase by one cycle increases the value of the maintenance cost by about 25%. The adopted warranty period has the least impact. These are certain dependencies that should be taken into account when considering the obtained results.

In order to obtain a broader picture of the model's usefulness, a multi-family residential, office and service building was analyzed. The results of the analyses allow us to conclude that, despite the higher price of materials, the most durable solutions, which are the most expensive to purchase, turn out to be the most advantageous in the entire life cycle of the building. It was found that when using solutions with a lower durability, the share of calculated building maintenance costs in the total costs incurred in 30 years of use is about 5% higher than when using solutions with higher durability.

In the extension of the method, the use of NPV for calculations was proposed. From the obtained results, it can be concluded that the most durable materials turn out to be the most profitable to apply, taking into account the entire life cycle of the facility. On the basis of the calculated values, it was found that when using solutions with a lower durability, the share of calculated building maintenance costs in the total costs incurred in 30 years of maintenance is always higher than when using solutions with higher durability. However, this difference is smaller than in the case of calculations in accordance with the Polish regulation and decreases with the increase in the discount rate. Extending the life cycle of a building to 80 years is primarily associated with increasing the number of product life cycles, which largely translates into the obtained results. In the case of windows and external doors, assuming an 80-year life cycle, the cheapest solution is the least durable solution (11 product life cycles were assumed). Intermediate solutions are the most expensive, and the most durable option is the second most economically profitable option. Other elements, i.e., internal doors, roof and floors, remain the most reasonable from the costing point of view for the choice of the most durable solutions, and the least durable appears to be the most expensive ones.

The formula proposed in the regulation is intended for specific purposes (comparison of tenders) and contains many generalizations and simplifications. However, in the authors' opinion, it can be used to estimate the maintenance costs of selected building elements.

The changes proposed by the authors may enable the model given in the Polish regulation, which is used nowadays only in public procurement areas in Poland, to extend the implementation possibilities both for other areas in Poland as well as in other countries.

Author Contributions: Conceptualization, E.P.; methodology, E.P.; resources, J.G.; data curation, J.G..; writing—original draft preparation, J.G.; writing—review and editing, E.P. All authors have read and agreed to the published version of the manuscript.

Funding: This research received no external funding.

Institutional Review Board Statement: Not applicable.

Informed Consent Statement: Not applicable.

Data Availability Statement: No new data were created or analyzed in this study. Data sharing is not applicable to this article.

Conflicts of Interest: The authors declare no conflict of interest.

References

1. Directive 2014/24/UE of the European Parliament and the Council of 26th February 2014 on Public Procurement and Repealing. Available online: https://www.legislation.gov.uk/eudr/2014/24/contents (accessed on 10 October 2021).
2. Regulation of the Minister of Investment and Development of 11 July 2018 on the Method of Calculating the Costs of the Life Cycle of Buildings and the Method of Presenting Information on These Cost. Available online: https://isap.sejm.gov.pl/isap.nsf/DocDetails.xsp?id=WDU20180001357 (accessed on 10 October 2021). (In Polish)
3. Baryłka, A.; Baryłka, J. *Eksploatacja Obiektów Budowlanych. Poradnik dla Właścicieli Zarządców Nieruchomości*; Wydawnictwo CRB: Łomianki, Poland, 2016. (In Polish)
4. ISO 15686-5:2017-Buildings and Constructed Assets-Service Life Planning Part 5: Life cycle costing. Available online: https://www.iso.org/standard/61148.html (accessed on 10 October 2021).
5. EN 13306:2017-Maintenance-Maintenance Terminology. Available online: https://standards.iteh.ai/catalog/standards/cen/5af77559-ca38-483a-9310-823e8c517ee7/en-13306-2017 (accessed on 10 October 2021).
6. Burt, N. *Facilities Management—Good Practice Guide*; Facility Management Association of Australia Ltd.: Docklands, VIC, Australia, 2012. Available online: https://www.melbourne.vic.gov.au/SiteCollectionDocuments/good-practice-guide-facilities-management.pdf (accessed on 10 October 2021).
7. Jońska, B. *Zarządzanie Nieruchomościami Komercyjnymi*; C.H. BECK: Frankfurt, Germany, 2014. (In Polish)
8. Roper, K.; Payant, R. *The Facility Management Handbook*; AMACOM: New York, NY, USA, 2014.
9. Sullivan, G.; Pugh, R.; Melendez, A.P.; Hunt, W.D. Operations & Maintenance Best Practices—A Guide to Achieving Operational Efficiency (Release 3). 2010. Available online: https://www.osti.gov/biblio/1034595/ (accessed on 10 October 2021).
10. Kaklauskas, A.; Zavadskas, E.K.; Raslanas, S. Multivariant design and multiple criteria analysis of building refurbishments. *Energy Build.* **2005**, *37*, 361–372. [CrossRef]
11. Starynina, J.; Ustinovichius, L. A multi-criteria decision-making synthesis method to determine the most effective option for modernizing a public building. *Tech. Econ. Dev. Econ.* **2020**, *26*, 1237–1262. [CrossRef]
12. Nagar, A. Development of Fuzzy Multi Criteria Decision Making Method for Selection of Optimum Maintenance Alternative. *Int. J. Appl. Res. Mech. Eng.* **2012**, *1*, 206–211. [CrossRef]
13. Tupenaite, L.; Zavadskas, E.K.; Kaklauskas, A.; Turskis, Z.; Seniut, M. Multiple Criteria Assessment Of Alternatives For Built And Human Environment Renovation. *J. Civ. Eng. Manag.* **2010**, *16*, 257–266. [CrossRef]
14. Zavadskas, E.K.; Kaklauskas, A.; Gulbinas, A. Multiple criteria decision support web-based system for building refurbishment. *J. Civ. Eng. Manag.* **2004**, *10*, 77–85. [CrossRef]
15. Bucoń, R. Model supporting decisions on renovation and modernization of public utility buildings. *Open Eng.* **2019**, *9*, 178–185. [CrossRef]
16. Nowogonska, B. Preventive Services of Residential Buildings According to the Pareto Principle. *IOP Conf. Series: Mater. Sci. Eng.* **2019**, *471*, 112034. [CrossRef]
17. Ighravwea, D.E.; Sunday, A.O. A multi-criteria decision-making framework for selecting a suitable maintenance strategy for public buildings using sustainability criteria. *J. Build. Eng.* **2019**, *24*, 100753. [CrossRef]
18. Ji, A.; Xue, X.; Wang, Y.; Luo, X.; Zhang, M. An Integrated Multi-Objectives Optimization Approach On Modelling Pavement Maintenance Strategies For Pavement Sustainability. *J. Civ. Eng. Manag.* **2020**, *26*, 717–732. [CrossRef]
19. Peach, R.H.; Visser, J.K. Measuring human factors in maintenance: A literature review. *S. Afr. J. Ind. Eng.* **2020**, *31*, 104–114.

20. Zhong, S.; Pantelous, A.A.; Goh, M.; Zhou, J. A reliability-and-cost-based fuzzy approach to optimize preventive maintenance scheduling for offshore wind farms. *Mech. Syst. Signal Pract.* **2019**, *124*, 643–663.
21. Batinić, D.D.; Bukvić, A. Use of OEE in Optimization of Maintenance Schedule. Available online: https://www.wirenet.org/wire-journal-international) (accessed on 10 October 2021).
22. Al-Refaie, A.; Al-Shalaldeh, H.; Lepkova, N. Proposed Procedure for Optimal Maintenance Scheduling Under Emergent Failures. *J. Civ. Eng. Manag.* **2020**, *26*, 396–409. [CrossRef]
23. Kim, J.-M.; Kim, T.; Yu, Y.-J.; Son, K. Development of a Maintenance and Repair Cost Estimation Model for Educational Buildings Using Regression Analysis. *J. Asian Arch. Build. Eng.* **2018**, *17*, 307–312. [CrossRef]
24. Farahani, A.; Wallbaum, H.; Dalenbäck, J.-O. Optimized maintenance and renovation scheduling in multifamily buildings—A systematic approach based on condition state and life cycle cost of building components. *Constr. Manag. Econ.* **2018**, *37*, 139–155. [CrossRef]
25. Li, C.-S.; Guo, S.-J. Life Cycle Cost Analysis of Maintenance Costs and Budgets for University Buildings in Taiwan. *J. Asian Arch. Build. Eng.* **2012**, *11*, 87–94. [CrossRef]
26. Le, A.T.H.; Domingo, N.; Rasheed, E.; Park, K.S. Building Maintenance Cost Planning and Estimating: A Literature Review. In Proceedings of the 34th Annual Association of Researchers in Construction Management Conference, ARCOM 2018, Belfast, UK, 3–5 September 2018.
27. Kwon, N.; Song, K.; Ahn, Y.; Park, M.; Jang, Y. Maintenance cost prediction for aging residential buildings based on case-based reasoning and genetic algorithm. *J. Build. Eng.* **2019**, *28*, 101006. [CrossRef]
28. Fregonara, E.; Ferrando, D.G. The Stochastic Annuity Method for Supporting Maintenance Costs Planning and Durability in the Construction Sector: A Simulation on a Building Component. *Sustainability* **2020**, *12*, 2909. [CrossRef]
29. Otmani, A.; Bouabaz, M.; Al-Hajj, A. Predicting Maintenance and Rehabilitation Cost for Buildings Based on Artificial Neural Network and Fuzzy Logic. *Int. J. Comput. Intell. Appl.* **2020**, *19*, 2050001. [CrossRef]
30. Juszczyk, M.; Leśniak, A.; Zima, K. ANN Based Approach for Estimation of Construction Costs of Sports Fields. *Complexity* **2018**, *2018*, 1–11. [CrossRef]
31. Leśniak, A.; Wieczorek, D.; Górka, M. Costs of facade systems execution. *Arch. Civ. Eng.* **2020**, *66*, 81–95.
32. Konior, J.; Stachoń, T. Bayes Conditional Probability of Fuzzy Damage and Technical Wear of Residential Buildings. *Appl. Sci.* **2021**, *11*, 2518. [CrossRef]
33. Wieczorek, D.; Plebankiewicz, E.; Zima, K. Model Estimation of the Whole Life Cost of a Building with Respect to Risk Factors. *Technol. Econ. Dev. Econ.* **2019**, *25*, 20–38. [CrossRef]
34. Chen, C.; Tang, L. BIM-based integrated management workflow design for schedule and cost planning of building fabric maintenance. *Autom. Constr.* **2019**, *107*, 102944. [CrossRef]
35. Lee, M.; Lee, U.-K. A framework for evaluating an integrated BIM ROI based on preventing rework in the construction phase. *J. Civ. Eng. Manag.* **2020**, *26*, 410–420. [CrossRef]
36. Leśniak, A.; Górka, M.; Skrzypczak, I. Barriers to BIM Implementation in Architecture, Construction, and Engineering Projects—The Polish Study. *Energies* **2021**, *14*, 2090. [CrossRef]
37. Zima, K.; Plebankiewicz, E.; Wieczorek, D. A SWOT analysis of the use of BIM technology in the Polish construction industry. *Buildings* **2020**, *10*, 16. [CrossRef]

sustainability

MDPI

Article

Building Circularity Assessment in the Architecture, Engineering, and Construction Industry: A New Framework

Nuo Zhang *, **Qi Han** and **Bauke de Vries**

Information Systems in the Built Environment, Eindhoven University of Technology, P.O. Box 513, 5600 MB Eindhoven, The Netherlands; q.han@tue.nl (Q.H.); b.d.vries@tue.nl (B.d.V.)
* Correspondence: n.zhang1@tue.nl

Abstract: Circular Economy (CE) has proved its contribution to addressing environmental impacts in the Architecture, Engineering, and Construction (AEC) industries. Building Circularity (BC) assessment methods have been developed to measure the circularity of building projects. However, there still exists ambiguity and inconsistency in these methods. Based on the reviewed literature, this study proposes a new framework for BC assessment, including a material flow model, a Material Passport (MP), and a BC calculation method. The material flow model redefines the concept of BC assessment, containing three circularity cycles and five indicators. The BC MP defines the data needed for the assessment, and the BC calculation method provides the equations for building circularity scoring. The proposed framework offers a comprehensive basis to support a coherent and consistent implementation of CE in the AEC industry.

Keywords: building circularity assessment; material flow model; building circularity material passport; building circularity calculation method

Citation: Zhang, N.; Han, Q.; de Vries, B. Building Circularity Assessment in the Architecture, Engineering, and Construction Industry: A New Framework. *Sustainability* **2021**, *13*, 12466. https://doi.org/10.3390/su132212466

Academic Editors: Oleg Kapliński, Lili Dong, Agata Bonenberg and Wojciech Bonenberg

Received: 12 September 2021
Accepted: 29 October 2021
Published: 11 November 2021

1. Introduction

The world is experiencing a growing threat of waste, emissions, and other environmental changes. An agile transformation in the construction sector [1] is needed to mitigate these threats. Since 36% of CO_2 in the world originated from the architecture, engineering, and construction (AEC) activities [2], the AEC industry has huge potential in reducing waste and emissions in all the phases of design, construction, and demolishment.

Circular Economy (CE) is recently recognized as a possible solution to solve waste and emissions problems in the AEC industry. It contrasts with the "take-make-dispose" linear economy and refers to an industrial economy based on closed loops [3]. By retaining the added value in the loops as long as possible, CE saves material, eradicates waste, and creates opportunities for moving towards more sustainable development.

In the AEC industry, Countries and agencies have set goals to reduce waste and promote circularity. For example, the EU Waste Framework Directive requires all EU countries to have a minimum recycling rate of 70% for Construction and Demolition Waste (CDW) generated by 2020 [4]. The Netherlands has set the target to operate with the CE principle by 2050, and the construction sector is one of the five priority sectors [5].

Researchers investigated the application of CE in the AEC industry for different aspects. Butkovic et al. reviewed 96 published papers and summarized five aspects of CE according to the papers: waste management, reducing the impact on the environment, material & product design, building design, and others [6]. Researchers focus on all stages of buildings, including the design stage [7]. Sparrevik et al. proposed and compared several methods based on LCA for analyzing the environmental performance of buildings. They came to the conclusion that methods at lower systemic levels as well as higher systemic levels can benefit the circular development in the built environment [8]. However, the assessment of CE in the AEC industry is still not yet unequivocally defined and is especially lacking practical approaches [9].

Much research about Building Circularity (BC) focuses on the material flow during the whole lifecycle of a building, especially in the recycling process of building materials and components. Research shows that in the United States, only 6% to 7% of the metallic aluminum has been lost to the environment since 1900. About 68% to 69% are still in use, while about 25% have no clear destination [10]. BC assessment can provide insight into the material flow by indicating the material flow's completeness and tightness (i.e., the circularity rate). Various research has been executed on circularity assessment approaches in the AEC industry [11]. Almost all the existing methods used in the AEC sector are based on a CE general assessment framework, such as the Material Circularity Indicator (MCI) [12], which is also widely used in other industries. Research about indicators and metrics are conducted to measure the circularity of products and systems. For example, Mesa et al. reviewed the existing CE indicators and developed a new set of indicators for measuring the circularity performance of product families [13]. Parchomenko et al. executed a multiple correspondence analysis of 63 metrics and proposed a standardized visualization framework for CE metrics. This framework is provided for integrating existing complementary CE metrics and facilitating further CE metrics [14]. Ruiz-Pastor et al. measured the relationship between personal intrinsic factors and circularity in design [15]. Vinante et al. collected and analyzed CE assessment indicators at firm-level metrics and organized them in a framework which matches metrics and functions and facilitates user individuation [16]. R strategy factors are developed to 10R in CE assessment [17]. However, by its nature, a building is a large industrial product with many materials and components. Size is more prominent than for other industrial products, and its lifetime is longer than most other industrial products. The assessment of CE in the AEC industry, defined as Building Circularity (BC) assessment in this paper, is relatively underexplored.

This research aims to fill the following observed research gaps:

1. While different tools are created for BC assessment, many assessment methods are limited to the perspective of a specific cycle, such as the environmental or technical cycle, which leads to an incomprehensive assessment.
2. Comprehensive dedicated circularity models with indicators for the AEC industry according the material flow in building construction are lacking.
3. The application of R strategy factors is far from practical for BC assessment.

In this paper, Building Circularity (BC) assessment is explored, focusing on the AEC domain. The framework developed in this paper contains three main parts: a material flow model, a BC Material Passports (MP), and a BC calculation method. The material flow model is the core part of the framework that redefines the concept of BC assessment, containing three circularity cycles and five indicators. It contains different circularity loops with R strategy factors. And all the R strategy factors are defined. The MP and the detailed BC calculation methods are combined with the material flow model to store data and complete specific calculations. Therefore, a systematic review is executed, and a new framework is proposed. Section 2 introduces the systematic review approach used in this paper. Section 3 addresses the material flow model, followed by the discussion of material passports in Section 4. Section 5 shows the BC calculation method. In Section 6, conclusions are drawn on the contribution to BC assessment and the limitations of the proposed framework.

2. Systematic Review Exiting Building Circularity Assessment Methods

The literature review of existing BC assessment methods is conducted by searching, selecting, and reviewing papers relevant to CE assessments in the AEC industry. In this research, Scopus was used to search for relevant papers. Since CE is a new concept, related terms are used to find more results. With the search term 1 "circular economy "or "circularity", used in combination with term 2 "construction", and term 3 "indicator" or "assessment", 442 articles were found. Many of the papers found in this first step have no clear relationship with the topic of this work. Therefore, they were excluded in the second step of selection, leaving 38 relevant papers. Nine of these papers are duplicated, so 27 papers are reviewed in this work. These papers showed that most of the developed assessment methods have been published since 2016. Table 1 shows the stepwise process, including the keywords and the number of found papers.

Table 1. Searching process for the reviewed papers.

Term 1	Term 2	Term 3	Paper Found in the First Step	Remaining Paper in the Second Step
Circular economy	Construction	Indicator	95	9
Circular economy	Construction	Assessment	281	17
Circularity	Construction	Indicator	22	4
Circularity	Construction	Assessment	44	8
Total			442	38

Next, the search was expanded with other keywords associated with the CE in the AEC industry. The keywords chosen here are the words that have a similar meaning relevant to CE, such as "recycling", or that are related to BC assessment, such as "sustainable development" and "waste management". We found this extension necessary because research in these fields has a long tradition and is more mature. Besides these academic papers, other documents published by companies, for example, the description of the CE calculation method for a building from the company Madaster [18], were included. Finally, 60 pieces of research were selected and reviewed to develop the new BC assessment framework presented in the following sections. All the scientific articles analyzed in this work have been published before September 2021.

The selected papers are categorized into three main BC assessment aspects: (1) material flow model, (2) material passports, and (3) BC calculation method, altogether constituting a novel BC assessment framework. These three BC assessment aspects are presented in the following sections subsequently. Every section has the same structure. First, the existing situation is discussed using the selected literature. Second, a new model or method is proposed, addressing the AEC demands. Third, an analysis of the new proposed model/methods is presented, highlighting the differences with the existing situation.

3. Material Flow Model

3.1. Existing Material Flow Models

The existing material flow models review starts from the definition of CE in general and CE in the AEC industry. CE is a new concept, which caused extensive discussion in recent years. Saidani et al. [19] showed that a good definition of CE is still lacking, while the number of CE indicators has reached 55. CE, as a complex and fuzzy concept, is hard to summarize as a generic concept. However, a suitable CE concept is needed to guide the successful implementation. One of the most used CE definitions that originate from the Ellen MacArthur Foundation is "an economic and industrial model that is restorative by intent and design" [20]. "Taking a new systemic perspective, it replaces the concept of waste with the one of restoration and aims to decouple economic growth from the use of virgin resources." Collecting and analyzing 114 CE definitions, Kirchherr et al. [21] conclude that "CE is an economic system that replaces the 'end-of-life' concept with reducing, alternatively reusing, recycling and recovering materials in production/distribution and consumption processes.".

While there are the concepts of CE in general, the concept for CE in the AEC industry, Building Circularity (BC), requires a dedicated definition within the AEC context. A CE definition in the AEC industry was found in CB23 [22], which states that "Circular construction means the development, use, and reuse of buildings, areas, and infrastructure without unnecessarily depleting natural resources, polluting the environment and affecting ecosystems." Research is needed for CE development in the AEC industry to develop a more consolidated theory on CE in construction and provide planning mechanisms to assist decision making.

Life Cycle Assessment (LCA) is a method for assessing the environmental impacts associated with all stages of the lifecycle of a commercial product, process, or service, and it has already become an important method for sustainability and waste assessment. Weissenberger and Ortiz's work reviewed the development of LCA for the assessment of environmental impact for buildings. Their work shows that the development of LCA has already contributed significantly to sustainability in the AEC industry [23,24] and has the potential to contribute to BC.

Waste management is an important theme of BC [25,26]. The structure and content of the waste management model have been studied for many years. As early as 1994, Craven et al. [27] had developed a linear production and consumption model with the four steps of resources: Virgin–Production–Use–Waste. Meibodi et al. [28] suggested that recycling waste can simultaneously solve several problems, such as decreasing the landfill and cost-saving. Gharfalkar et al. [29] summarized existing "waste hierarchy" and proposed alternative definitions of them. The concept of waste management hierarchy developed by Gertsakis et al. [30] made systematic classification and assessment of the waste in the construction process, which could also be an essential part of the material flow management under BC. It shows that avoiding waste should proceed with recycling and disposal. The hierarchy proposed in the research contains reduce, reuse, recycle, treatment, and disposal, ranging from most desirable to least desirable.

Many other researchers proposed their alternative waste hierarchy structure by R strategies. The research by Gehin et al. [31] presented a 3R strategy: reuse–remanufacturing–recycle, and a framework that applies the 3R strategy in designing products in the early phase. Li [32] developed a 5R framework for a comprehensive CE assessment, in which the indexes of element, environment, economy, social, and management are included. The research by Yeheyis et al. [33] developed a 3R framework of reduce–reuse–recycle and a comprehensive strategy for implementing the framework for the decision-making of selection, classification, and management of materials. Bakker et al. [34] introduced three hierarchical design strategies for product life extension and recycling: prevention–reuse–recycling. Vermeulen et al. expanded the R strategies to 10R and redefined each of them [17].

Few studies focused on R strategies in the AEC industry. Ping et al. [35] used a case study in Taiwan to show their 5R strategy of CE in a construction project. In this research, 5R of rethink–reduce–reuse–repair–recycle is adopted. Table 2 shows an overview of the R strategies chosen by these different pieces of research.

Table 2. R strategies in the research of waste management.

R Strategy	Gertsakis et al., 2003 [30]	Gehin et al., 2007 [31]	Yeheyis et al., 2013 [33]	Bakker et al., 2014 [34]	Ping et al., 2021 [35]	Vermeulen et al., 2018 [17]
Refuse						√
Reduce	√		√		√	√
Reuse	√	√	√	√	√	√
Repair					√	√
Refurbish						√
Remanufacture		√				√
Repurpose						√
Recycle	√	√	√	√	√	√
Recover energy						√
Remine						√
Treatment	√					
Disposal	√					
Prevention				√		
Rethink					√	

Publication reports increased in recent years on the topic of BC assessment models. We will discuss the recently published papers about BC assessment, followed by the influential BC assessment models in the AEC industry from MCI, Madaster, and CB′23 in more detail.

The papers reviewed here assessed the circularity of buildings under different dimensions and different types of buildings. Zimmermann et al. proposed a framework including scenarios for preservation and renovation to evaluate circular economy strategies in existing buildings [36], while Gravagnuolo et al. focused on the environmental impacts of historic buildings conservation under the CE perspective [37]. Wolf et al. analyzed the environmental impact of buildings according to the CE principles [38]. Eberhardt et al. made a comparison of allocation approaches to solving the problems of allocating benefits and burdens between systems [39]. Antonini et al. studied the indicators of reversibility and durability in the BC assessment [40]. Abadi pointed out a development direction of the BC assessment model, including twelve indicators [41]. Nuñez-Cacho analyzed the development of a CE measurement scale and created a framework in seven dimensions [42], and Charef summarized 64 indicators and 5 entities [43]. Hossain reviewed the existing research and concluded that the environmental dimension and policy framework are only studied by a small percentage of research. Moreover, the economic and social dimensions of CE in the AEC industry are ignored in the existing studies. Further comprehensive evaluation strategies adopting the CE principle in the AEC industry which considers the full life cycle of buildings and the multiple dimensions are needed [44].

The MCI model, "calculating a Material Circularity Indicator for manufactured products and companies", is currently used to make assessments about the circularity of products in general [12]. The platform from Madaster functions as "a missing link in the transition to a circular economy: a central platform where the identity, quality, as well as the location of materials in buildings can be registered." The model is designed to "objectively measure the circularity level of both technical and biological lifecycles, and to determine a single Madaster score" [45]. CB′23 is a model established for "providing insight into the degree of circularity of a material, product, structure or area" [22].

In MCI, there are two cycles in the circular process: the technical cycle—the recycled properties of materials and the biological cycle—the environmental influence. Material source in input and output, and the efficiency of the recycling process are used as indicators for the material flow; the lifetime and the functional units achieved during the product lifetime are used as indicators for utility. By using these indicators, it calculates the circularity of products and gets a score as the assessment [12]. MCI provides scores of Circularity, Value capture, Recycled content and Reuse index [46]. Madaster only focuses on the technical cycle [18]. Since MCI is the most popular circularity model for all industries, the model from Madaster borrowed the basic concepts from MCI and made some adjustments. In the model, three scores for the three-phase of buildings are calculated separately: the production phase, the usage phase, and the demolishment phase. CB′23 claims the inclusion of the biological cycle. This model first classifies the materials and then calculates the circularity of each material with several different indicators.

3.2. New Material Flow Model

A new material flow model is introduced in this paper. As mentioned before, currently, there is no clear BC definition. When combining CE's definition [21] with the activities' characteristics in the AEC industry, the new definition of BC can be derived. BC is a building property that describes the circular capability, including its construction activities to create environmental quality, economic prosperity, and social equity by repair, reuse, refurbish, remanufacture, and recycle.

In line with the definition, the BC is assessed through all four aspects of the environment, economy, society, and technology, shown in Figure 1. The model contains both the technical and biological cycles and adds the social and economic cycles. The circularity calculation of the biological cycle builds on the existing LCA methods and data, while the circularity calculation of the economic cycle takes the existing Life Cycle Cost (LCC) method and data as the starting point. The social aspect is hard to quantify and not further elaborated here.

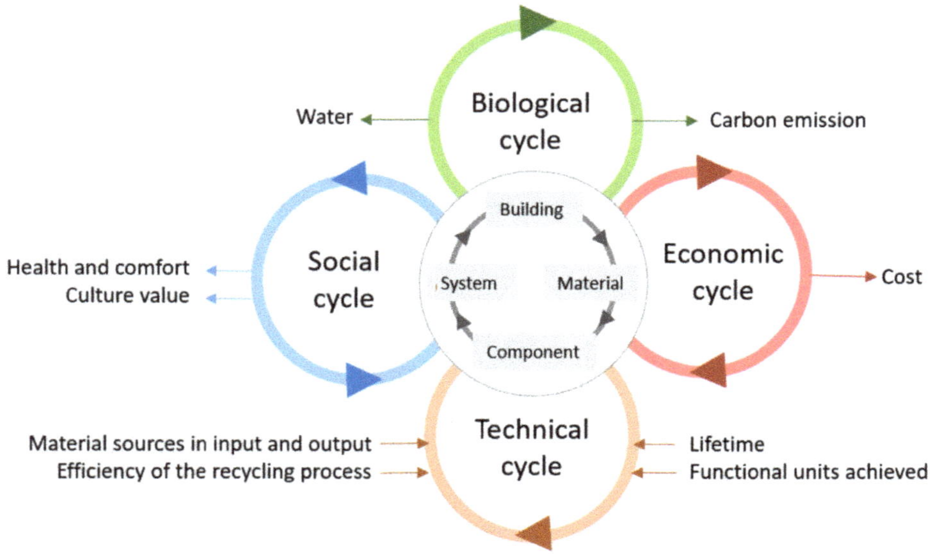

Figure 1. Circular cycles in BC.

Figure 2 shows the new material flow model and the different indicators that influence BC in different phases during the buildings' whole technical, economic, and biological lifecycle. The materials undergo the process of "rapid renewable material or virgin material" in the production phase; "part components", "components", and "systems" in the construction phase; "buildings" in the usage phase; and "landfill or burning" in the demolishment phase in the linear economy [47]. In the building lifecycle, there are other, new phases—recycling that includes reuse, repair, refurbish, remanufacture, and recycle (shown in Figure 2). Material source in input and output is influenced by the length of loops in the recycling process, and all the circularity loops here are closed. There are several kinds of loops here. In the shortest loop, the recycling of the systems is conducted directly, while in the longer loop, the systems are further disassembled into components and circulated. Material recycling exists in the longest loop. In the actual recycling of construction, there may be a mixture of each loop.

Five indicators are included in the loops. "Material source in input" is the input of material entering the production phase and recycling process. "Material source in output" is the output of material in the demolishment and recycling phase. In the loop, the resource can get lost in the "efficiency of the recycling process", which is associated with the concept of "disassembly". "The functional units archived" of a product system is a quantified description of the performance requirements that the product system fulfills. It means how many people use these buildings during their lifetime. "The lifetime" is the duration of materials, components, or buildings being used. All of the indicators are shown in Figure 2.

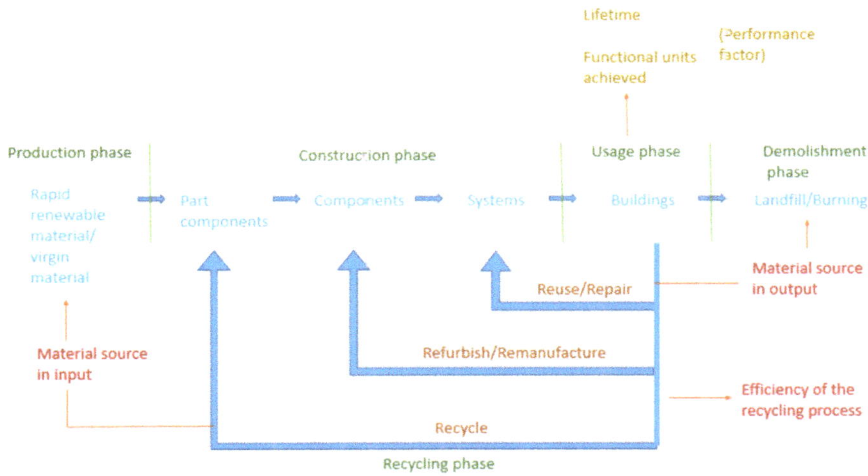

Figure 2. The new material flow model.

A new concept, "Factor", is defined here, referring to properties of the recycling phase in R strategy that describe the source and destination of material sources in the material flow. The factors are shown in Figure 3, including reuse, repair, refurbish, remanufacture, and recycle. All five material flow factors of R strategy are defined: Reuse—the building/component/system is used in other buildings directly without any process; Repair—the building/component/system is changed partly, and the renewed one is the same as the old one; Refurbish—most of the building/component/system is changed, and the renewed one is almost the same as the old one; Remanufacture—the building/component/system are completely remanufactured and used again for the same purpose; Recycle—the material/component/system are used again for a different purpose. To calculate these factors more systematically, the circularity level of these factors should be defined. For example, the circularity level of reuse should be the highest among all these factors. Therefore, "recycle" should be less circular than "reuse". It is also a more BC-friendly choice to use rapidly renewable material rather than other virgin materials, but it is not as circular as reused materials. At the end of the recycling process, the materials finally become waste and enter the demolishment phase. We can assign weights to these different factors for calculation purposes.

Figure 3 shows the relationship between the used BC indicators and the factors in material flows. All five indicators are used for the calculation of the BC. Among them, the three BC indicators (Material source in input, Material source in output, Efficiency of the recycling process) are indicators influenced by the factors, while the other two indicators (Functional units achieved, Lifetime) have no relationship with the factors.

3.3. Material Flow Model Analysis

The application scopes of MCI, Madaster, CB'23, and new material flow model differ, as shown in Table 3 [12,18,22]. While MCI is established for all industries, the other three models are created for the AEC industry only. The material flow model contains the three cycles in the assessment: the technical cycle, the biological cycle, and the economic cycle. The cycle comparison is shown in Table 3.

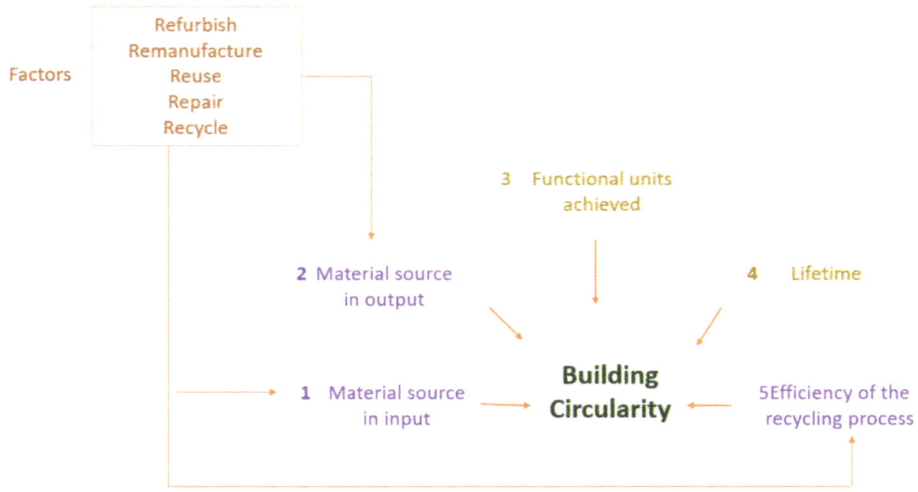

Figure 3. Factors and Indicators (1–5) for the new BC model.

Table 3. Cycle comparison between MCI, Madaster, CB′23, and the material flow model.

Item		MCI [12]	Madaster [18]	CB′23 [22]	Material Flow Model
Application domain		All industries	AEC industry	AEC industry	AEC industry
Cycles	Technical cycle	Yes	Yes	Yes	Yes
	Biological cycle	Yes	No	Yes	Yes
	Economic cycle	No	No	No	Yes

Table 4 shows the comparison of indicators used in these models. For example, the model of Madaster ignores the indicators for the efficiency of the recycling process and functional units achieved [48]. The material flow model includes the indicator of functional units achieved. This indicator refers to the average number of people who use the building every day during the whole life cycle of the building.

Table 4. Indicators comparison between MCI, Madaster, CB′23 and the material flow model.

Indicators Included in the Technical Cycle	MCI [12]	Madaster [18]	CB′23 [22]	Material Flow Model
Material source in input	Yes	Yes	Yes	Yes
Material source in output	Yes	Yes	Yes	Yes
Efficiency of the recycling process (Disassembly)	Yes	No	Yes	Yes
Lifetime	Yes	Yes	No	Yes
Functional units achieved	Yes	No	No	Yes

Madaster uses the same R strategy factors as MCI. Despite using different words, the factors defined in MCI and CB′23 are similar, such as "reuse materials" and "secondary material from reuse". There are a few differences between these two models. CB23 divides the material by input and output and defines waste scenarios for loss, burning, and landfill. In contrast, MCI defines material for refurbishment and remanufacture. The material flow model includes five defined factors: recycling, refurbishment, remanufacture, reuse, and repair. However, it does not include waste scenarios for loss, burning, and landfill, because these are not part of the recycling phase (see Figure 2).

R strategy factors in the model are used to calculate the indicators in the BC assessment models. Although the material flow model factors are the same for input and output, the values are different. The input value shows the material flow before the building/system/component was installed/constructed, while the output value denotes the estimated material flow after recycling.

4. Material Passports

4.1. Existing Material Passports

BC assessment needs data that provide detailed information on materials and components. Data used for the material flow models mentioned in Section 3 are reviewed here. Building as Material Banks [49] is a type of Material Passport (MP) that provides a European data set that focuses on the circularity information of the materials and components, such as the amount of material used from recycling. The model from Madaster uses two local Material Databases (MD): the Netherlands Institute for Building Biology and Ecology (NIBE) and Nationale Milieudatabase (NMD) [50,51]. MD such as NIBE and NMD and MP such as BAMB all provide information on materials and components. The distinction is that MD does not record information for materials and components in building projects, while MP follows the development of the materials and components during their whole lifecycle, including the recycling process. MD and MP can exist at different levels: building, system, component, and material. The higher levels contain aggregated data from the lower levels. The following screenshots are adapted from the official website of these databases and translated from Dutch to English if needed.

NIBE is a Dutch MD, providing the data of components, including a brief description and the data of properties. This MD collects data from various companies. It includes the lifetime of the components, and it divides the amount of material in the waste scenario into five kinds of categories: landfill, burning, reuse, recycling, and others. Figure 4 shows a screenshot of NIBE [50], including a figure, shadow costs, weight, lifetime, and the waste scenario. The shadow cost is an estimated measurement of environmental impacts, while the waste scenario divided the material into five categories by percentages. NIBE also provides the average distance for transport from the factory to the construction site. In the example shown in Figure 5 [50], the lifetimes of the materials foamed concrete, exterior wall paint, and mortar are 50 years, 35 years, and 75 years, respectively. The exterior wall paint and mortar are 100% burning and landfill, respectively, while the foamed concrete is 99% recycling.

Figure 4. Data provided in NIBE [50].

Foamed concrete

Product Features:

Weight	50.0	kg
Lifespan	50	years
Transport distance to factory	150	km

Waste scenario:

Landfill	1.0	%
Burning	0.0	%
Recycling	99.0	%
Reuse	0.0	%
Others	0.0	%

Exterior wall paint

Product Features:

Weight	0.4	kg
Lifespan	35	years
Transport distance to factory	150	km

Waste scenario:

Landfill	0.0	%
Burning	100.0	%
Recycling	0.0	%
Reuse	0.0	%
Others	0.0	%

Mortar

Product Features:

Weight	47.1	kg
Lifespan	75	years
Transport distance to factory	150	km

Waste scenario:

Landfill	100.0	%
Burning	0.0	%
Recycling	0.0	%
Reuse	0.0	%
Others	0.0	%

Figure 5. Data provided in NIBE for foamed concrete, exterior wall paint, and mortar [50].

NMD provides some data of the whole components, such as the component name and code. At the same time, it also provides some data of each material in the component, such as the material weight and lifetime. NMD divides the material in the waste scenario proportionally into three categories: burning, landfill, and recycling. Figure 6 is the screenshot of NMD [51]. Both the NIBE and NMD material databases provide environmental data that calculate the material's environmental impact by the shadow cost, which is an estimated price for the environment's intangible assets. For example, in the NMD, for curtain walls with the ID 21.03.007, the shadow price of Global Warming Potential (GWP) is 0.13169154, and that of Acidification is 0.00057094. The sum of these shadow prices of material is used to assess the environmental impact.

Product features

Explanation of product	No explanation available
Datacategorie	3
Productcode	11.01.001
Product name	Sand
Elementcode	11.01
Element name	Soil replenishments
Unit	m3
Productlevensduur	1000
Transport distance to construction site [km]	50
Means of transport code	900t SBK 900t Transport, freight, lorry, unspecified (or project National Environmental Database SBK version 2.2 (ecoinvent 3.4))
Scaling	n.v.t.

Construction + replacement; production, transport, and disposal

Code	Profile production	Part	Number	Unit	Construction waste	Lifespan	Code	Great	Spend	Recy
294	SBK 294 Sand (NVLB: B1 fill sand / B2 industrial sand)	addition	200.0	kg	0.05	1000	038	0.01	0.0	0.99

Figure 6. Data provided in NMD [51].

MP developed in BAMB are sets of data describing the defined characteristics of materials in products providing percentages for recovery and reuse, as shown in Figure 7 [52]. BAMB also contains information about the manufacturers. Currently, BAMB contains MP for a selection of materials, part components, and components. Moreover, it provides data on material health [49].

Other researchers have developed a database or MP for specific purposes. Heeren et al. [53] provided a database framework that contains a coding system and data structure with additional climate indicators. Honic et al. [54,55] developed an MP using Building Information Modeling (BIM) for design optimization. MP has shown its potential in storing data and facilitating analysis [56,57].

Figure 7. Data provided in BAMB [49].

4.2. New Material Passport (MP)

The new BC MP presented in this paper focuses on the circularity features of materials and components. The BC MP records the recycling information of materials and components in the project. Figure 8 shows an example of the newly proposed BC MP. It contains all the data needed for the proposed material flow model (see Figure 2), including five parts: (1) The basic information—the name, ID, size, and manufacturer; (2) The product feature—the weight, lifetime, functional units achieved and a brief description of the material or component; (3) The circularity feature—the data of five factors in percentages for input and output; (4) The environmental feature; (5) The economic feature.

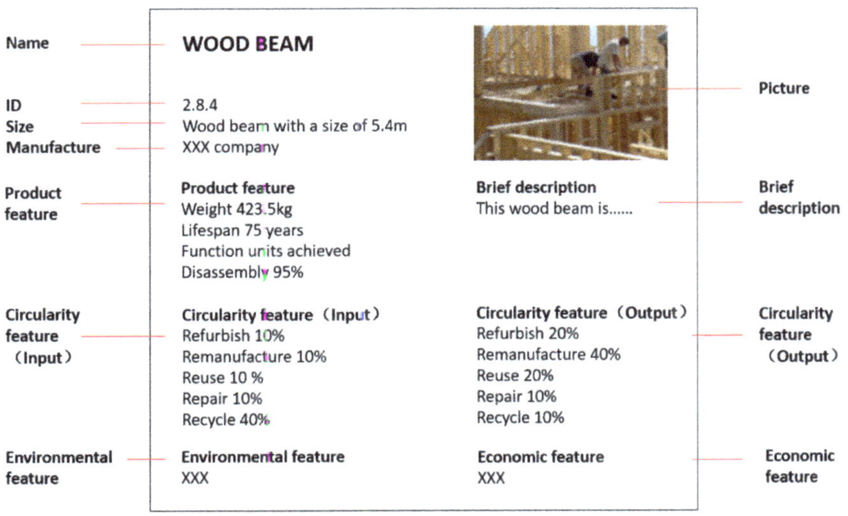

Figure 8. An example of the new BC MP.

The economic data contain the costs during the production and construction of the material or component, while the environmental data contain the data that show the environmental performance of the material or component, such as the carbon emissions. The data collected contain information both for materials and components, as sometimes the components information is collected directly, while in other cases only the materials information is available. Consequently, the calculation process may start at either materials or components.

4.3. New Material Passport Analysis

Tables 5 and 6 show the comparison between the existing MD and MP and the proposed BC MP for the indicators and the factors (see Figure 3) [49–51]. The existing MD contain several recycling data but fail to provide all the data needed for the material flow. They focus on recycling in output but do not give detailed information. For example, in the MD of NIBE, there are no data for material source in input. The only way to get the material source in input is to interpret information from the brief description of the material or component, such as "This wood beam comes from the rapidly renewable forest". However, this is not quantified data and is inaccurate. Furthermore, functional units achieved of the components and materials are also missed in all the existing MD. Last but most importantly, although the existing MD provide the lifetime of the materials and components, the data are not useful. The lifetime of a large number of materials and components are all set to be 1000 years in this database. In the BC MP, the material sources, both in input and output, are provided, along with functional units achieved and lifetime. It also provides data on the efficiency of the recycling process (disassembly). Almost all the existing MD only provide the data of reuse and recycle, while in the proposed BC MP, all required recycling features are described by a percentage for each factor. Table 5 provides an overview of the indicators covered by the different MD databases. Indicators of material source in input and output are calculated by the factors in Table 6.

Table 5. Indicators in MD and MP.

Class	Name	Material Source in Input	Material Source in Output	Functional Units Achieved	Lifetime	Efficiency of the Recycling Process
MD	NIBE [50]	×	√	×	√	×
	NMD [51]	×	√	×	√	×
MP	BAMB [49]	√	√	×	√	√
	BC MP	√	√	√	√	√

Table 6. Factors in MD and MP.

Class	Name	Refurbish	Remanufacture	Reuse	Repair	Recycle
MD	NIBE [50]	×	×	√	×	√
	NMD [51]	×	×	×	×	√
MP	BAMB [49]	×	×	√	×	√
	BC MP	√	√	√	√	√

As shown in Figure 2, the building construction process includes material–part component–component–system–building, which requires data at different levels. Both NIBE and NMD focus on the part component level. NIBE provides a group of data for the part component, while NMD provides several groups of data for each material in the part component. It means that from NIBE, data of the whole part component can be collected, but the data of each material in the part component are missing. Detailed data of each material in the part component from NMD can be collected, but the data of the part component need to be calculated. Both BMAB and BC MP focus on more levels—material and part component—and their data are provided at these different levels. In the new BC MP, the detailed data of each material in the component and for the whole component are provided.

The mentioned MD are based on the coding from the Nf/sfb database, but there are also differences. Since NIBE focuses on the part component level, it uses only codes with two digits that indicate the part component system, such as "21" for exterior walls. In NMD, the codes consist of three groups with seven digits in total. For example, "23.01.003" is the third type of concrete floor that can be found in NMD. The "23", "01", and "003" codes indicate the system, the material, and the sequence number, respectively. In the new BC MP, a material is coded with seven digits, while a part component only uses the first two digits. Both material and part component codes follow the Nf/sfb coding system, which allows for linking to the existing MD databases.

5. Building Circularity Calculation Method

5.1. Existing Building Circularity Calculation Methods

Among all the existing BC calculation methods, the method of MCI is one of the most used methods. MCI is a method for all industries, and it is recognized as a basic method [12]. Many other methods are revised from MCI and tailored to a specific industry. Two main indicators from the technical cycle calculate the circularity rate: the linear flow index and the utility factor built as a functional of the utility of a product. The former is determined by the material source of the components, which refers to the percentages of virgin material, recycled material, reused material, and waste of the materials and components. The latter is determined by the lifetime and functional units achieved during the lifecycle of the building.

$$MCI = 1 - LFI \times X \tag{1}$$

$$LFI = \frac{V + W}{2M + \frac{W_F - W_C}{2}} \tag{2}$$

$$X = \left(\frac{L}{L_{av}}\right) \times \left(\frac{U}{U_{av}}\right) \tag{3}$$

In Equation (1) "*MCI*" is Material Circularity Indicator, "*LFI*" is Linear Flow Index, and *X* is the utility factor of a product. *LFI* in Equation (2) is determined by "*V*" as the mass of virgin feedstock used in a product and "*W*" as the mass of unrecoverable waste associated with a product. "*M*" is the mass of a product, "W_F" is the mass of unrecoverable waste generated when producing recycled feedstock for a product, and "W_C" is the mass of unrecoverable waste generated in the process of recycling parts of a product. Utility factor *X* in Equation (3) is a functional of the utility of a product, where "*L*" is the actual average lifetime of a product, "L_{av}" is the actual average lifetime of an industry-average product of the same type, "*U*" is the actual average number of functional units achieved during the use phase of a product, and "U_{av}" is the actual average number of functional units achieved during the use phase of an industry-average product of the same type.

The BC calculation method from Madaster is similar to the MCI method, but it ignores some indicators, such as the functional units. In the method from CB'23, it classifies the materials into several categories and calculates the circularity rate of each of them. The equation used for the circularity rate calculation of one specific material is:

$$V_X = \sum (M_i \times M_{vi}) \sum M_i \tag{4}$$

where "M_i" is the mass of a (partial) object (*i*) and "M_{vi}" is the mass percentage of primary (virgin) raw materials in a (sub) object. "V_X" represents the percentage of primary raw materials. The share of primary raw materials is calculated per (sub) object.

5.2. New Building Circularity Calculation Method

A new method is developed in this paper for accurate calculation based on a more comprehensive understanding of the BC concept and the indicators and factors that influence the circularity of buildings, shown in Figures 9 and 10 with their corresponding symbols. It borrow from existing research in CE indicators, which focus on products and components as references, such as MCI, Total Restored Products, Circular Economy Index, and Longevity indicator [12,58–60]. The BC calculation method follows the proposed material flow model, which contains three cycles: the technical cycle, the environmental cycle, and the economic cycle (i.e., the social cycle is left out in this paper). From the technical cycle, all five indicators (CRmi, CRmo, Fu, Ls, DS) are considered for the calculation of BC, and for the calculation of material source in input and output, all five factors (Rifb, Rimu, Riui, Ripr, Rici) are used (see Figure 9). The calculation of indicators in the technical cycle is based on the MCI. The BC calculation follows the order of material–component–system–building (see Figure 2). First, the circularity of the material is calculated. The circularity of the component is the sum of all the materials' circularity that it consists of. A definition of every symbol (in bold) used in Figures 9 and 10 is provided, together with its calculation equation. The calculations of the environmental cycle and economic cycle are based on existing LCA and LCC research. The proposed method focuses on the calculation from the material level.

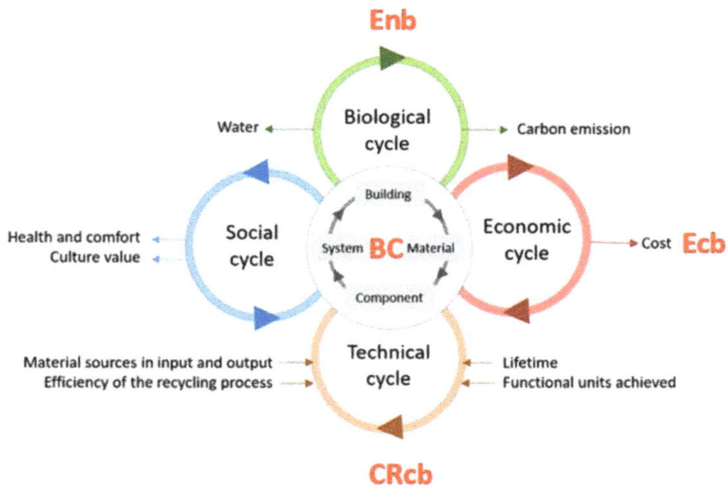

Figure 9. Circularity cycles in the BC calculation method.

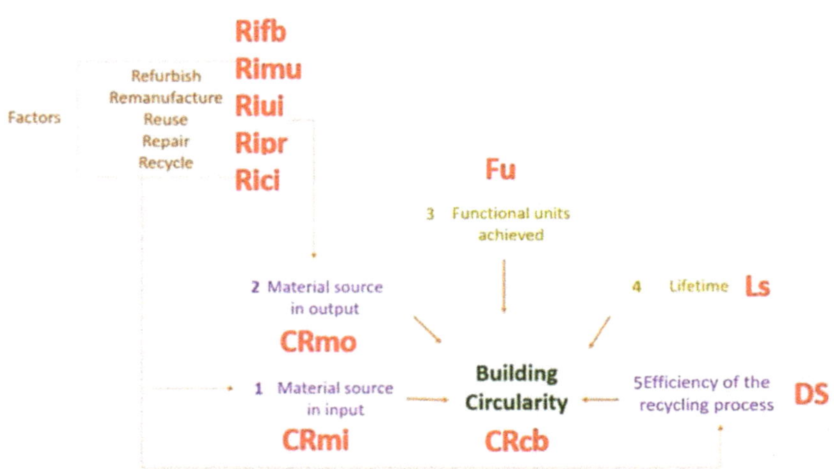

Figure 10. Symbols of indicators and factors in the BC calculation method.

Building circularity (BC): "BC" is the building circularity. "CR_{cb}" is the technical circularity rate. "E_{nb}" is the environmental impact. "E_{cb}" is the economic impact. "m" is the number of materials. "n" is the number of components. "l" is the number of systems. BC is calculated as presented in Equation (5).

$$BC = \sum_{1}^{L}\sum_{1}^{N}\sum_{1}^{M} f(CR_{cb}, E_{nb}, E_{cb}) \tag{5}$$

Environmental impact (Enb): The handbook on lifecycle assessment shows the basic assessment method of environmental impact [61]. The calculation of Enb employs the current LCA method, using the shadow cost [62]. The calculation needs both environmental data from an existing LCA database and the project data of the building. The proposed BC MP provides the project data of each material and component, as shown in Figure 8, such as the weight.

Economic impact (Ecb): Ecb is calculated by employing the existing LCC method, which extracts economic data from the LCC database and the project data from the proposed BC MP, as shown in Figure 8.

Technical circularity rate (CRcb): The five indicators result in a CR_{cb} score between 1 (fully circular) and 0 (fully linear). "CR_{mi}" is the material source in input. "CR_{mo}" is the material source in output. "DS" is disassembly (Efficiency of the recycling process). "LS" is the lifetime. "F_u" is functional units achieved. CRcb is calculated as shown in Equation (6):

$$CR_{cb} = (CR_{mi} + CR_{mo} \times DS)/2 \times LS \times F_u \tag{6}$$

Material source in input (CRmi): "CR_{mi}" is the technical circularity rate in the material input. During the production of the components, "R_{ici}" is the percentage of recycled materials, "R_{imu}" is the percentage of remanufacturing materials, "R_{ifb}" is the percentage of refurbishing materials, "R_{ipr}" is the percentage of repair materials, "R_{iui}" is the percentage of reused materials. "a, b, c, d", "e" is the weight for each of the factors. CR_{mi} is calculated as shown in Equation (7):

$$CR_{mi} = R_{ici} \times a + R_{imu} \times b + R_{ifb} \times c + R_{ipr} \times d + R_{iui} \times e \tag{7}$$

The weights "a, b, c, d, e" are developed for different factors in the R strategy, and they show the circular level of each factor. For example, the R strategy of reuse has the highest circular level, as it means that the demolished one can be used again directly and does not need additional manufacturing. Now, the values of "a, b, c, d, e" are not decided yet; further research is needed to solve this.

To show the implication of Equation (7), we make the following assumptions: weight factor (a) for recycling percentage is 0.5, weight factor (b) for remanufacturing percentage is 0.6, weight factor (c) for refurbishing percentage is 0.7, weight factor (d) for repair percentage is 0.8, and weight factor (e) for reuse is 1. Under these assumptions, the circularity rate in the material input of a component (CR_{mi}) that is produced with 100% reused material (R_{iui}) calculates to 1, while, in case a component is produced with 100% recycled material (R_{ici}), the CRmi calculates to 0.5. In case of a component that is 50% made of reused material and 50% made of recycled material, the circularity rate in the material input (CR_{mi}) calculates to 0.75, showing the sensitivity of the CR_{mi} value to the weights that are used given the (local) standards and conditions.

Material source in output (CRmo): "CR_{mo}" is the technical circularity rate in the material source in output. The calculation of it is similar to the circularity rate in the material source in input. "R_{oci}" is the percentage of recycled materials, "R_{omu}" is the percentage of remanufacturing materials, "R_{ofb}" is the percentage of refurbishing materials, "R_{opr}" is the percentage of repair materials, and "R_{oui}" is the percentage of reused materials. "a, b, c, d, e" is the weight for the factors. CR_{mo} is calculated as shown in Equation (8):

$$CR_{mo} = R_{oci} \times a + R_{omu} \times b + R_{ofb} \times c + R_{opr} \times d + R_{oui} \times e \tag{8}$$

The weight factors "*a, b, c, d, e*" in the calculation for output are the same as in the calculation for input. The implications of the applied weight factors on the calculated CRmo value are also similar to what was discussed for Material source in input (CRmi).

Disassembly: The Design for Disassembly (*DS*) is adopted from the industrial engineering sector. The disassembly factor of a building shows the efficiency of the recycling process in the demolishment of the building. DS is calculated considering two aspects: first, the wear and tear of the material (DS_a), which is determined by material properties; second, the connection type with other materials (DS_b), which is determined by the project. Data for the calculation of the first aspect (DS_a) are extracted from MP directly. For the second aspect (DS_b), the disassembly is calculated following the methods developed by Durmisevic (2002). *DS* is calculated as shown in Equation (9):

$$DS = DS_a \times DS_b \tag{9}$$

Lifetime: The lifetime "*LS*" is calculated from the design value of the building/system/component and the actual life time until being demolished. "LS_{dv}" is an average lifetime of a kind of building/system/component, which can be extracted from the MP. "LS_{au}" is the lifetime of the building/system/component, and the designer could provide a figure for it. *LS* is calculated as shown in Equation (10):

$$LS = LS_{au}/LS_{dv} \tag{10}$$

For every material/component/system, their own life time is considered as LS_{au}. For example, in case of a beam in the building, the actual usage time of the beam can be 20 years, while the lifetime of the (temporary) building is only ten years. During the demolishment process of the building, the beam is moved to another building that exists for ten years more. After the demolishment of the second building, the beam is landfilled. In that case, for the BC calculation of the beam, the lifetime is 20 years.

Functional units achieved: The functional units achieved "F_u" is calculated from the number of users of the building during its lifetime. "F_{au}" is the actual functional units (i.e., actual number of building users) achieved, which means the average user of the building every day during the lifetime of the building. The designer could give a figure for it. "F_{dv}" is the average functional units achieved (i.e., average number of building users) for a specific building type, which means the average user every day of a specific building type. F_u is calculated as shown in Equation (11):

$$F_u = F_{au} / F_{dv} \tag{11}$$

5.3. Building Circularity Calculation Method Analysis

Several problems exist in the current assessment methods, including issues related to the circularity cycles, the indicators, and the factors in BC assessment. The proposed BC calculation method in Section 5.2 can overcome these problems. The most prominent problems in the current BC methods are discussed in more detail in the following sections, with reference to the proposed calculation method in Section 5.2.

5.3.1. Problems Related to the Circularity Cycles

Studies of BC assessment in the AEC industry focus on specialized areas and fail to develop a holistic framework [63,64]. Equally important as the technical cycle, the biological cycle and the economic cycle are integral parts of the proposed BC calculation method that follows the proposed material flow model (see Figure 2). Consequently, the proposed BC assessment is more comprehensive.

5.3.2. Problems Related to Indicators

For the technical cycle, many indicators in the current methods need to be calculated more systematically. They are either completely ignored or have to be entered manually by the user. For example, in some models, the indicator for the efficiency of the recycling process is missing [65]. As a consequence, this method can only be used in a building with components that can be disassembled without any loss of material.

To illustrate the implication of the shortcomings, we present Table 7, showing the circularity calculation of a beam under different settings. The example proves that ignoring the disassembly indicator influences the BC results significantly. In Setting 1, the disassembly factor is ignored. In comparison, setting 2 in the proposed material flow model is followed, which considers the disassembly and has a DS value of 50%. The results show that due only to the inclusion of the DS (Disassembly) factor, the calculated BC is significantly lower.

Table 7. Different settings for the efficiency and factors of the recycling process.

Indicator		Setting 1 (Jianli et al.) [65]	Setting 2 (Material Flow Model)	Setting 3 (Madaster) [48]	Setting 4 (Material Flow Model)
CRmi	Rici	0%	0%	0%	0%
	Rimu		0%		0%
	Rifb		0%		0%
	Ripr		0%		0%
	Riui	0%	0%	0%	0%
CRmo	Roci	0%	0%	**100%**	**100% (weight: 0.4)**
	Romu		0%		0%
	Rofb		0%		0%
	Ropr		0%		0%
	Roui	100%	100%	0%	0%
DS		–	**50%**	100%	100%
LS		100%	100%	100%	100%
Fu			100%		100%
BC		**50%**	**25%**	**50%**	**20%**

Bold text highlights the differences between setting 1 and 2, 3 and 4.

5.3.3. Problems Related to the Factors

In the current BC calculation models, the number of considered R strategy factors is less than in the traditional models of waste management. Moreover, the R strategy factors for material source in input and output are all considered equally circular.

Using the example of the beam presented in Table 7, the implications can be illustrated using the current model and the proposed material flow model. Setting 3 is the current Madaster model [48], which considers the recycling materials and the reused materials equally circular. Setting 4 is the proposed material flow model, with the weight of the recycling material set to 0.4, assuming that the recycling material is less circular than the reusing material. The result of the BC calculation according to setting 3 is 50%, while for setting four, the results are lower, namely 20%, as shown in Table 7. It is evident that providing weights for different factors can help give more insights into the effect of different recycling strategies and generate more accurate BC values.

6. Conclusions

6.1. Discussion of Contributions

In this research, we proposed a new framework for BC assessment, including a material flow model, an MP, and a BC calculation method. It contributes to improving the BC assessment on three aspects: the assessment cycles, the assessment indicators, and the assessment R strategy factors. This effort could be highly valuable both for different stakeholders and future research.

This work promotes the BC assessment to a more comprehensive degree, based on accurate content. By improving the assessment model, policymakers can make more detailed regulations to promote circularity in the AEC industry. This work can also benefit building designers and engineers with convenient, comprehensive, and easy to understand results about the BC value of their design and help them improve the circularity of their design. Further, the proposed BC assessment methods can be used to determine the circular level of a building as part of the building quality assessment.

It also promotes the development of BC assessment in the academic field. The BC assessment framework provides a wide perspective and facilities the future research both in holistic and specific research. In more detail, the framework contributes to existing literature as follows. First, while CE influences in the AEC industry are related to the technical cycle, the environmental cycle, the economic cycle, and the social cycle, some existing models fail to include them all. In this research, the proposed material flow model contains three cycles for a comprehensive assessment. Second, some of the important indicators are lost in the current assessment methods. The material flow model proposed here includes significant indicators for more accurate calculation results. Third, in the current BC assessment models, R strategy factors are far from well-researched, and they are assumed to be equally weighted. This research redefines the R strategy factors in a consistent manner and provides calculation methods for each of them. Distinguishing different R strategy factors can encourage users to choose a recycling strategy that is more effective and efficient.

6.2. Limitations and Further Research

The main limitation of this work is that the proposed material flow model and assessment framework are built based on the available economic and environmental assessment metrics and fail to add the entire new, not yet widely accepted calculation metrics of the economic and environmental cycle according to the CE. In the future, a more comprehensive calculation model may be developed following these developments. Secondly, different stakeholders participate in the design process, and their different views on BC assessment should be considered. While currently the application of the BC framework is most dominant in the building design stage, future research may adjust the BC assessment method to accommodate the requirements of other stakeholders after building construction while maintaining a unified standard. Thirdly, the weights for R strategies factors in the circularity calculation of material source in input and output require more research to determine their value for a specific building project. Finally, a standardized data collection of the recycling properties of materials and components is needed. This research proposed a new data structure to support the data collection. However, the data collection of recycling properties is still in the early development phases, and further study should focus on the data collection approaches in practice. A comprehensive and reliable data collection on material and components is paramount to a standardized BC assessment in the AEC industry.

Author Contributions: Conceptualization, N.Z., Q.H., B.d.V.; methodology, N.Z., Q.H., B.d.V.; writing—original draft preparation, N.Z.; writing—review and editing, Q.H., B.d.V.; supervision, Q.H., B.d.V. All authors have read and agreed to the published version of the manuscript.

Funding: This research received no external funding.

Institutional Review Board Statement: Not applicable.

Informed Consent Statement: Not applicable.

Data Availability Statement: Not applicable.

Conflicts of Interest: The authors declare no conflict of interest.

References

1. IPCC. *Climate Change 2014: Synthesis Report. Contribution of Working Groups I, II and III to the Fifth Assessment Report of the Intergovernmental Panel on Climate Change*; UN: New York, NY, USA, 2014; ISBN 9789291691432.
2. Mousa, M.; Luo, X.; McCabe, B. Utilizing BIM and Carbon Estimating Methods for Meaningful Data Representation. *Procedia Eng.* **2016**, *145*, 1242–1249. [CrossRef]

3. Lazarevic, D.; Valve, H. Narrating expectations for the circular economy: Towards a common and contested European transition. *Energy Res. Soc. Sci.* **2017**, *31*, 60–69. [CrossRef]
4. Construction and Demolition Waste. Available online: https://ec.europa.eu/environment/topics/waste-and-recycling/construction-and-demolition-waste_en (accessed on 11 September 2021).
5. Netherlands Enterprise Agency. *A Circular Economy in The Netherlands by 2050*; Government of the Netherlands: The Hague, The Netherlands, 2019; 72p.
6. Lovrenčić Butković, L.; Mihić, M.; Sigmund, Z. Assessment methods for evaluating circular economy projects in construction: A review of available tools. *Int. J. Constr. Manag.* **2021**, 1–10. [CrossRef]
7. Eberhardt, L.C.M.; Birkved, M.; Birgisdottir, H. Building design and construction strategies for a circular economy. *Archit. Eng. Des. Manag.* **2020**, 1–21. [CrossRef]
8. Sparrevik, M.; de Boer, L.; Michelsen, O.; Skaar, C.; Knudson, H.; Fet, A.M. Circular economy in the construction sector: Advancing environmental performance through systemic and holistic thinking. *Environ. Syst. Decis.* **2021**, *41*, 392–400. [CrossRef]
9. Antwi-Afari, P.; Ng, S.T.; Hossain, M.U. A review of the circularity gap in the construction industry through scientometric analysis. *J. Clean. Prod.* **2021**, *298*, 126870. [CrossRef]
10. Chen, W.Q. Recycling rates of aluminum in the United States. *J. Ind. Ecol.* **2013**, *17*, 926–938. [CrossRef]
11. Heisel, F.; Nelson, C. RhinoCircular: Development and Testing of a Circularity Indicator Tool for Application in Early Design Phases and Architectural Education. In Proceedings of the 2020 AIA/ACSA Intersections Research Conference: CARBON, online, 30 September–2 October 2020; pp. 1–6.
12. Ellen MacArthur Foundation. *Ellen MacArthur Foundation Circularity Indicators: An Approach to Measuring Circularity*; Ellen MacArthur Foundation: Cowes, UK, 2015; p. 12.
13. Mesa, J.; Esparragoza, I.; Maury, H. Developing a set of sustainability indicators for product families based on the circular economy model. *J. Clean. Prod.* **2018**, *196*, 1429–1442. [CrossRef]
14. Parchomenko, A.; Nelen, D.; Gillabel, J.; Rechberger, H. Measuring the circular economy—A Multiple Correspondence Analysis of 63 metrics. *J. Clean Prod.* **2019**, *210*, 200–216. [CrossRef]
15. Ruiz-Pastor, L.; Chulvi, V.; Mulet, E.; Royo, M. The relationship between personal intrinsic factors towards a design problem and the degree of novelty and circularity. *Res. Eng. Des.* **2021**. [CrossRef]
16. Vinante, C.; Sacco, P.; Orzes, G.; Borgianni, Y. Circular economy metrics: Literature review and company-level classification framework. *J. Clean. Prod.* **2021**, *288*, 125090. [CrossRef]
17. Vermeulen, W.J.V.; Reike, D.; Witjes, S. Circular Economy 3.0: Getting Beyond the Messy Conceptualization of Circularity and the 3R's, 4R's and More. In *CEC4Europe Factbook*; Circular Economy Coalition for Europe: Vienna, Austria, 2018; pp. 1–6.
18. Madaster Services B.V. *Madaster Circularity Indicator Explained*; Madaster Services, B.V.: Laren, The Netherlands, 2018; 17p.
19. Saidani, M.; Yannou, B.; Leroy, Y.; Cluzel, F.; Kendall, A. A taxonomy of circular economy indicators. *J. Clean Prod.* **2019**, *207*, 542–559. [CrossRef]
20. MacArthur, E. *Towards the Circular Economy, Economic and Business Rationale for an Accelerated Transition*; Ellen MacArthur Foundation: Cowes, UK, 2013.
21. Kirchherr, J.; Reike, D.; Hekkert, M. Conceptualizing the circular economy: An analysis of 114 definitions. *Resour. Conserv. Recycl.* **2017**, *127*, 221–232. [CrossRef]
22. Platform CB'23. *Guide for Measuring Circularity (Platform CB'23, 2020c): Core Method for Measuring Circularity in the Construction Sector*; Platform CB'23: The Hague, The Netherlands, 2020.
23. Weißenberger, M.; Jensch, W.; Lang, W. The convergence of life cycle assessment and nearly zero-energy buildings: The case of Germany. *Energy Build.* **2014**, *76*, 551–557. [CrossRef]
24. Ortiz, O.; Castells, F.; Sonnemann, G. Sustainability in the construction industry: A review of recent developments based on LCA. *Constr. Build. Mater.* **2009**, *23*, 28–39. [CrossRef]
25. Mesa, J.A.; Fúquene, C.E.; Maury-Ramírez, A. Life cycle assessment on construction and demolition waste: A systematic literature review. *Sustainability* **2021**, *13*, 7676. [CrossRef]
26. Hossain, M.U.; Thomas Ng, S. Influence of waste materials on buildings' life cycle environmental impacts: Adopting resource recovery principle. *Resour. Conserv. Recycl.* **2019**, *142*, 10–23. [CrossRef]
27. Craven, D.J.; Okraglik, H.M.; Eilenberg, I.M. Construction Waste and a New Design Methodology. *Sustain. Constr.* **1994**, *16*, 89–98.
28. Meibodi, A.B.; Kew, H.; Haroglu, H. Most popular methods for minimizing in-situ concrete waste in the UK. *N. Y. Sci. J.* **2014**, *7*, 111–116.
29. Gharfalkar, M.; Court, R.; Campbell, C.; Ali, Z.; Hillier, G. Analysis of waste hierarchy in the European waste directive 2008/98/EC. *Waste Manag.* **2015**, *39*, 305–313. [CrossRef]
30. Gertsakis, J.; Lewis, H. Sustainability and the Waste Management Hierarchy. *EcoRecycle Vic.* **2003**, 16.
31. Gehin, A.; Zwolinski, P.; Brissaud, D. A tool to implement sustainable end-of-life strategies in the product development phase. *J. Clean Prod.* **2008**, *16*, 566–576. [CrossRef]
32. Li, W. Comprehensive evaluation research on circular economic performance of eco-industrial parks. *Energy Proc.* **2011**, *5*, 1682–1688. [CrossRef]

33. Yeheyis, M.; Hewage, K.; Alam, M.S.; Eskicioglu, C.; Sadiq, R. An overview of construction and demolition waste management in Canada: A lifecycle analysis approach to sustainability. *Clean Technol. Environ. Policy* **2013**, *15*, 81–91. [CrossRef]
34. Bakker, C.; Wang, F.; Huisman, J.; Den Hollander, M. Products that go round: Exploring product life extension through design. *J. Clean Prod.* **2014**, *69*, 10–16. [CrossRef]
35. Ping Tserng, H.; Chou, C.M.; Chang, Y.T. The key strategies to implement circular economy in building projects-a case study of Taiwan. *Sustainability* **2021**, *13*, 754. [CrossRef]
36. Zimmermann, R.K.; Kanafani, K.; Rasmussen, F.N.; Andersen, C.; Birgisdóttir, H. LCA-Framework to evaluate circular economy strategies in existing buildings. *IOP Conf. Ser. Earth Environ. Sci.* **2020**, 588. [CrossRef]
37. Gravagnuolo, A.; Angrisano, M.; Nativo, M. Evaluation of Environmental Impacts of Historic Buildings Conservation through Life Cycle Assessment in a Circular Economy Perspective. *Aestimum* **2020**, *2020*, 241–272. [CrossRef]
38. De Wolf, C.; Hoxha, E.; Fivet, C. Comparison of environmental assessment methods when reusing building components: A case study. *Sustain. Cities Soc.* **2020**, *61*, 102322. [CrossRef]
39. Malabi Eberhardt, L.C.; Van Stijn, A.; Rasmussen, F.N.; Birkved, M.; Birgisdottir, H. Towards circular life cycle assessment for the built environment: A comparison of allocation approaches. *IOP Conf. Ser. Earth Environ. Sci.* **2020**, 588. [CrossRef]
40. Antonini, E.; Boeri, A.; Lauria, M.; Giglio, F. Reversibility and durability as potential indicators for circular building technologies. *Sustainability* **2020**, *12*, 7659. [CrossRef]
41. Abadi, M.; Sammuneh, M.A. Integrating circular economy and constructability research: An initial development of a lifecycle "circularity" assessment framework and indicators. In Proceedings of the 36th Annual ARCOM Conference, Leeds, UK, 7–8 September 2020; pp. 516–525.
42. Nuñez-Cacho, P.; Górecki, J.; Molina-Moreno, V.; Corpas-Iglesias, F.A. What gets measured, gets done: Development of a Circular Economy measurement scale for building industry. *Sustainability* **2018**, *10*, 2340. [CrossRef]
43. Charef, R.; Lu, W. Factor dynamics to facilitate circular economy adoption in construction. *J. Clean Prod.* **2021**, *319*, 128639. [CrossRef]
44. Hossain, M.U.; Ng, S.T.; Antwi-Afari, P.; Amor, B. Circular economy and the construction industry: Existing trends, challenges and prospective framework for sustainable construction. *Renew. Sustain. Energy Rev.* **2020**, *130*, 109948. [CrossRef]
45. Platform—Madaster. Available online: https://madaster.com/platform/ (accessed on 11 June 2020).
46. Ellen MacArthur Foundation. Circularity Calculator. Available online: http://www.circularitycalculator.com/ (accessed on 15 October 2021).
47. Durmisevic, E.; Ciftcioglu, Ő.; Anumba, C.J. *Knowledge Model for Assessing Disassembly Potential of Structures*; Delft University of Technology: Delft, The Netherlands, 2003.
48. Heisel, F.; Rau-Oberhuber, S. Calculation and evaluation of circularity indicators for the built environment using the case studies of UMAR and Madaster. *J. Clean Prod.* **2020**, *243*, 118482. [CrossRef]
49. Materials Passports—BAMB. Available online: https://www.bamb2020.eu/topics/materials-passports/ (accessed on 10 June 2020).
50. Milieuclassificaties van Bouwproducten. Available online: https://www.nibe.info/nl/milieuclassificaties (accessed on 10 June 2020).
51. Database—Nationale Milieudatabase. Available online: https://milieudatabase.nl/database/ (accessed on 10 June 2020).
52. Smeets, A.; Wang, K.; Drewniok, M.P. Can Material Passports lower financial barriers for structural steel re-use? *IOP Conf. Ser. Earth Environ. Sci.* **2019**, 225. [CrossRef]
53. Heeren, N.; Fishman, T. A database seed for a community-driven material intensity research platform. *Sci. Data* **2019**, *6*, 23. [CrossRef] [PubMed]
54. Honic, M.; Kovacic, I.; Rechberger, H. BIM-Based Material Passport (MP) as an Optimization Tool for Increasing the Recyclability of Buildings. *Appl. Mech. Mater.* **2019**, *887*, 327–334. [CrossRef]
55. Honic, M.; Kovacic, I.; Sibenik, G.; Rechberger, H. Data- and stakeholder management framework for the implementation of BIM-based Material Passports. *J. Build. Eng.* **2019**, *23*, 341–350. [CrossRef]
56. Honic, M.; Kovacic, I.; Aschenbrenner, P.; Ragossnig, A. Material Passports for the end-of-life stage of buildings: Challenges and potentials. *J. Clean Prod.* **2021**, *319*, 128702. [CrossRef]
57. Atta, I.; Bakhoum, E.S.; Marzouk, M.M. Digitizing material passport for sustainable construction projects using BIM. *Resour. Conserv. Recycl.* **2021**, *43*, 103233. [CrossRef]
58. Pauliuk, S. Critical appraisal of the circular economy standard BS 8001:2017 and a dashboard of quantitative system indicators for its implementation in organizations. *Resour. Conserv. Recycl.* **2018**, *129*, 81–92. [CrossRef]
59. Di Maio, F.; Rem, P.C. A robust indicator for promoting circular economy through recycling. *J. Environ. Prot.* **2015**, *6*, 1095–1104. [CrossRef]
60. Franklin-Johnson, E.; Figge, F.; Canning, L. Resource duration as a managerial indicator for Circular Economy performance. *J. Clean Prod.* **2016**, *133*, 589–598. [CrossRef]
61. De Bruijn, H.; van Duin, R.; Huijbregts, M.A.J.; Guinee, J.B.; Gorree, M.; Heijungs, R.; Huppes, G.; Kleijn, R.; de Koning, A.; van Oers, L.; et al. (Eds.) *Handbook on Life Cycle Assessment*; Springer: Dordrecht, The Netherlands, 2002; p. 7. [CrossRef]
62. ISO. ISO 14040:2006—Environmental Management—Life Cycle Assessment—Principles and Framework. Available online: https://www.iso.org/standard/37456.html (accessed on 11 September 2021).

63. Akhimien, N.G.; Latif, E.; Hou, S.S. Application of circular economy principles in buildings: A systematic review. *J. Build. Eng.* **2021**, 38. [CrossRef]
64. Lei, H.; Li, L.; Yang, W.; Bian, Y.; Li, C.-Q. An analytical review on application of life cycle assessment in circular economy for built environment. *J. Build. Eng.* **2021**, *44*, 103374. [CrossRef]
65. Zhai, J. BIM-Based Building Circularity Assessment from the Early Design Stages A BIM-Based Framework for Automating the Building Circularity Assessment from Different Levels of a Building's Composition and Providing the Decision-Making Support on the Design of the Circular Building from the Early Design Stages. Master's Thesis, Eindhoven University of Technology, Eindhoven, The Netherlands, September 2020.

sustainability

MDPI

Article

Sustainable Water Management in a Krakow Housing Complex from the Nineteen-Seventies in Comparison with a Model Bio-Morpheme Unit

Wojciech Bonenberg [1], Stanisław M. Rybicki [2], Grażyna Schneider-Skalska [3] and Jadwiga Stochel-Cyunel [4,*]

1 Faculty of Architecture, Poznan University of Technology, 60-965 Poznan, Poland; wojciech.bonenberg@put.poznan.pl
2 Faculty of Environmental Engineering and Energy, Cracow University of Technology, 31-155 Kraków, Poland; srybicki@pk.edu.pl
3 Faculty of Architecture, Cracow University of Technology, 31-155 Kraków, Poland; gschneid@pk.edu.pl
4 Doctoral School, Cracow University of Technology, 31-155 Kraków, Poland
* Correspondence: j.stochel@doktorant.pk.edu.pl

Abstract: Cities grow through the addition of new housing structures, but the existing tissue is also modernized. Krakow, like any city with a historical origin, has typologically varied housing tissue. A large area of the city is occupied by multi-family panel-block housing estates which are being revitalised and the scope of this revitalization should include sustainable design elements. This paper determines the potential for implementing integrated water management, that utilizes rainwater in an existing basic urban unit that is a housing estate from the nineteen-seventies, located in Krakow (Poland), in conjunction with the Bio-Morpheme—the fractal reference model unit. The parameters of the Bio-Morpheme were established by earlier research as the optimum for a housing unit with regards to the circular economy and improving water use efficiency. The study covers the need to improve the quality of the housing environment, linked with the presence of natural elements, including a water reservoir, in the direct vicinity of the development. The analyses explored the potential to employ integrated water management with rainwater reuse in a basic urban unit (Krakow-Morpheme) and then compared the findings with the outcomes obtained by the proposed Bio-Morpheme complex. The results indicate that the potential to achieve a lower demand of water from the water supply system and to lower wastewater production were obtained, with a simultaneous opportunity to lay out an open water reservoir into the Krakow-Morpheme urban interior for improvement of the health value and well-being of inhabitants.

Keywords: water management; housing environment quality; multi-family settlements; Bio-Morpheme

Citation: Bonenberg, W.; Rybicki, S.M.; Schneider-Skalska, G.; Stochel-Cyunel, J. Sustainable Water Management in a Krakow Housing Complex from the Nineteen-Seventies in Comparison with a Model Bio-Morpheme Unit. *Sustainability* **2022**, *14*, 5499. https://doi.org/10.3390/su14095499

Academic Editors: Andreas N. Angelakis and Miklas Scholz

Received: 28 January 2022
Accepted: 30 April 2022
Published: 3 May 2022

Publisher's Note: MDPI stays neutral with regard to jurisdictional claims in published maps and institutional affiliations.

1. Introduction

Ongoing urbanization can be observed across all continents. According to United Nations data, more developed regions saw growth in urbanized areas of up to 79.1% in 2020 (compared with 54.8% in 1945) and this is expected to rise to 86.6% in 2050 [1]. Development and technical infrastructure are becoming more concentrated, while the amount of biologically active areas is decreasing. There is also visible spatial, economic, societal and climate change [2,3]. Citizens continue to pursue opportunities in cities [4]. Estimates indicate that in 2050, as much as 70% of the global population will live in cities, use urban infrastructure, and expect a good quality of life with access to water of a suitable quality [5,6]. The growth of cities and changes in their urban structure necessitate a new outlook on high-density housing areas and blue infrastructure planning. Citizens also expect higher-quality living environments [7–9].

Poland is a country where, using the United Nations typology, the largest cities are in the medium-sized category with populations of between one and five million and between 500,000 and one million: Warsaw has a population of 1,790,658, and the Silesian Conurbation has a population of 2,072,200. The second group includes, among others, Wroclaw—642,869, Poznan—534,813 and Krakow, which was reported to have a population of 779,115 on 31 December 2019 [10].

Constantly growing housing areas form a majority of the structure of contemporary cities. Climate change, linked with dry spells and torrential rainfall events, affects housing areas that become inundated with rainfall, while the overall water resources are becoming smaller. The microclimate deteriorates and temperatures increase, due to significant amounts of impervious surfaces and a lack of water reservoirs that regulate temperature and humidity in the environment. In Poland, the current water exploitation index (WEI) [11], which describes the ratio of the amount of water consumed to the amount of (available) renewable water resources, is estimated at 20%, which is significantly below the 17.7% threshold, which indicates that Poland experiences a permanent water shortage. It should be noted that the available surface water resources in Poland are unevenly distributed; therefore, local water unavailability and the risk of local water deficits show considerable variability.

The necessity to store and utilise rainwater in housing areas has the potential to even out the water balance, while also improving microclimate [12]. The potential use of rainwater in the water balance is tied with the existing and desired development structure. Rainwater can be managed by allocating and harvesting its fractions, such as roof run-off, footpath and bicycle path run-off, road and parking area run-off and rainwater falling on green/recreational areas [13]. The storage and use of rainwater in the water-management system of housing areas would require classifying water type depending on source and level of pollution, collection methods and purpose, as well as treatment for ensuring compliance with domestic and global quality standards [14].

Contemporary European cities, including Krakow, are characterised by a diverse range of housing development types, ranging from historical areas to a significant amount of large-scale, multi-family housing estates successively supplemented by smaller complexes built by real-estate development companies since 1989 [15]. Changes in the Polish political and economic system since 1989 have resulted in unfavourable processes in the structure of residential areas. Housing complexes which developed very quickly in open green areas, often gated, have strengthened the urban sprawl phenomenon, created spatial chaos and worsened the quality of the environment [16–19].

Large multi-family housing complexes built with industrialised methods in the 1960s and 1970s are part of the housing structure of Polish cities, such as Warsaw (e.g., residential complexes—Ursynów, Bródno), Kraków (e.g., housing estates—Mistrzejowice, Azory, Ugorek), Gdańsk (e.g., housing estates—Zaspa, Przymorze) and Wrocław (e.g., housing estates—Bartoszowice, Huby) [20,21]. It is estimated that 60% of housing developments built in Polish cities from 1966 to 1995 are large-panel buildings. Currently, around 10 million Poles live in this type of block (i.e., over $\frac{1}{4}$ of the country's population) [22]. After World War II, large multi-family estates also emerged in other European countries. In most countries, where the stock of such facilities is significant, the modernization and rehabilitation path is chosen to varying degrees and with different objectives. In Germany, significant modernization of large estates in Berlin (Gropiusstadt, Markisches Viertel and Thermometersiedlung), as well as in Dresden and Cottbus, has occurred. This also applies to other countries, such as Sweden and the Netherlands [20,21,23].

Housing-estate developments that consist of large multi-family housing complexes from the nineteen-sixties and seventies occupy the largest parts of Krakow. Both the number and quality of multi-family housing estates in the city support the assumption that they can be a suitable object of sustainable revitalization and the application and promotion of sustainable water management.

Some housing estates are planned for revitalization as a part of the municipal revitalization plan. The updated version of this plan indicated that one of its operational goals included the Krakow urban municipality housing block development revitalization program [24,25]. The largest number of these housing estates is in the northern belt, where the biggest housing complexes were built, such as Mistrzejowice, Bieńczyce, Prądnik, Azory and Krowodrza. The estates were a response to housing needs, but their spatial structure fulfilled the postulates of the Athens Charter of 1933, which provided access to light, sun and greenery. This is the reason for the long distances between buildings, allowing for a large amount of green space to be left, and the inclusion of school and kindergarten gardens. Large green areas also resulted from decision-making strategies (state ownership) and a low motorization rate. With the passage of time, the appreciation of urban space in these estates has increased [20,21].

Revitalization is based on academic studies, such as a multi-criteria analysis of nineteen block housing estates located in the limits of the Krakow Urban Municipality [26]. A function-spatial diagnosis of the Olsza II and Ugorek housing estates was also taken into account [27]. The diagnosis indicates that block housing estates are highly rated by their residents and have a large quantity of free areas with great potential for the rainwater management process in various forms [24,27]. Currently, the city of Krakow is carrying out the project "A study of the housing environment quality of Krakow's housing complexes". The project is planned to conclude towards the end of 2022.

Action towards improving the quality of the housing environment in cities is included both in decisions by Krakow's municipal authorities and in new initiatives undertaken in the European Union, as part of the New European Bauhaus [28]. Improving the human living environment, and an emphasis on revitalization based on the application of contemporary, sustainable structures, including, and perhaps most importantly, blue infrastructure, occupy a crucial place in the European program. Previous plans by the Krakow municipality only marginally concerned this field and revitalization strategies focused on the societal and technical issues of panel-block development. However, the document that includes a program of technical inspections of panel-block residential buildings located within the city limits of Krakow Urban Municipality proposes the use of rainwater to create open reservoirs among a variety of space-revitalization tools [24].

Revitalization actions by the city of Krakow, that target large housing estates, require an investigation of the degree to which these housing estates, and the housing development typology, facilitate the application of modern forms of comprehensive water management. It is also crucial to investigate the potential for balancing gains within a single complex.

The effects of comprehensive water management that include rainwater management must be approached holistically and account for the significance of water, understood as water in the environment, which includes the housing environment, in its environmental and economic, climate, and recreational and aesthetic aspects. The proposed idea of BioMorpheme seems to be a very good solution, especially with regards to the location of facilities in built-up areas, where the rainwater-drainage capacity is currently too low to connect new facilities. As the granting of a building permit usually depends on the "available" capacity of the sewage system collecting rainwater, the presented idea, which significantly reduces the volume of rainwater discharged, is a proposal that allows the standards to be met without significant investments in the rainwater sewage system.

Environmental and economic aspects refer to the correct shaping of water management, rational and efficient resource use, enhancing biodiversity, maintaining continuity of the ecological corridor and lowering greenery maintenance expenditures [29]. It also refers to improving the health of residents and their wellbeing, lowering medical treatment expenditures and providing opportunities for direct contact with the aquatic environment [9,30]. It further refers to the correction of microclimates in housing complexes, lowering air temperature, raising humidity, changing ionization, and affecting the air turbulence and movement that contribute to purification [14].

Recreational and aesthetic aspects concern both physical recreation and psychological effects. The potential for introducing various types of recreation associated with water, and the forms which these may take, are determined by an area's capacity, the scale of the water feature, the degree of development of the bank and the diverse needs of various local community groups. The aesthetic portion of these aspects is the shaping of a beautiful setting, including the form of reservoirs or watercourses, the design of their banks and the sounds produced (standing or flowing water), and the wealth of aquatic greenery [31].

The principles of contemporary sustainable design include the above-mentioned aspects and comprehensive water management in the housing environment is found both in the 3R principle—reduce, reuse, recycle, and in the principles of respect for people and respect for site [32]. Thus far, the revitalization activities of large housing complexes undertaken by the city of Krakow mainly concern communication and technical issues. The aim of revitalization should also include improvements to the quality of the housing environment in the complex, as well as to small housing units. [33]. Taking into account the microclimate, the health of residents, the condition of green areas and the level of aesthetics, including contact with nature, it is advisable to investigate the possibility of integrated water management on a selected example, with particular emphasis on the use of rainwater.

Referring to the above statements, our main research goals are:

- determining the possibility of implementing integrated water management using rainwater in a housing estate built in the 1970s in Krakow (Poland) on the example of a selected, typical housing complex;
- checking whether in the indicated housing complex it is possible to obtain water savings from the network by supplying rainwater to it and at the same time introducing an open reservoir or watercourse into the urban interior;
- checking whether the proposed method of shaping the urban structure of a large housing estate, consisting of the separation of smaller spatial units, allows for better use of rainwater than at present. The object of reference for the research is a typical building in the city and its surroundings;
- comparing the findings with the results of a reference unit. Checking the potential of the Bio-Morpheme complex solutions, by which the rationalization of water management has led to lower demand for water from the municipal water supply system and the amount of sewage produced, accompanied by introducing an open water reservoir into its urban interior layout [34].

2. Materials and Methods

In application to the research process, conceptual methods in urban-hydrological analyses had a cascade structure (Top-Down Approach). The conceptual approach was located at the top of this cascade and acted as a "decision gate", in terms of strategies for further urbanization towards sustainable water management in Krakow.

The aim of the presented method is its intention to answer the question as to whether the implementation of the idea of spatial structuring, inspired by the form and function of Bio-Morpheme, makes sense in Krakow and whether it could be applied to the modernization of Krakow's "Large Housing Estates".

In the event that this model does not work, other ways of solving the problem would have to be sought. The calculations showed that, based on general hydrological data for Krakow, the model is serviceable for the urban renewal of Krakow's "Large Housing Estates". Detailed implementation requires moving to lower levels of the methodological cascade and, for evident reasons, requires more precise methods, which are expensive and risky at the conceptual stage.

In theoretical terms, the general analysis method (metamodeling) is the logic gate of the planning decision-making system for the revitalization of the "Large Housing Estates" in Krakow (due to their specificity and common hydrological and urban characteristics). The gateway operation consists of rejecting or accepting possible further research on the selected urban units of Krakow (by principles included in the publication).

This approach corresponds to methods of planning risk assessment at the conceptual stage of decision-making. This type of approach is used in risk management theory, and its implementation in spatial planning and water management in Krakow is an original research achievement.

The study of the possibility of introducing integrated water management in a housing complex from the nineteen-seventies was a preliminary investigation which aimed to show whether it is possible to achieve the desired goals in the Krakow-Morpheme unit. The adapted calculation simple method was adequate for the initial study, and a positive result would allow continuing work using the more precise calculation methods appropriate for specialists from the field of hydrology research, especially the methods described in "Applied Hydrology" by Ven Te Chow [35].

The adopted fast and simple method has provided advantages of work and features of originality at the same time, constituting an easy and effective method of proceeding, which is accessible and understandable to a large group of architects, planners, and town planners. It increases the chances of its extensive application at the stage of preliminary identification of the possibility of introducing integrated water management in revitalized multi-family housing complexes in Krakow and other cities with similar housing structures.

The similarities, in terms of the typology of buildings and urban configurations, in the large housing estates of Krakow from the nineteen-seventies mean that it is appropriate to start the research on one selected example. Such an example was considered to be the Mistrzejowice complex created in the years 1969-1982. It is a complex of housing estates for approximately forty thousand inhabitants and consists of four estates with typical five- and eleven-storey buildings. Out of four units, the Bohaterów Września housing estate was selected, a fragment with five-storey linear buildings. It was decided to distinguish the basic spatial unit, a morpheme, taking into account the principles of residence. This is conducive to maintaining a small scale and identification with the area; it facilitates contact with residents and the promotion of sustainable solutions and facilitates logistics, enabling corrective actions to be carried out in stages [33]. The Morpheme, named Krakow-Morpheme, is an area comparable to the reference Bio-Morpheme [34].

The hydrological analyses of the existing housing development areas, including the problems of area saturation (development intensity) and the possibility of loading the existing infrastructure (among other water and sewage grids), encourage the transformation and division of development areas into more tiny structural elements, which aspire to self-sufficiency. At the same time, the necessity of improving the climate and health of places of residence requires the introduction of blue-green engineering solutions. In assumption, the Bio-Morpheme unit, in both structural and functional terms, is a fractal part of the city organism pursuing a circular economy. The functional-spatial parameters of the Bio-Morpheme make it suitable to maximum harvesting and use of rainwater, while minimizing the consumption of tap water and maximizing the use of energy from renewable sources, so that it becomes as autonomous a unit as possible (see Table 1). It assumes high efficiency of rainwater harvesting from the roof, while using the roof surface for energy production by covering it with photovoltaic cells. However, to maintain the favourable microclimate conditions of the unit (air cooling, limiting the heat island phenomenon), it was predicted to introduce vertical gardens and an open water reservoir.

In the Krakow-Morpheme, the balancing of rainwater from roofs, bicycle paths, roadways, parking lots and pavements was adopted as well as checking the possibility of using rainwater to achieve savings in the water supply network, and creating external reservoirs/watercourses to improve the microclimate and small retention.

The subject of investigation, a complex comprised of two typical residential buildings from the nineteen-seventies located in the Mistrzejowice housing estate in Krakow, has green areas, pavements and parking spaces, and an urban interior open on two sides. (See Figure 1).

Table 1. Summary of the housing environment parameters of the Krakow-Morpheme and the Bio-Morpheme, which are essential for integrated resource management and increase in resident well-being.

Research Field	Element Expectations	Effect Features	Results	Occurrence of the Features in Krakow-Morpheme	Occurrence of the Features in Bio-Morpheme
The architectural and urban structure of housing development [13,29–34]	The form of the residential unit and land development preferences–partially open forms, proportions of the neighborly interior	1:3 - the proportions of the building height to the width of the open interior	sense of comfort and security	YES	YES
	Scale and proportionsROMANIApreferences—a human scale of development	From 3 to 5 storeys	sense of comfort and a human scale	YES	YES
	Concept of housing development preferences—striving to live in a group called neighborhood	200–500 inhabitants - the size of the development unit favours the creation of a neighborly space	sense of comfort	YES	YES
	Function in the residential unit preferences—an attractive recreational program in the neighborhood space	adequate space for recreation resulting from the proportions of the urban interior	increase of well-being	YES	YES
Research Field	**Element Expectations**	**Effect Features**	**Results**	**Occurrence of the features in Krakow-Morpheme**	**Occurrence of the features in Bio-Morpheme**
Water in residential environment [12–14,24,29,31–35]	Water management Preferences - water retention in the morpheme, - rainwater reusing, storage and recycling in site	Introduction solutions of blue-green engineering	Increase of water safety	NO	YES
	The required infrastructure and water use efficiency Preferences: - Sufficient amounts of water to meet the living needs of the inhabitants; - Striving to save water from water supply systems without reducing comfort and without affecting health of residents negatively	Reuse of rainwater and greywater in the building	Reduction of no less than 40% of the Monthly consumption of tap-water	NO	YES
Research Field	**Element Expectations**	**Effect Features**	**Results**	**Occurrence of the features in Krakow-Morpheme**	**Occurrence of the features in Bio-Morpheme**
Healthy residential environment (In aspect of water use) [8,9,12,14,24,29–35]	Structure of the land development preferences—the need for contact with elements of nature, including an open water reservoir—to ensure physical and mental health	Introduction solutions of blue-green engineering	Increase of Well-being	NO	YES
	Climatic comfort preferences—the need to stay in a healthy environment ensuring climatic comfort (including adequate humidity and water availability)	Introduction solutions of blue-green engineering including the open water reservoir	Lowering the perceived temperature, improving ventilation and humidity	NO	YES

Figure 1. Housing Complex from the nineteen-seventies in Krakow (Poland)—Krakow-Morpheme.

For the purposes of this study, the plan was minimally simplified, which had no effect on the final results.

Territory—the area's size is 100 × 90 m = 9000 m^2
Development features—two multiple-core-type buildings, each consisting of four segments with a height of five storeys
Urban form type—linear development
Number of residential units—sixty in each building, 120 in total
Number of residents—assumed to be in compliance with the number estimated in the nineteen-seventies, with 220 residents in each building, making a total of 440
Objective—balancing run-off from roofs, bicycle paths, roads, parking spaces and paved surfaces; determining rainwater-utilization potential with the intent of increasing savings on potable water and the creation of external water bodies/courses to improve the microclimate and small retention
Climate conditions—standard for Krakow (Poland)
Characteristics of the area selected for comparison—Bio-Morpheme [34]

Characteristics of the Bio-Morpheme

- The spatial structure of the area is based on the cooperation of infrastructures, according to Ken Yeang's typology
- The Bio-Morpheme is, as far as possible, autonomous in terms of the use and storage of rainwater within the blue infrastructure.
- The Bio-Morpheme is assumed to be in the form of a building quarter, with a side length of 100 m, partially open, with a four-storey residential development enriched with service functions, pedestrian and vehicular traffic, with favourable interior

proportions and of an interior size enabling the realization of a domestic program for residents and the introduction of a pond (see Figure 2).

Figure 2. The Bio-Morpheme reference unit.

Table 1 summarizes the features of Morpheme important for the integrated management of resources (including water) and the simultaneous improvement of the comfort and well-being of the inhabitants.

Table 2 summarises the area of the surfaces that receives precipitation in the Bohaterów Września housing estate in Krakow and in the reference complex (the Bio-Morpheme). These values were used as data for rainwater harvesting volume in further sections.

Table 2. Technical parameters of elements of a catchment area for the Krakow-Morpheme housing unit and the Bio-Morpheme.

Surface	Krakow-Morpheme		Bio-Morpheme	
	Surface Area [m^2]	Percentage of the Krakow-Morpheme Total Surface Area [%]	Surface Area [m^2]	Percentage of the Bio-Morpheme Total Surface Area [%]
Roofs of residential building and service areas	1680	18.67	2610	26.10
Parking and road way	2040	22.67	1630	16.30
Cycle paths	0	0	746	7.46
Permeable pavements	1026	11.40	1055	10.55
Green belt	262	2.91	777	7.77
Greenery: trees and bushes	3992	44.35	2862	28.62
Surface-water reservoir	0	0	320	3.20
Total surface area	9000	100	10,000	100

The calculations were based on actual data for Krakow (monthly average values) collected over a period of ten years, which enabled the calculation of the volume of rainwater that could potentially be harvested for consumer use. The paper presents the proposals for rainwater management as well as greywater usage. These proposals reflect solutions that limit water consumption and reduce the amount of wastewater generated within the housing complexes. For this study the following specific assumptions were made concerning using water 'not-from-the-tap': specific contaminants of household wastewater streams, as well as constituents of rainwater, are limiting factors in the general idea of water reclamation and it is assumed that the applied technical solutions should be as uncomplicated as possible and cannot be economically unfavourable. Calculations were made as follows:

i. Greywater, i.e., wastewater from bathing, handwashing and laundry, will be collected and treated in simple installations (coalescent filter and pump if necessary), built-in bathroom structures of each flat, and used for toilet flushing; an excess volume of this water will be discharged to the sewerage; as these three streams create more wastewater than necessary for toilet flushing. This very process, flushing, will be a limiting factor for this way of water saving.
ii. Rainwater from the roof will be collected in a rainwater harvesting tank, common for the entire Bio-Morpheme, and will then be pumped to a specific water line in the plumbing system to support each washing machine with water. The volume of the tank is assumed to be for three days of operation, i.e., 15 cubic meters. It would be equipped with a filtration unit (sieve) to remove particles, UV radiation for disinfection purposes, and quality control via a simple turbidimeter and pH meter online. Location of the tank and its dimensions will depend on the internal design and arrangements of the building. Calculations were made for a steady state based on real rainfall data for Krakow [34].

Such water management requires proper internal plumbing of buildings, so the Bio-Morpheme concept seems to be an efficient idea for buildings and their associated planning and design. Water reclamation potential has been summarized in Table 3, comparing greywater use with the harvested rainwater potential use.

Table 3. Possibility of using greywater to substitute drinking water in Bio-Morpheme unit, authors' own, based on [36].

Purpose of Water Use	Percentage Share of Water Consumption in Daily Water Consumption	Purpose of the Use	Whether this Can Be Substituted by Rainwater
Dishwashing	3%	Toilet flushing	No
Laundry	22%		Yes
Bathing	17%		No
Washing hands	1.7%		No

The technical solution of the Bio-Morpheme would operate per the following ways/procedures of water use:

a. Rainwater collected from the roofs of residential and commercial areas (see Figure 2), after basic water treatment processes, would be directed to a tank located in the underground part of the building (the capacity of the tank would be based on dimensions of an average three-day quantity of rainwater harvested from these collection areas, i.e., roofs). Rainwater requires minimum treatment (dust removal) and, due to its low hardness, is suited for laundries, differing from greywater streams in this feature. After such treatment, a separate internal distribution system would supply water to washing machines in individual residential units. Calculations warned that in winter months and some periods of summer a three-day period would not cover demand for laundry water. In such periods, without precipitation, washing machines would operate with tap water delivery through a three-way valve. As the authors' observations confirmed that evaporation significantly reduces the volume of rainwater harvested when daily rainfall is below 1 mm/day, further calculations omit rainfall from such days.

b. Car parks would be equipped with water collection/infiltration systems, such as Wavin Q-Bic, so it was assumed that this type of rainwater runoff would not be taken into water balance.
c. Access roads, pavements and bicycle paths would be drained with the use of a stormwater drainage system compliant with the DWA-M 153 standard. This would ensure that pre-treatment conditions would be in accordance with the Regulation of the Minister of Maritime Affairs and Inland Navigation of 12 July 2019. Water collected from this part of the area would be directed to the ornamental/recreational element in the form of a pond for the Bio-Morpheme or a watercourse, a reservoir with an elongated shape adapted to the development layout of the Krakow-Morpheme. The reservoirs would have an average depth of 0.70 m. This depth was chosen for safety reasons, as a deeper pond requires protection barriers which could significantly degrade the landscaping of the entirety of the Bio-Morpheme. An analysis of the area size of this element is presented later in the paper.

The main hydro-ecological aim of the proposal is to find meaningful ways of utilizing rainwater harvested from roofs, which is of relatively good quality and requires little pre-treatment (removal of suspended solids), to reduce the consumption of water supplied by the municipal waterworks. A functional analysis of water use indicates that it would be most efficient to use this water in washing machines. In this way, it would be possible to collect rainwater that is relatively frequently available from the environment. In the Bio-Morpheme, 'excess' rainwater, which would not be used for washing machines, along with water from paved areas, i.e., roads and bicycle paths, would be directed to a specially constructed pond. Only a portion of rainwater, which exceeds the needs for washing, and also water excess over the capacity of this pond, would generate a runoff to the municipal rainwater drainage system (sewer). It is not possible to build such a pond in the reference estate; therefore, unused rainwater would be discharged into the sewer system, increasing the overflow effect. For this reason, regardless of the increase in comfort by stabilising the air humidity (pond), the Bio-Morpheme would significantly save water. Another article by this team focuses on the problem of using grey water from households for reuse [34]

3. Results

3.1. Estimations of Savings in Water Supply Needs Resulting from the Morpheme Dimensioning

The calculation of water savings for inhabitants was based on the assumption that due to specific climate conditions, in the months of January and February, rainwater harvesting for washing purposes would be negligible. The calculations were done for housing complexes shown in the figures above; characteristics are as presented in Tables 1 and 2, and shown in the Figures 1 and 2 above.

The calculation formula is described by the Equation (1) [34]:

$$Q = \varphi \times \Psi \times q \times F\,[\mathrm{L}] \tag{1}$$

In which:
φ—drain lag coefficient [dimensionless]
Ψ—run-off coefficient [dimensionless]
q—rainfall intensity [$L/m^2 \times s$]
F—catchment area

As the components of the morpheme are characterised by different values of the run-off coefficients, a proxy coefficient was calculated according to the Equation (2):

$$\Psi z\,\frac{\Psi 1 * F1 + \Psi 2 * F2 + \ldots + \Psi i * Fi}{F1 + F2 + \ldots + Fi} \tag{2}$$

where symbols Ψ_1, Ψ_2,.. Ψ_i and F_1, F_2,..F_i refer to the values of run-off coefficients and catchment area for elementary surfaces, as shown in Table 1.

Formulae for RWH volume calculations used surface areas, as presented in Figure 1 (above), and a specific run-off coefficient for impermeable roads, paths and pavements (0.95 was applied, based on own measurements on tarmac in the yard at Cracow University of Technology). This coefficient is generally similar to values from literature [37–39] and was confirmed by real tests. Intensive water usage is visibly higher in the summer, due to the climate.

Water evaporation from the pond was calculated based on standardized formulae, with the following parameters:

$$g_h = \Theta A(x_s - x) \quad (3)$$

$$g_h = \text{amount of evaporated water per hour [kg/hour]} \quad (4)$$

$$\Theta = (25 + 19v) = \text{evaporation coefficient [kg/(m}^2 \times \text{hour)]} \quad (5)$$

$$v = \text{velocity of air above the water surface [m/s]} \quad (6)$$

$$A = \text{water surface area [m}^2\text{]} \quad (7)$$

$$x_s = \text{humidity ratio in saturated air at the same temperature as the water surface [kg/kg] (kg H}_2\text{O in kg dry air)} \quad (8)$$

$$x = \text{humidity ratio in the air [kg/kg] (kg H}_2\text{O in kg dry air)} \quad (9)$$

$$x = 0.62198 p_w/(p_a - p_w)$$

$$(p_w = \text{partial pressure of water vapour in moist air [Pa, psi]})$$

The proper assessment of rainwater volume is very difficult and carries the risk of inaccuracies due to the significant uneven rainfall incidence over time. This is clearly a key issue in the dimensioning of RWH devices. Common practice is that the initial design is based upon long-term average rainwater precipitation; however, tests and observations performed by the authors in the city of Krakow (Poland) have shown that this problem should not be underestimated. Data collected by the authors shows that the maximum observed monthly precipitation was as high as 403.1 mm (2010), while the lowest was observed in November 2010 (0.6 mm) [34]. During relatively low precipitation periods, only household wastewater may be the source of 'non-tap water' (water which is not sourced from the municipal waterworks). By contrast, during periods of heavy rainfall, the inflow of this water is much higher than the actual demand. The monthly precipitation value tended to vary between years, and years with precipitation significantly lower than average are usually named 'dry years', while years when precipitation is higher than average are called 'wet years', The impact of the characteristics of the year on the real unevenness of precipitation is shown in Figure 3. Rainfall volume data were obtained by the Cracow University of Technology (CUT) directly from the Institute of Meteorology and Water Management—National Research Institute. Data concerning rainfall intensity were measured by the CUT meteorological station, located in Cracow (Kraków) 24 Warszawska St., as monthly values in millimeters.

3.2. Possible savings in tap water

Calculations of the possible replacement of tap water with harvested rainwater and grey water [40] were performed for daily average water demand and use in the EU [36]. The determinations of rainwater usage and grey water application were based on monthly averages but converted to daily volumes by averaging the number of rainy days. The number of inhabitants in a Bio-Morpheme complex (230 people) and in the complex under study (440 inhabitants) justified the application of such a calculation scheme. The model calculations were conducted for three scenarios of inhabitants' daily water consumption in the units [36]:

a. 140 L/person × day, as stipulated in Polish design standards [10];
b. 100 L/person × day which corresponds to the average water consumption in Polish cities [10,37];

c. 90 L/person × day, which corresponds to the target value of daily average water use without leakage of pipelines [26,38].

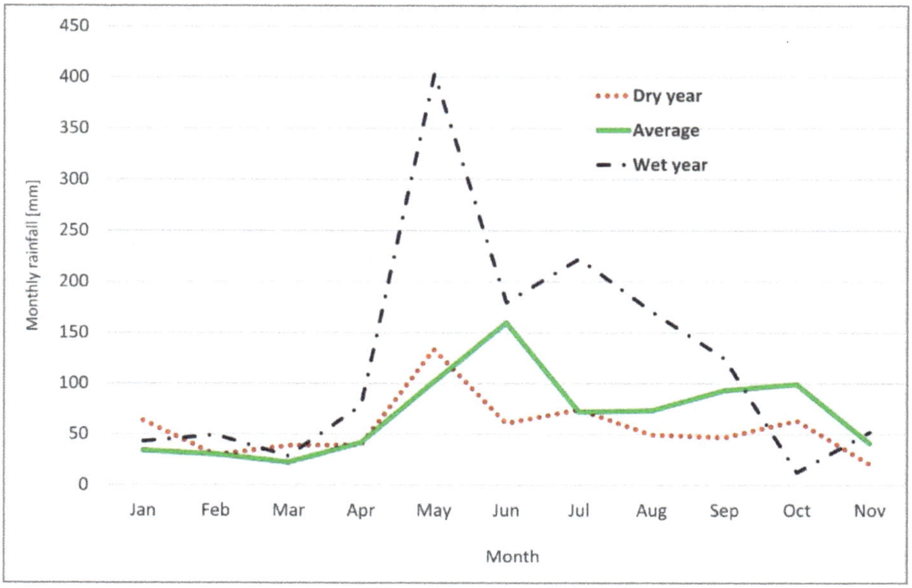

Figure 3. The average monthly rainfall amounts for Kraków—variability of rainwater harvesting potential—data for Krakow (Poland) in the period from 2010 to 2018.

The authors assumed that tap-water replacement should be done in an economically feasible way. Based on the authors' previous investigation [34], it was assumed that toilet flushing generates wastewater that is not usable within a Bio-Morpheme complex without highly efficient treatment, while direct consumption is non-recoverable from the point of view of water management. The calculations for the Krakow complex did not take into consideration problems associated with the remodelling of the internal water/wastewater system installed inside the building, as the main aim of the paper is to compare possible water savings between the innovative Bio-Morpheme concept and a 'conventional' design (reference buildings). The typical structure of household water consumption is as follows (ordered in descending volume of water consumption):

- Toilet flushing 28.1%
- Laundry 22%
- Bathing 17%
- Unaccounted (losses) 13%
- Meal preparation 11%
- Dishwashing 3%
- Other needs (e.g., lawn watering) 2.2%
- Direct consumption 2%
- Hand washing 1.7%

Water used for the purposes typed in bold can be reused as greywater, mainly for toilet flushing, which represents 44% of the consumption of water that has reuse potential but with a narrow field of application, i.e., toilet flushing reduces this input in water saving to a maximum of 28% of the total water consumption. These data clearly prove that the proper use of rainwater can significantly reduce the consumption of tap water.

4. Discussion

Table 4 presents a summary of results of calculations for the 140 L/cap × day scenario, for Bio-Morpheme-specific water management, proving that this concept may lead to an average of 47% of water savings on annual water consumption. The same calculations for the Krakow-Morpheme have shown less optimistic predictions with water savings at slightly over 30%. It should be noted that the amount of rainwater to be stored in a pond tends to vary significantly between months.

Table 4. Prediction of possible overall water savings by implementing Bio-Morpheme-specific water management compared with the conventional design concept.

Measuring Time Period	RWH Effective [L/d]	Grey Water [L/d]	Tap Water [L/d]	Savings [L/d]	Savings as Percentage of Total Consumption
			Krakow-Morpheme		
Monthly max	5894	17,248	41,553	23,142	38%
Month. aver.	3833	17,248	40,428	21,081	34%
Monthly min	2699	17,248	38,358	19,947	32%
			Bio-Morpheme		
Monthly max	9156	9048	18,959	18,204	57%
Month. aver.	5955	9048	17,197	15,003	47%
Monthly min	4193	9048	13,996	13,241	41%

The difference in tap water savings between the two design concepts undoubtedly lies in better use of RWH. This issue is illustrated in Figure 4, which shows the results of calculations relating to the use of rainwater (without the use of grey water), with various fluctuations.

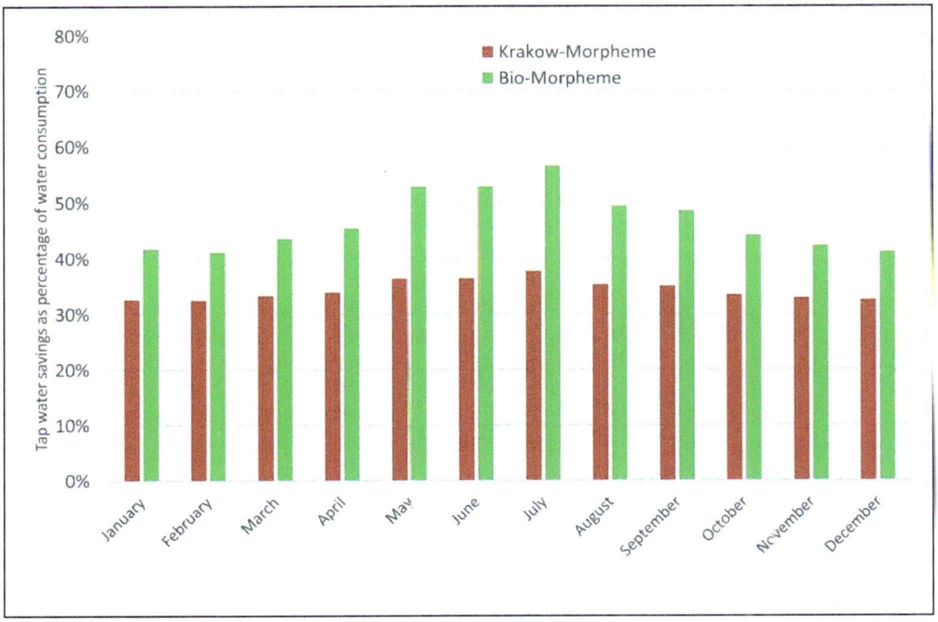

Figure 4. Comparison of tap-water savings for the Krakow housing complex—Krakow-Morpheme and Bio-Morpheme.

Rainwater collected from precipitation on roads and impervious pavements is not suitable for use by residents and will be directed to the open pond as presented in Figure 1, where it will be stored with possible loss due to natural evaporation. Evaporation will obviously increase the comfort of the inhabitants through a favourable change in humidity. At the same time, this reservoir will act as a buffer reservoir on the rainwater sewage system, preventing the drainage system of these waters from being overfilled during periods of heavy rain. This solution can protect a Bio-Morpheme complex from flooding. The housing complex in Krakow does not have such a feature and is thus unprotected; the maximum feasible area of a constructed water evaporation system would not exceed 204 m^2 (See Figure 5).

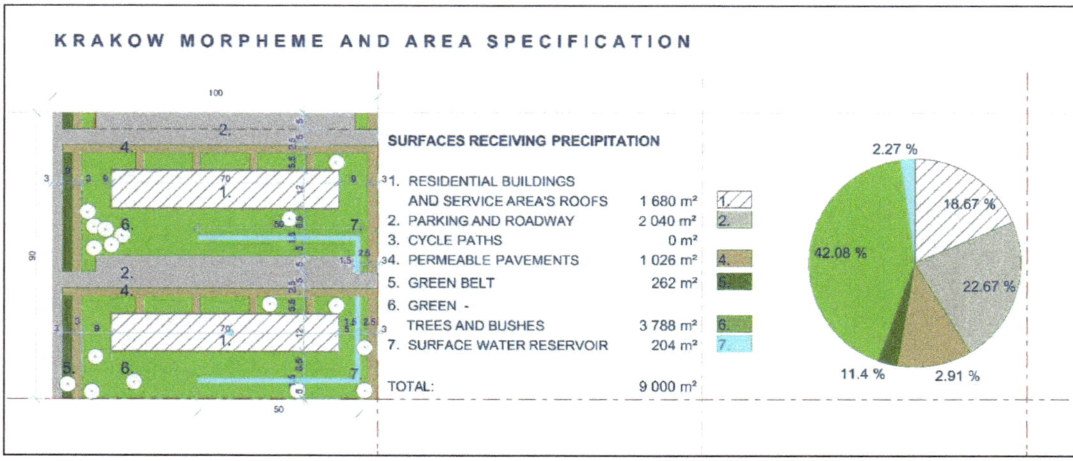

Figure 5. Krakow-Morpheme with water reservoirs.

Table 5 summarizes the area of the surfaces that receive precipitation in the Bohaterów Września housing estate in Krakow and the reference complex (the Bio-Morpheme). These were used as data for rainwater possible harvest volume designation, and to determine the size of the water reservoir and the possibility to introduce it into the Krakow-Morpheme area.

Table 6 presents the results of calculations of the amount of harvested rainwater, water use for domestic purposes and the volume of evaporated pond water and water used for lawn watering (i.e., not creating outflow to the municipal sewer system).

Figure 6 illustrates the results of calculations of water loss from the reservoir due to evaporation, inner-yard cleaning, and lawn irrigation, and indicates that in the period of the highest rainfall and high air temperature, this reservoir, as a part of the Bio-Morpheme, will reduce the total quantity of water discharged into the municipal rainwater sewage system by as much as 3%. It is clear that evaporation calculations are always general and the actual effectiveness of this process depends on current meteorological conditions.

Comparison in terms of water and sewage management

Krakow housing complex—Krakow-Morpheme:

- Average savings of 34% on potable water from municipal system (grey water included)
- Less impact of the reservoir on surface load of sewerage (if constructed)

Bio-Morpheme:

- Average savings of 47% on potable water from the municipal system
- Average reduction of 32% in load on surface-water sewer system during periods with highest rainfall (May-August)

Table 5. Summary of the area of surfaces that receive precipitation in the Bio-Morpheme and in the reference complex (the Bohaterów Września housing estate in Krakow).

Surface	Krakow-Morpheme		Krakow-Morpheme with Integrated Water Management		Bio-Morpheme	
	Surface Area [m^2]	Percentage of the Krakow-Morpheme Total Surface Area [%]	Surface Area [m^2]	Percentage of the Krakow-Morpheme Total Surface Area [%]	Surface Area [m^2]	Percentage of the Bio-Morpheme Total Surface Area [%]
Roofs of residential buildings and service areas	1680	18.67	1680	18.67	2610	26.10
Greenery: trees and bushes	3992	44.35	3788	42.08	2862	28.62
Permeable pavements	1026	11.40	1026	11.40	1055	10.55
Green belt	262	2.91	262	2.91	777	7.77
Parking and road way	2040	22.67	2040	22.67	1630	16.30
Cycle paths	0	0	0	0	746	7.46
Surface-water reservoir	0	0	204	2.27	320	3.20
Total surface area	9000	100	9000	100	10,000	100

Table 6. Specific values of rainwater generated on roads, parking and impermeable paths vs. water evaporated and used on lawn watering.

Specific Value [m^3/month]	Krakow-Morpheme		Bio-Morpheme	
	Water Harvested	Water Evaporated/Used	Water Harvested	Water Evaporated/Used
Monthly maximum	437	41	254	111
Average	292	113	170	40
Monthly minimum	200	0	116	0

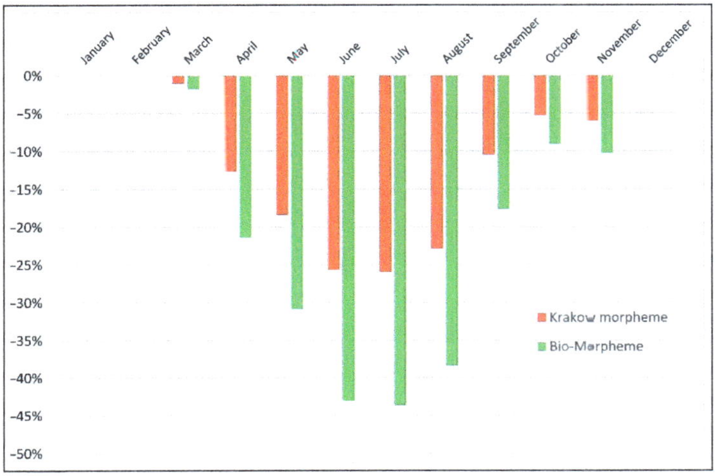

Figure 6. Comparison of the relative reduction of rainwater flow to the municipal sewer system between the Krakow housing complex—Krakow-Morpheme and Bio-Morpheme.

5. Summary and Conclusions

In the literature references on the subject, there are no studies of the autonomous Bio-Morpheme type unit called the Krakow-Morpheme or the Bio-Morpheme reference example in terms of spatial, environmental and social structure. There are numerous studies based on numerical data and studies on typologically different buildings [37,39], while contemporary publications, including groups of works devoted to rainwater harvesting, are mostly focused on the issue of the proper determination of the volume of collected rainwater [35,36,41], quantitative issues [42] or life cycle assessment (LCA) of these systems [43]. Similarly, papers on the use of greywater, in terms of design, are limited to indoor installations [44], or methods of treating these waters prior to circulation [45], usually referring to pre-treatment methods. There is an apparent lack of available references relating to conditions for the simultaneous use of rainwater and greywater at the design stage of buildings.

The studies on the completed housing complexes also do not consider the problem of the independence of the basic unit, instead, they present examples of diverse, most often mixed housing in climatic zones other than Krakow.

Current design models for urban-water management systems and corresponding infrastructure using centralised designs, such as the "sponge city", often fail from the perspective of cost-effectiveness and inability to adapt to future changes [46].

In cities such as Krakow, it is difficult to implement a centralised model of integrated water management, due to organisational and economic reasons, and the requirement for a large amount of underground construction work and high investment costs.

The presented Bio-Morpheme concept is innovative in terms of the integration of architectural factors and hydrological and water supply issues being introduced as early as the building design stage. A new value is also the possibility of using it in shaping new residential areas and as an element of an independent urban seal, especially due to the clearly lower values of rainwater runoff to the municipal sewage system than in conventional solutions.

Our approach is distinguished by its decentralised nature. It is adapted to the local conditions of compact residential complexes in Krakow. Dividing the investment process into small investment projects allows these to be implemented in places where they are most needed, and to divide large investment costs into smaller projects implemented successively.

Our concept is based on the principle of the flexible decentralization of integrated water management within compact morphemes directly connected to residential blocks. An important feature is that the morphemes are independent of each other and the shutdown of one morpheme does not interfere with the operation of the others.

To sum up, the advantages of this solution are as follows:

- Reduction of the distance between source and demand. This affects the efficiency of the system—this approach has been validated by recent studies [47,48];
- Reducing the cost. Traditional surface-water management systems for large areas must take into account the varied topography and existing underground infrastructure networks, roads, and existing buildings, significantly increasing investment costs. While small areas (morphs) with similar topographic characteristics reduce the costs associated with infrastructure and the pumping of water and wastewater in a given area. This is confirmed by the studies of Salvan, L. et al. [49] and Yifru, B. A. et al. [50];
- Better utilisation of landforms, which are more homogeneous on the urban microscale;
- Achieving Water Security and Resiliency. Diversifying water sources is critical for achieving water security and resilience. Morphemes provide additional water sources that are not used in existing centralized systems. Morphemes can be incorporated into centralized systems and, as a result, can improve water security. In addition, the morpheme system reduces the amount of water drawn from the centralized water source. Morphemes absorb disturbances and can independently adapt to the desired state. This is important when using rainwater, as confirmed by studies [51,52].

The conclusions from the conducted research, and concerning the above summary, are as follows:

1. It is asserted that the analysis results for the residential building from the nineteen-seventies are promising and the results indicate the feasibility of introducing the proposed solutions into the undergoing revitalization of Krakow's housing estates, as well as in other sites with similar morphological features. (The two-building housing complex provides the potential for supporting the water-supply system and constructing an open-water reservoir).
2. Solutions proposed in the article significantly increase the real feasibility of the idea of water reuse and RWH, all the more so because the pond proposed by the authors is a kind of water buffer reducing the nuisance of torrential rainfall, which is not taken into account in the quoted items [35–39,41–43].
3. The conducted research has development potential and should initiate further works. The site development of housing estates from the nineteen seventies leaves little free space between buildings to construct open water reservoirs. It should be investigated whether it is possible to introduce watercourses, along with buildings, in adjacent green areas, with possible wider sections, which could contribute to the interior's cross-ventilation and improve the microclimate, while the reservoir could be located outside the developed area in the recreational green zone. Introducing watercourses along buildings makes them visible from apartments (Figure 5). It is proposed to extend the investigation and indicate how the model in question could perform:
    a. in a complex of two buildings with a height of eleven storeys (the second type of development employed in housing estates from the nineteen-sixties and seventies alongside five-story buildings),
    b. in housing complexes built by real-estate developers in the nineteen-nineties and the beginning of the twenty-first century.

In these configurations, it is suggested to investigate the technical potential of reservoir and watercourse placements in relation to existing infrastructure and spatial layouts

To sum up, the research showed that under conditions of decreasing availability of water in the environment, it is possible to implement the precepts of modern water management in the development of existing housing blocks in Krakow. These precepts include multiple reuses of water resources within a complex; implementing a circular economy, ranging from the scale of individual households and the complex to the city as a whole; the use of the water supply system and sewage system in combination with quantitative and qualitative water protection in the environment; the use of any available water sources to produce a high-quality housing environment and fostering features beneficial to health; securing the engagement of the city's residents, institutions, organisations and industrial facilities in implementing the adopted technical solutions, with a particular focus on social groups that inhabit the various housing estates and complexes [34].

Author Contributions: Conceptualization, W.B., S.M.R., G.S.-S. and J.S.-C.; methodology, W.B., S.M.R., G.S.-S. and J.S.-C.; software, W.B., S.M.R., G.S.-S. and J.S.-C.; validation, W.B., S.M.R., G.S.-S. and J.S.-C.; formal analysis, W.B., S.M.R., G.S.-S. and J.S.-C.; investigation, W.B., S.M.R., G.S.-S. and J.S.-C.; resources, W.B., S.M.R., G.S.-S. and J.S.-C.; data curation, W.B., S.M.R., G.S.-S. and J.S.-C.; writing—original draft preparation, W.B., S.M.R., G.S.-S. and J.S.-C.; writing—review and editing, W.B., S.M.R., G.S.-S. and J.S.-C.; visualization, W.B., S.M.R., G.S.-S. and J.S.-C.; supervision, W.B., S.M.R., G.S.-S. and J.S.-C.; funding acquisition, J.S.-C. All authors have read and agreed to the published version of the manuscript.

Funding: This research and the APC were funded by NCBiR and Cracow University of Technology as part of the project "ROAD TO EXCELLENCE—a comprehensive university support programme", implemented under the Operational Programme Knowledge Education Development 2014–2020 co-financed by the European Social Fund; agreement no. POWR.03.05.00-00-Z214/18.

Institutional Review Board Statement: Not applicable.

Informed Consent Statement: Not applicable.

Data Availability Statement: Not applicable.

Acknowledgments: The research (work) was carried out as part of the project "ROAD TO EXCELLENCE—a comprehensive university support programme" implemented under the Operational Programme Knowledge Education Development 2014–2020 co-financed by the European Social Fund; agreement no. POWR.03.05.00-00-Z214/18.

Conflicts of Interest: The authors declare that they have no known competing financial interests or personal relationships that could have appeared to influence the work reported in this paper.

References

1. United Nations Department of Economic and Social Affairs. *World Urbanization Prospects: The 2018 Revision*; United Nations: New York, NY, USA, 2019.
2. Carter, J.G.; Cavan, G.; Connelly, A.; Guy, S.; Handley, J.; Kazmierczak, A. Climate change and the city: Building capacity for urban adaptation. *Prog. Plan.* **2014**, *95*, 1–66. [CrossRef]
3. Competitive Cities and Climate Change. OECD Conference Proceedings, Milan, 9–10 OCTOBER 2008. Available online: https://www.oecd.org/cfe/regionaldevelopment/50594939.pdf (accessed on 1 January 2022).
4. Roser, M.; Ortiz-Espina, E. World Population Growth. OurWorldInData.org. 2019. Available online: https://ourworldindata.org/world-population-growth (accessed on 12 January 2022).
5. Ritchie, H.; Roser, M. Urbanization, Our World in Data.org. Available online: https://ourworldindata.org/urbanization/ (accessed on 12 December 2021).
6. Corburn, J. *Towards the Healthy City. People, Places, and the Politics of Urban Planning*; The MIT Press: Cambridge, MA, USA, 2009.
7. Iojă, C.I.; Badiu, D.L.; Haase, D.; Hossu, A.C.; Niță, M.R. How about water? Urban blue infrastructure management in Romania. *Cities* **2021**, *110*, 103084. [CrossRef]
8. Miró, A.; Hall, J.E.; Rae, M.; O'Brien, C.D. Links between ecological and human wealth in drainage ponds in a fast-expanding city, and proposals for design and management. *Landsc. Urban Plan.* **2018**, *180*, 93–102. [CrossRef]
9. Summers, J.K.; Smith, L.M.; Case, J.L.; Linthurst, R.A. A Review of the Elements of Human Well-Being with an Emphasis on the Contribution of Ecosystem Services. *Ambio* **2012**, *41*, 327–340. [CrossRef]
10. Demographic Yearbook of Poland. GUS Poland. 2020. Available online: www.stat.gov.pl (accessed on 15 December 2021).
11. Berkowska, E.; Gwiazdowicz, M. Deficyt Wody w Polsce. 2020, Biuro Analiz Sejmowych, Infos 1(267). Available online: http://orka.sejm.gov.pl/WydBAS.nsf/0/A8BC86EF5BC4E446C12584E900399A4B (accessed on 10 December 2021).
12. Manteghi, G.; Bin Limit, H.; Remaz, D. Water bodies an urban microclimate: A review. *Mod. Appl. Sci.* **2015**, *9*, 1–12. [CrossRef]
13. Haupt, P. Natural elements of composition in the space of housing complexes. *Sr. Mieszk./Hous. Environ.* **2018**, *24*, 89–98. [CrossRef]
14. Rahman, S.; Khan, M.T.R.; Akib, S.; Bin Che Din, N.; Biswas, S.K.; Shirazi, K.M. Sustainability of Rainwater Harvesting System in terms of Water Quality. *Sci. World J.* **2014**, *2014*, 721357. [CrossRef]
15. Solarek, K. *Urban Design in Town Planning. Current Issues and Dilemmas from the Polish and European Perspective*; Faculty of Architecture WUT: Warszawa, Poland, 2018.
16. Bradecki, T.; Swoboda, J.; Nowak, K.; Dziechciarz, K. *Study of Contemporary Housing Estates Implemented in Poland in 2003–2016*; Wydawnictwo Politechniki Śląskiej: Gliwice, Poland, 2019.
17. Stachura, E. Studies on current housing conditions in Poland: Urban scale and commonly used areas attributes in housing estates. *Tech. Trans.* **2013**, *110*, 125–135. Available online: https://repozytorium.biblos.pk.edu.pl/resources/30751 (accessed on 15 December 2021).
18. Ogrodowczyk, A. Spatial aspects of housing policy transformation in Poland after 1989—Example from Lodz. *J. Econ. Manag.* **2015**, *19*, 137–154.
19. Polanska, D.V. Urban policy and the rise of gated housing in post-socialist Poland. *GeoJournal* **2014**, *79*, 407–419. [CrossRef]
20. Gronostajska, B. *Kreacja i Modernizacja Przestrzeni Mieszkalnej. Teoria i Praktyka na Przykładzie Wybranych Realizacji Wrocławskich z lat 1970–1990*; Oficyna Wydawnicza Politechniki Wrocławskiej: Wrocław, Poland, 2007.
21. Ostańska, A. *Wielka Płyta. Analiza Skuteczności Podwyższania Efektywności Energetycznej*; PWN: Warszawa, Poland, 2016.
22. Tofiluk, A.; Knyziak, P.; Krentowski, J. Revitalization of Twentieth-Century Prefabricated Housing Estates as Interdisciplinary Issue. *IOP Conf. Ser. Mater. Sci. Eng.* **2019**, *471*, 112096. [CrossRef]
23. Gruszecka, K.; Gzell, S.; Rembarz, G. *Osiedle: Reurbanizacja*; Urbanista: Warszawa, Poland, 2009.
24. Sprawozdanie z Monitorowania Miejskiego Programu Rewitalizacji Krakowa, Urząd Miasta Krakowa 2020, Kraków, Poland. Available online: https://www.bip.krakow.pl/plik.php?zid=282124&wer=0&new=t&mode=shw (accessed on 9 September 2021).
25. Program Rewitalizacji Zabudowy Blokowej Osiedli na Terenie Gminy Miejskiej Kraków. Aktualizacja MPRK. Wydział ds. Przedsiębiorczości i Innowacji, Krakow, Poland. 2020. Available online: https://www.bip.krakow.pl/?dok_id=51896 (accessed on 19 December 2021).

26. Wielokryterialna Analiza Dziewiętnastu Osiedli Zabudowy Blokowej Położonych na Terenie Gminy Miejskiej Kraków. Urząd Miasta Krakowa, Krakow, Poland. 2011. Available online: http://krakow.pl/data/rewitalizacja/27322_0.pdf?_ga=2.172258083.1298109433.1642345210-1628345790.1630074405 (accessed on 10 December 2021).
27. Kobylarczyk, J.; Schneider-Skalska, G.; Haupt, P.; Racoń-Leja, K.; Sumlet, W.; Tor, P. *Diagnoza Funkcjonalno-Przestrzenna Osiedli: Olsza II i Ugorek*; CUT Press: Cracow, Poland, 2015.
28. The New European Bauhaus. Available online: https://europa.eu/new-european-bauhaus/index_pl (accessed on 20 December 2021).
29. Verma, P.; Singh, P.; Singh, R.; Raghubanshi, A.S. *Urban Ecology. Emerging Patterns and Social-Ecological Systems*; Elsevier Science Publishing Co Inc.: New York, NY, USA, 2020.
30. Vlahov, D.; Galea, S. Urban health: A new discipline. *Lancet* **2003**, *362*, 1091–1092. [CrossRef]
31. Georgiou, M.; Morison, G.; Smith, N.; Tieges, Z.; Chastain, S. Mechanisms of Impact of Blue Spaces on Human Health: A Systematic Literature Review and Meta-Analysis. *Int. J. Environ. Res. Public Health* **2021**, *18*, 2486. [CrossRef] [PubMed]
32. Vale, B.; Vale, R. *Green Architecture: Design for a Sustainable Future*; Thames and Hudson: London, UK, 1991.
33. Schneider-Skalska, G. *Designing a Healthy Housing Environment. Selected problems*; Lambert Academic Publishing: Saarbrucken, Germany, 2011.
34. Rybicki, S.; Schneider-Skalska, G.; Stochel-Cyunel, J. Bio-Morpheme as innovative design concept for "Bio City" urban structure in the context of water saving and human health. *J. Clean. Prod.* **2022**, *in press*.
35. Chow, V.T.; Maidment, D.R.; Mays, L.W. *Applied Hydrology*; McGraw-Hill, Inc.: New York, NY, USA, 1988; 540p, ISBN 0070108102.
36. Eurostat Statistics Explained. Water Statistics. Available online: https://ec.europa.eu/eurostat/statistics-explained/index.php?title=Water_statistics (accessed on 20 January 2022).
37. Pinzon, T.M. Modelling and Sustainable Management of Rainwater Harvesting in Urban Systems. Ph.D. Thesis, Universitat Autònoma de Barcelona, Barcelona, Spain, 2012.
38. Well, F.; Ludwig, F. Blue–green architecture: A case study analysis considering the synergetic effects of water and vegetation. *Front. Archit. Res.* **2020**, *9*, 191–202. [CrossRef]
39. Godyn, I.; Grela, A.; Stajno, D.; Tokarska, P. Sustainable Rainwater Management Concept in a Housing Estate with a Financial Feasibility Assessment and Motivational Rainwater Fee System Efficiency Analysis. *Water* **2020**, *12*, 151. [CrossRef]
40. Ward, S.; Memon, S.A.; Butler, D. Performance of a large building rainwater harvesting system. *Water Res.* **2012**, *46*, 5127–5134. [CrossRef]
41. McGrane, S.J. Impacts of urbanisation on hydrological and water quality dynamics, and urban water management: A review. *Hydrol. Sci. J.* **2016**, *61*, 2295–2311. [CrossRef]
42. Quinn, R.; Melville-Shreeve, P.; Butler, D.; Stovin, V. A critical evaluation of the Water Supply and Stormwater Management Performance of Retrofittable Domestic Rainwater Harvesting Systems. *Water* **2020**, *12*, 1184. [CrossRef]
43. Yan, X.; Ward, S.; Butler, D.; Daly, B. Performance assessment and life cycle analysis of potable water production from harvested rainwater by a decentralized system. *J. Clean. Prod.* **2018**, *172*, 2167–2173. [CrossRef]
44. Juan, Y.-K.; Chen, Y.; Lin, J.-M. Greywater Reuse System Design and Economic Analysis for Residential Buildings in Taiwan. *Water* **2016**, *8*, 546. [CrossRef]
45. Campos Rodrigues, K.; Salomão Rael de Morais, L.; Martins de Paula, H. Green/sustainable treatment of washing machine greywater for reuse in the built environment. *Clean. Eng. Technol.* **2022**, *6*, 100410. [CrossRef]
46. Zevenbergen, C.; Fu, D.; Pathirana, A. Transitioning to Sponge Cities: Challenges and Opportunities to Address Urban Water Problems in China. *Water* **2018**, *10*, 1230. [CrossRef]
47. Karsu, O.; Kara, B.Y.; Akkaya, E.; Ozel, A. Clean Water Network Design for Refugee Camps. *Netw. Spat. Econ.* **2021**, *21*, 175–198. [CrossRef]
48. Sitzenfrei, R.; Wang, Q.; Kapelan, Z.; Savić, D. Using Complex Network Analysis for Optimization of Water Distribution Networks. *Water Resour. Res.* **2020**, *56*, e2020WR027929. [CrossRef] [PubMed]
49. Salvan, L.; Abily, M.; Gourbesville, P.; Schoorens, J. Drainage System and Detailed Urban Topography: Towards Operational 1D-2D Modelling for Stormwater Management. *Procedia Eng.* **2016**, *154*, 890–897. [CrossRef]
50. Yifru, B.A.; Chung, I.; Kim, M.; Chang, S.W. Assessment of Groundwater Recharge in Agro-Urban Watersheds Using Integrated SWAT-MODFLOW Model. *Sustainability* **2020**, *12*, 6593. [CrossRef]
51. Canales, F.; Gwoździej-Mazur, J.; Jadwiszczak, P.; Struk-Sokołowska, J.; Wartalska, K.; Wdowikowski, M.; Kaźmierczak, B. Long-Term Trends in 20-Day Cumulative Precipitation for Residential Rainwater Harvesting in Poland. *Water* **2020**, *12*, 1932. [CrossRef]
52. Werbelo, L.; Brown, R. Working towards sustainable urban water management: The vulnerability blind spot. *Water Sci. Technol.* **2011**, *64*, 2362–2369. [CrossRef]

sustainability

Article

Market Regeneration in Line with Sustainable Urban Development

Justyna Borucka [1,*], Piotr Czyż [1], Giorgio Gasco [2], Weronika Mazurkiewicz [1], Dorota Nałęcz [1] and Marcin Szczepański [3]

1 Faculty of Architecture, Gdańsk University of Technology, 80-233 Gdansk, Poland
2 Faculty of Art, Design and Architecture or Department of Architecture, Bilkent University, Ankara 06800, Turkey
3 Faculty of Civil and Environmental Engineering, Gdańsk University of Technology, 80-233 Gdansk, Poland
* Correspondence: justyna.borucka@pg.edu.pl

Abstract: This article presents the study of the optimal design solutions for regeneration of marketplaces. It examines the design variants for the revitalisation of the marketplace, in particular, investment in their modernisation in order to find the most optimal model for transforming these public spaces to have a significant impact on the city's development. The research is a comparative analysis of the implementation of regeneration design models on the marketplace within the Oliwa district of Gdansk (Poland).The data for the case study design models includes analysis based on various optimisation criteria, taking into account the urban and economic aspects of the city landscape when selecting a specific space revitalisation design model. The implementation of regeneration investment includes a number of complex processes that must be sustainable and so require rational social and spatial planning, as well as proper organisation in terms of cost and time.

Keywords: economic analysis; marketplace regeneration; optimisation; space syntax; spatial analysis; urban design

Citation: Borucka, J.; Czyż, P.; Gasco, G.; Mazurkiewicz, W.; Nałęcz, D.; Szczepański, M. Market Regeneration in Line with Sustainable Urban Development. *Sustainability* **2022**, *14*, 11690. https://doi.org/10.3390/su141811690

Academic Editors: Oleg Kapliński, Lili Dong, Agata Bonenberg and Wojciech Bonenberg

Received: 30 June 2022
Accepted: 7 September 2022
Published: 17 September 2022

1. Introduction

Public markets are vital elements of public space in modern cities. They are some of the most truly unique and city-defining elements of the landscape of a city.

All over the world, marketplaces have and still do play an important role because of various historical, cultural and social issues.

As authors have mentioned [1–7], in many parts of the world, those marketplaces are still vivid spots and key connecting points in the city space (e.g., Beşiktaş Fish Market in Istanbul, New Market Pazarii Ri in Tirana, New Market in Stazione, Termini in Rome).

In Europe, there is a visible change in the understanding of the role of the market in the contemporary city, as well as important changes in what constitutes market communities. Markets are no longer just places of trade, but also cultural spaces. Undoubtedly, public markets play an important role in the socio-economic structure of modern cities [8,9]. In addition, cities, their districts and different public spaces including public and local markets, like other market brands, carry cultural meaning, cache and recognisable symbols of a given city or neighbourhood. Furthermore, markets also play an economic role in defining the city's or neighbourhood's identity whilst being its recognisable symbol [10].

In Poland, the existence of traditional markets has been an integral part of the lives of urban communities, but their role is seen only as a space of trade, often used only for two days of the week [11,12]. At the same time, there is a growing demand for local centres in every neighbourhood, and a lack of meeting places for residents. Transforming existing markets into vital public space, therefore, can also be important for the local community. The transformed marketplace can create a space for multiple purposes: place for trade, meeting place, cultural venues and restaurant zones.

This case study focuses on the case of the Oliwa Marketplace in Gdansk. It is a pilot regeneration, which will serve as an example of market regeneration strategy in Poland. It constitutes a prototype for future research to be used in other locations.

It is part of a research and implementation program carried out by a research team consisting of three units represented successively by: the City Initiative Association (Stowarzyszenie Inicjatywa Miasto)—Project Leader, the Faculty of Architecture of the Gdańsk University of Technology and the Academy of Fine Arts in Gdańsk.

The results of the study will form a comprehensive strategy for the transformation of market areas in Poland.

The purpose of the present study is to find the most optimal model for transforming these public spaces. This article discusses optimisation whilst choosing a final design solution which is part of the market regeneration plan, using different variants/models of spatial design for the case study. The "Polanki" Marketplace (Targowisko "Polanki") case study, located in the Gdańsk-Oliwa district, Poland, is closely related to the ongoing research project concerning the regeneration of urban markets [13] (pp. 71–78). The study's main challenge was to search for the micro scale optimal design/solution proposal and that is why the site of the marketplace has been taken into consideration and tested. For the purposes of our study, we specifically focused on the process of the most effective use of the square—hence, the deliberate delimitation of the area.

Firstly, the article considers optimisation issues: optimisation in general and optimisation in the specific context of spatial and social costs and time factors of implementation.

Secondly, it describes the overall goal of the project, conducted within the framework of the GOSPOSTRATEG program financed by the National Centre for Research and Development in Poland, and the purpose of the research [14,15].

Thirdly, it focuses on a particular case study—the pilot implementation plans for the Oliwa Marketplace—with the aim of defining the regeneration processes of this marketplace, and optimising the criteria adopted throughout the study.

Finally, it discusses the analysis of the case study design models, leading to a final conclusion defining the spatial and economic boundaries of the optimal design model and its recommendations for implementation.

The proposed models are the result of 2 years of analysis and research conducted at the marketplace [15]. Project designs were preceded by several detailed studies and additional workshops with market space users (buyers and sellers).

The purpose of optimisation is to arrive at the best possible solution (selected from a specific set) on the basis of adopted criteria, which are expressed using mathematical algorithms, i.e., objective functions. At the point of result and decision making, multi-criteria optimisation brings much better results, and so the authors chose that model in the research. Multi-criteria decision problems can be classified into the following groups:

Multiple-purpose decision problems wherevalues do not have a predetermined number of options with problem-specific feature values.

The problem of multi-attribute decisions, for which there are a limited number of decision options and models [16–18].

In the research, we use optimisation in the context of spatial and social implementation factors and in the context of cost and time implementation factor issues. The models subjected to a detailed analysis presented in the article are the result of deeper studies, representative examples of extensive design analyses carried out during the grant implementation, which was preceded by separate studies (i.e., workshops with the market space users) on demography and neighbouring objects.

The purpose of this paper is to choose the most optimised version for market distortions using the three above-mentioned criteria.

2. Stages of the Optimisation Process

2.1. Spatial

Space syntax is a method [19,20] commonly used to show the relationship between the topological dependencies of space in cities and the presence of people in these spaces. The most important factor influencing the analysis of space is still its configuration, i.e., the relationships taking place within it. The theory suggests that the way elements of space are stacked together does in fact influence the behaviour of its users [21]. It is usually used as a tool to show the relationship between the syntax of different spaces in cities and their usage [22–25]. This method overlaps with observations made in space and is based on intuitive exploration. As has been proven in many previous studies (see, e.g., https://spacesyntax.com/project/trafalgar-square/ accessed on 26 June 2021), the mathematical method of space syntax coincides with observations made in space. For this reason, it was decided to use this method as representative.

In our case, we needed to evaluate the spatial configuration options for changing the functions of the existing market.

The goal of the study was to choose the best spatial option that would fulfil a number of criteria: it should be convenient for stall holder trading, it would best integrate the entire quarter (in which the district town hall stands) and through which we would create a connection with the market hall standing in the square. Due to the fact that the space syntax method has fractal features, the entire market space was divided into smaller spaces (like in a city) and analysed on the basis of the longest axial lines situated inside of each space.

2.2. Economy (Cost and Time)

In practice, the essential elements of optimisation and, at the same time, project management investment are as follows: the budget for the implementation of a given project, the resources assigned to it and its implementation time. Very often, the search for the optimal solution comes down to taking into account two criteria: minimising costs and the time of project implementation. It should also be emphasised that the shortest completion time of a construction investment is not usually the same as the lowest cost. In the curve of total costs (kc) (Figure 1), one can surmise, inter alia, two important factors from the point of view of optimisation [26].

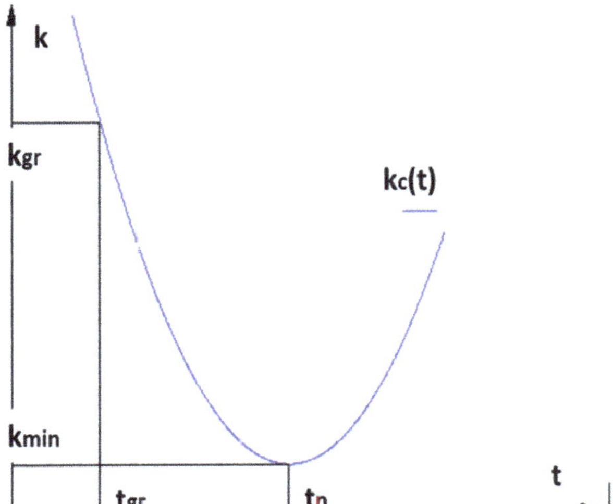

Figure 1. Dependence of project cost over time of implementation (after Fridgeirsson and Roston, 2017 [26]).

tn represents the normal time of project implementation, which corresponds to the lowest cost of its implementation (kmin),

tgr represents the limit time, i.e., the shortest possible time for implementation of the project, which corresponds to the limit cost (kgr).

Appropriate analysis of various variants of related projects as well as the corresponding costs and execution times makes it possible to determine the total cost curve. When we exceed the time for normal project implementation, we observe an increase in investment cost. This gives a clear sign to the investor that the optimal solution he or she is looking for is within the range between normal and cut-off time. The choice of the best solution depends in this case on the investor's preference [27].

It should also be noted that the optimisation of practical projects in the field of architecture and construction is complex, which means that the time needed to solve investment problems grows exponentially with the size of the investment [28–30]. There are also numerous scope limitations using optimisation methods, including heuristic and accurate ones, to find the optimal solution. [26].

One of the basic optimisation problems is scheduling tasks. The cause of this problem is the interdisciplinary nature of architectural and construction investment projects and their high specificity [31–34]. One of the tools that can be used to solve problems in the field of construction, architecture and scheduling is costs criteria [35–38] or multi-criteria analysis framework and optimisation [39,40]. These algorithms ensure that the best possible solution is found, but at the same time, it should be noted that the results obtained by applying any optimisation methods are strictly dependent on the input parameters used. The above standards must be considered at the stage of infrastructure design, but also maintained throughout the period of its operation and take into account the problems of effective financial spending [41–44]. In the analysed example, the authors focused mostly on space syntax, the criterion of time and cost of implementation depending on various development variants.

It is critically important also to remember that research design was aimed specifically at a micro study of a very specific area in the city. The aim of the study was dictated by the regulations related to the research grant under which the regeneration of the market square was carried out. In essence, the research grant provided the scope but also the limitations to our study that we had to keep in mind. In the above-mentioned GOSPOSTRATEG research grant for the square case study, the issues deemed to be most crucial were specifically defined as economic, spatial and social issues. One of the key guidelines was then the finances and duration of the project, and those considerations had to play a central role. Such a framework was necessary to implement the changes, hence, the decision that it would constitute one of the key guidelines during the optimisation process.

3. Purpose of Research: Case Study Description

Marketplaces, i.e., separate areas or buildings (square, street, market hall) with permanent or seasonal retail outlets or devices, are intended for trade and play an important role. They have a significant impact on the quality of life in cities. What also matters is their attractiveness in the eyes of the inhabitants. The key issue of marketplace management is at the heart of issues related to sustainable development—it combines economic, environmental and social issues. All this has been underlined in the research of the Gospostrateg project.

The aim of this research is to analyse the variants of pilot projects for the modernisation and regeneration of the Oliwa Market [14].

In particular, the initial marketplace designs and the programme for their implementation: type, scope and method of construction works were analysed in order to select the most sustainable development investment, three models for implementation were selected.

The article presents these three selected model-variants for which detailed design solutions were proposed. During the studies in the earliest stages of this project, there were more than only three model designs, but those presented in the article have been elaborated in detail as the most representative ones.

It is assumed that the results of the analysis are to identify a solution that takes into account the spatial modifications (to redesign the space) of the market and the implementation of the spatial concept. Essentially, this would ensure the functionality of a given layout and increase the number of buyers and sellers, as well as stress the importance of this location on the city map as an important public and city-forming space.

In order to indicate the optimal solution, quantitative data (cost and time of implementation of activities for each of the three selected models) and qualitative data (determined based on space syntax expertise) are analysed. This complex analysis will arrive at a rational and effective solution for the urban layout of the public space. It takes into account numerous varied criteria such as economic, technical, functional, environmental and social aspects.

To this end, the site of the Oliwa Market required regeneration and changes to its spatial organisation. Specifically, it was necessary to arrange the space by replacing the surfacing to designate zoning, that is, marking the places for trade in the "drawing" through floor surfaces. The aim of the work was to obtain an orderly space for trade on trading days when the market was filled with farmers and artisans selling their goods.

The condition of the Oliwa Market in 2018 required immediate intervention. Besides socio-economic problems, this space was neglected in terms of infrastructure and space quality as well as aesthetics. On market days (Wednesdays and Saturdays), the number of users was constantly declining. Outside of market days, the area was used as an illegal parking lot. It was an example of public space in Gdańsk, which, like many others in Poland, had lost its importance, was slowly degrading and was vulnerable to the prospect of transforming the area into another function altogether. The place, despite its past, undoubtedly constituted one of the most important elements for building the identity of the district, but was no longer an interesting local district centre.

The social factor (such as age and occupation of the space users of the market places) was also the subject of studies before the design process was undertaken. Based on that work, a number of conclusions were arrived at. It has been proved that approximately 25–30% of buyers at city marketplaces are people of retirement age. At some marketplaces, already more than half of the sellers are people of retirement age. At the same time, there are not many new, young sellers. If the current trends remain unchanged, the percentage of people in 2030 will be as high as 60. A disrupted generational change, resulting from the lack of interest in running trade fairs by young people, may lead to a demographic collapse in the next few years. Currently, this is the greatest threat facing trade in marketplaces. Marketplace monitoring data which was conducted during the whole process of transformation of the elaborated public place clearly shows this observation [45–49]. Monitoring data constitutes also the preliminary design guidelines for the design models.

The case concerns a specific place study (covering a marketplace site which has been a subject of the implementation pilot design conducted and financed under the research grant) constituting a prototype which, in the future, can potentially be used in other locations.

The wider context of the research has been underlined and expanded in the introduction, where the explanation of the particular site has been elaborated. However, the study searches for the micro scale optimal design proposal (preceded already by extensive demographic research and numerous partial design studies), that is why the site of the marketplace and the final three models have been taken into consideration and tested.

4. Projects Stage Model Description

Suggestions for solving the current problem within the research mainly involved developing a comprehensive operational strategy for socio-economic activation, enriching the cultural offer and transforming the programme concept. In addition, the project proposed improving the aesthetics of degraded exhibition areas and providing redevelopment models. Those model variants serve as the basis of the research. The subject of the analysis is a variation of a three-stage task:

Stage I: consecutive demolition of the existing surfaces and the removal of the barracks, pavilions and kiosks in the marketplace;

Stage II: construction of new surfaces for the shopping square;

Stage III: construction of small architecture elements, including waste and storage shelters together with another similar structure.

Three design variants were developed and then analysed.

4.1. Description of the Designed Condition—Option 1/Model 1

The design assumption in Model 1 was:

Trade is to take place in planned sales points within 3 × 3 m squares designated by surfacing divisions, where, for the duration of the market trading, tents will be set up. There are planned 80 cm spacing breaks in the layout of sales points due to the restrictions introduced during the COVID-19 pandemic. Many of the retail outlets allow for the possibility of direct sales from cars. On other days, the square space will be used as a culture-forming place—an active public space intended for various events within the business plan of the market operator(Figure 2).

Figure 2. Oliwa Marketplace site plan. Proposed development of the square—option 1/Model 1, drawing by the authors.

In this option on fair days, there will be a number of cars parked in the square. A similar system has been operating in the market for the past 30 years. Whilst it is more convenient for sellers, the aesthetic level and comfort for buyers is very low. There is no space for additional greenery because of the space needed for cars to manoeuvre and park. The complex scope of construction works for the first option includes, respectively: demolition works, earthworks, surface works and elements of small architecture. As part of the surface works, it is planned to build a square with a granite paved surface area of 1060 m^2 and create 35 m^2 of landscaped areas. As part of the small architecture works, the plan proposes the creation of a small wooden shed.

4.2. Description of the Designed Condition—Option 2/Model 2

The design assumption in Model 2 was:

Trade is to take place in the planned sales points within 3 × 3 m squares designated by surfacing divisions and only for the duration of trading will tents be set up. There are planned 80 cm spacing breaks in the layout of sales points due to the restrictions introduced during the COVID-19 pandemic. Some of the retail outlets assume direct sales from cars. On other days, the square space will be used as a culture-forming place—an active public space intended for various events within the business plan of the market operator (Figure 3).

Figure 3. Oliwa Marketplace site plan. Proposed development of the square—option 2/Model 2, drawing by the authors.

The commercial square will be separated by a system of plant pots with vegetation, as shown on the land development plan, between these, steel ropes with fasteners will be stretched to delineate utility separation. A waste and storage shed will be located in the northeast corner of the plot at Schopenhauer Street. There are at least two parallel internal pavements proposed in the fair space that will assure safe circulation of pedestrians on the plot. Only ten trade places are provided for cars. The other two parking places are for supply vans for the trade hall. There is also a place for parking food trucks in the north of the market. Thus, in addition to trade, a small gastronomic zone in the area is also possible.

The complex scope of construction works for the second option includes, respectively: demolition works, earthworks, surface works and elements of small architecture. As part of the surface works, it is planned to build a square with a granite paved surface with an area of 846 m^2, recreate the nearby pavement on the right with an area of 44 m^2, build a gravel surface path with an area of 190 m^2 and create 35 m^2 of landscaped areas and 10 m^2 of rain gardens. As part of the works related to small architecture, the plan is to include the following structures: a wooden shed, 21 sales boxes, a wooden platform, 11 decorative pots and 4 wooden benches.

4.3. Description of the Designed Condition—Option 3/Model 3

The design assumption in Model 3 was:

The market trade and sales activities are to take place in planned 3 × 3 m sales points designated by surfacing divisions, and only for the duration of trading will tents be set up. There are planned 80 cm spacing breaks in the layout of sales points due to the restrictions introduced during the COVID-19 pandemic. None of the retail outlets assume direct sales from cars. Cars are completely excluded from the area. On other days, the square space will be used as a culture-forming place—an active public space intended for various events within the business plan of the market operator(Figure 4).

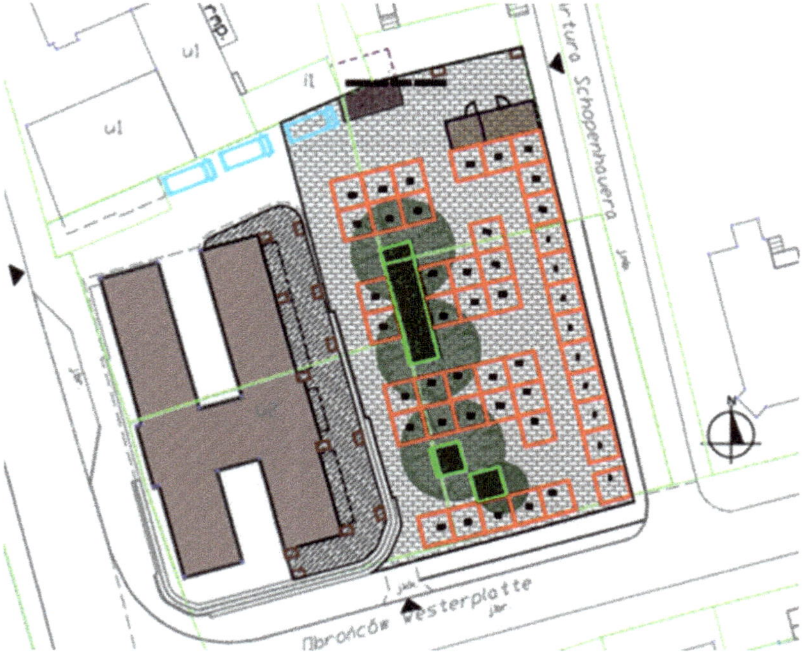

Figure 4. Oliwa Marketplace site plan. Proposed development of the square—option3/ Model 3, drawing by the authors.

The commercial square will be separated by a system of plant pots with vegetation, marked on the land development plan, between these, steel ropes with fasteners will be stretched to delineate utility separation. A waste and storage shed will be located in the northeast corner of the plot. Several internal pavements which assure safe circulation of pedestriansare proposed in the market space.

The complex scope of construction works for the third option includes, respectively: demolition works, earthworks, surface works and elements of small architecture. As part of the surface works, the plan is to build a square with a granite paved surface with an area of 1035 m^2, construct a path with a gravel surface with an area of 180 m^2 and create 60 m^2 of landscaped areas. As part of the works related to small architecture, it is planned to construct: a wooden shed, 45 sales boxes, a wooden platform, 14 decorative pots and 6 wooden benches.

5. Criteria Adopted for Optimisation, Selection of the Optimal Variant

5.1. Social, Spatial (Environmental, Security and Functional) Criteria

Each of the three spatial development design proposals were transformed into a model (a fictive urban system), which was then converted into an axial map and tested via spatial analysis. The model treats closed buildings, green spaces and stall spaces as inaccessible. All open spaces, stairs and passages through buildings were treated as convex spaces. In the end, the parameters of Connectivity, Global Choice and Global Spatial Integration were illustrated (Figures 5–7).

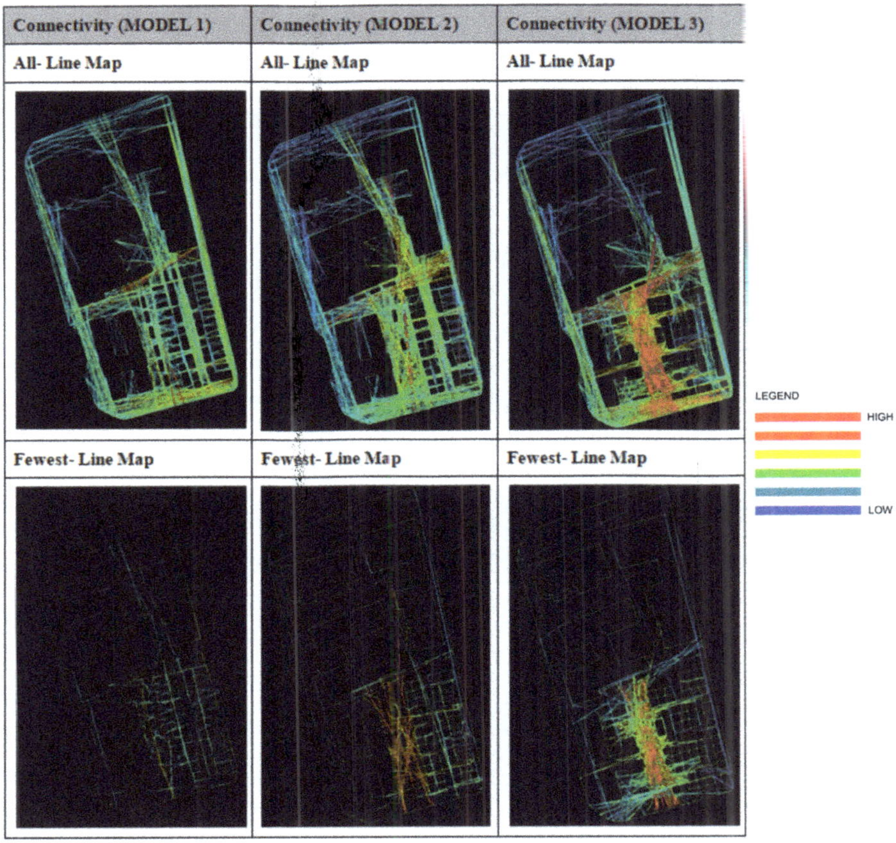

Figure 5. Oliwa Marketplace space syntax analysis. Connectivity parameter; comparative statement for three model designs, drawing by the authors.

These three parameters are the basic parameters that describe the relationship between the spaces.

Connectivity measures the number of spaces immediately connecting a space of origin [19] (p. 103). Conventional measures of place network connectivity capture only the metric characteristics of streets related to physical connectivity without considering the geometric characteristics related to visual connectivity. Our bodies interact with the built environment through a system of metric distances while our minds interact with the built environment through a system of visual distances [21,37].

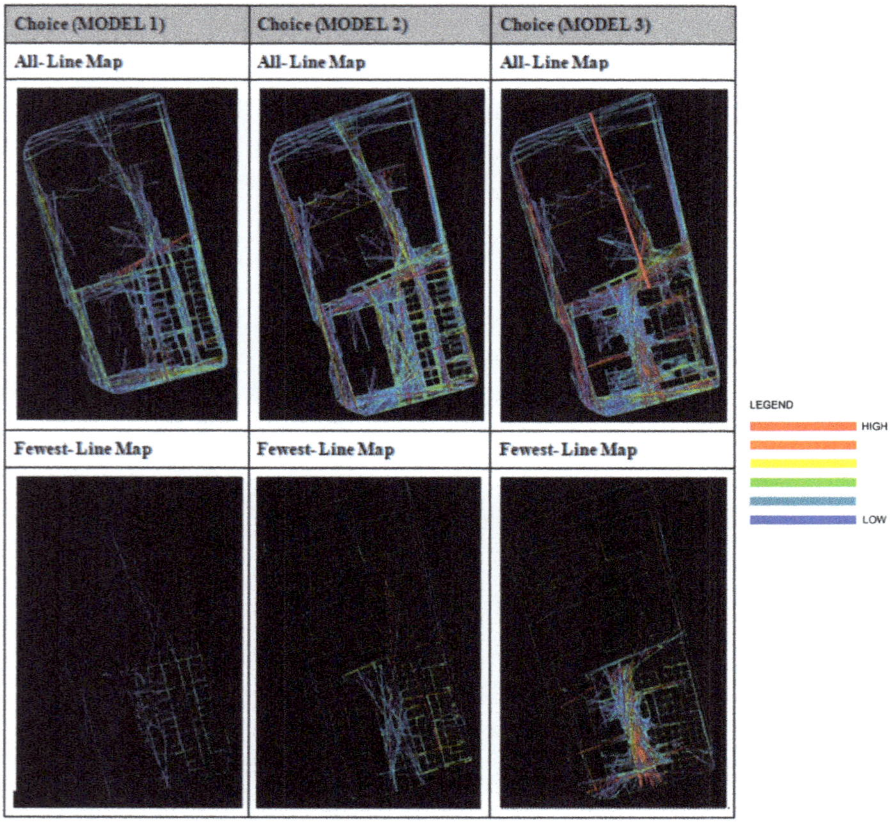

Figure 6. Oliwa Marketplace space syntax analysis. Choice parameter; comparative statement for three model designs, drawing by the authors.

The parameter of Choice, as Hillier et al. say: "measures how likely an axial line or a street segment is passed through using the shortest routes from all spaces to every other spaces in the entire system or within a predetermined distance (radius) from each segment" [50] (p. 237). The parameter of Integration is a parameter normalised by measurement of distance from a space of origin to all others in a given system under study. In general, it calculates how close the origin space is to all other spaces and can be seen as the measurement of relative asymmetry (or relative depth) [19] (pp. 108–109).

The parameter of Integration measures the potential for meetings in a space for social exchanges [20]. The potential of meetings is determined by the presence of people within the space. The concept of Integration is sometimes used interchangeably with the concept of accessibility [51]. The algorithm used calculates the shortest distance—expressed in a topological way—between individual points on the map.

Summarising, the parameter of Connectivity shows how many other spaces cross a space of origin (the more red lines there are, the more connections there are to this space), the parameter of Choice shows how busy the space is and the parameter of Integration shows how central the space is. To find more information about space syntax analysis attribute summaries, please refer to Appendix A.

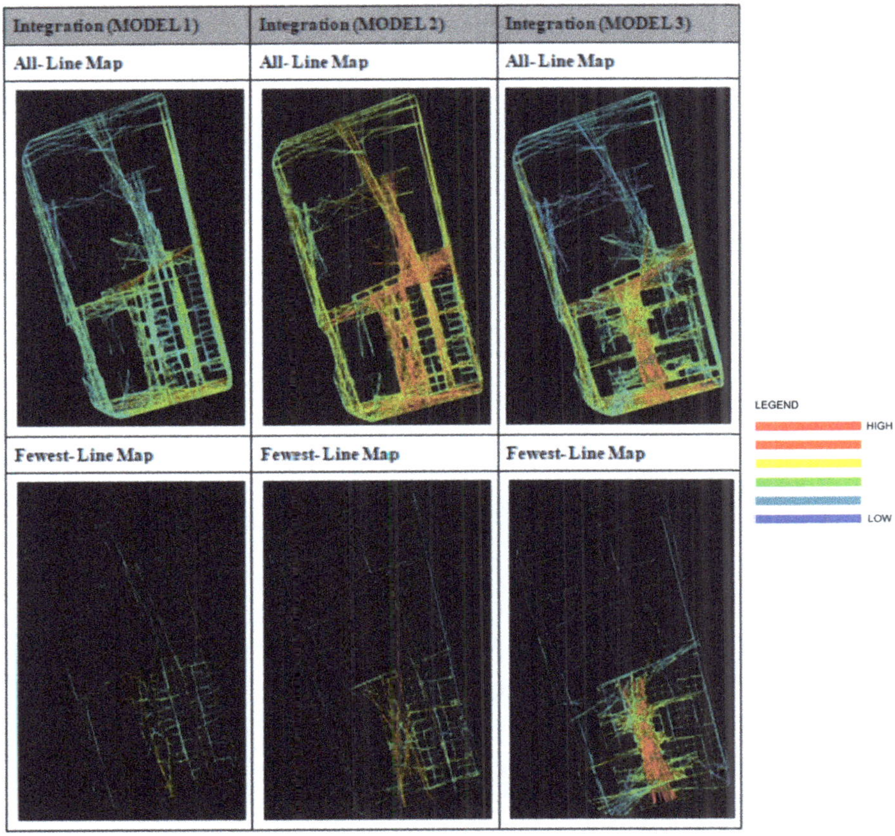

Figure 7. Oliwa Marketplace space syntax analysis. Integration parameter; comparative statement for three model designs, drawing by the authors.

Connectivity of these three models shows that the most connected is the third model. The decision to open the side entrances to the market hall and connect it with the open-air market resulted in an increase in the connectivity parameter in such a way that the middle part of the quarter has the most number of connections with other spaces in the entire system.

The Choice parameter shows how often a given space will be used to get to all other spaces in the system. It is clearly visible that the third system would favour choosing the passage between the northern and southern part of the quarter and moving through the hall and the open-air market.

The Integration parameter shows which spaces are the most central in the entire system, and thus the probability of users appearing in them. None of the spaces in the first model are well integrated with other spaces in the system. The second model clearly shows very good integration of both the north–south passage and the square in front of the hall. In the third system, the northern spaces are less integrated with the market space, while the hall and the marketplace are very well integrated with each other.

5.2. Cost Criteria

For each variant, the cost of work had to be properly calculated and determined. To do that, all project documentation was prepared, on the basis of which a bill of quantities was prepared. Based on these, three investment cost estimates were prepared.

The following assumptions were made for the calculation:

The detailed method used the rates and prices from the first quarter of 2021 of the Sekocenbud system price list, current market prices of materials and information obtained from material producers. The following works were taken into account: demolition works, edges, resistors, earthworks, surface works, landscaping and elements of small architecture. On the basis of the cost estimate prepared, taking into account the above assumptions, the cost of carrying out the works to be undertaken under the three proposed variants was determined. The results are presented in the statement of cost estimates (including VAT) for the implementation of the scope of work provided for in options 1, 2 and 3, below. The data are as follows:377,086.66 monetary units for option 1 (Model 1), 476,594.42 monetary units for option 2 (Model 2) and 568,762.22 monetary units for option3 (Model 3).

Option 3 has the highest cost of works planned to be carried out under the given solution (Figure 8).

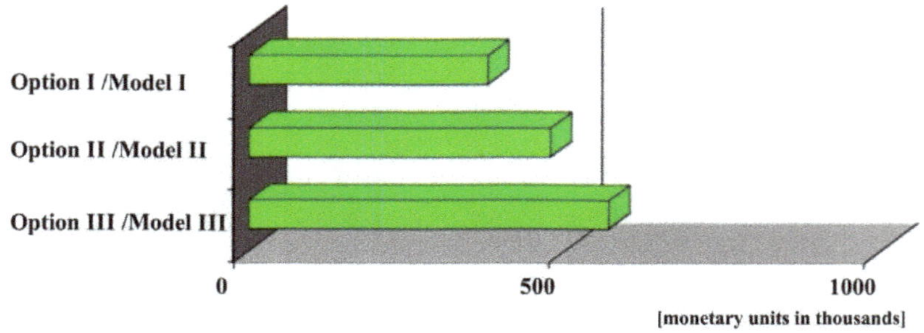

Figure 8. Costs of works envisaged for implementation under options 1, 2 and 3.

The cost of performing the scope of work covered by option 2 represents 83.80% of the cost that would be incurred by implementing option 3. The cost of carrying out work according to option 1 represents almost 66.30% of the cost of performing the scope of work provided for in option 3 (Figure 9).

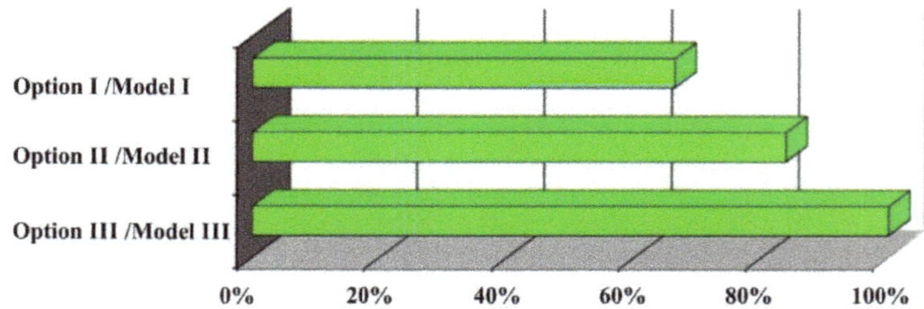

Figure 9. Percentages of the costs of works envisaged for implementation under options 1, 2 and 3.

5.3. Time Criteria

In order to determine the time of the execution of works, construction schedules were prepared for each of the three proposed variants.

The following assumptions were made within the schedules: the scope of works results from the project documentation prepared for the three considered variants, the works are carried out in the shortest possible time (using the method of uniform work, in some cases parallel and subsequent execution) and the basis for determining the tangible expenditure of labour and equipment is obtained from the construction cost estimates prepared using the detailed calculation method.

The data below presents the statement of cost estimates (excluding VAT) for the implementation of the scope of work provided for in options 1, 2and 3.

The summary of the implementation times for the three proposed variants are as follows:

There are a total of 438 work shifts for option 1 (Model 1), 423 work shifts for option 2 (Model 2) and 467 work shifts for option 3 (Model 3), where work shifts are the times in units.

The time related to the implementation of the works covered by option 3 is the longest—467 working shifts (Figure 10). This is primarily due to the greatest range of activities to be carried out.

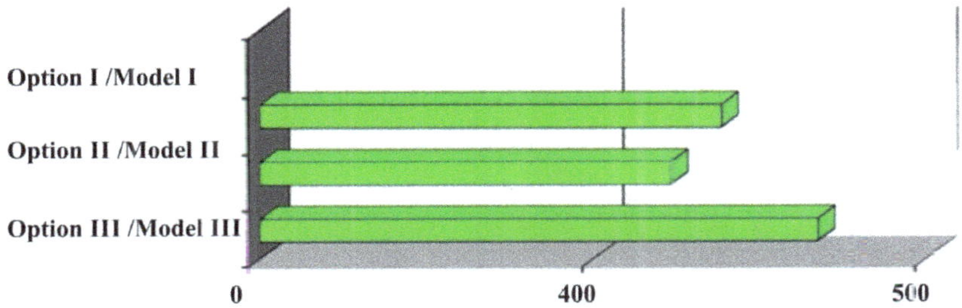

Figure 10. Duration of the works envisaged for implementation under options 1, 2 and 3.

6. Conclusions

Remark: As proven, the first design model, which focuses on the habits of sellers and their points of view, is the most economical solution. However, this model, which favours parking cars on the square, is not the right solution in the context of creating a user-friendly space. On the other hand, considering only the spatial aspect of the space and its functioning and connection with the neighbourhood, measured by the space syntax analysis, the third model appears to be the best solution. Finally, we conclude that Model 2 is the most optimal, and that by balancing the costs versus spatial solutions, it is possible to effectively improve the current management structures of marketplaces and, as a result, transform them as living public spaces. We conclude that the balance between cost and design assumptions informs implementation in various conditions, allowing for the changing nature of local markets to thrive and bring them closer to the inhabitants (in a more sustainable way).

The consideration and analysis of the selected models to implement investment in market regeneration (Figure 11) executed and presented in the article justifies the methods, conclusions and statements presented below:

1. The implementation of regeneration investment must take into account the balance of profits and losses and creating optimal solutions for the main project. Social and spatial aspects are not the only elements that make up a sustainable implementation model. Cost and time criteria are critical in determining the final optimal implementation (Table 1). It is imperative that the final investment decision, and all related practical aspects, undergo broad analysis. It must take into account criteria of a different nature, such as social and environmental considerations.

2. Given the complex nature of the interactions of various criteria, and the multiplicity of possible decisions and actions, it is essential that the process of solution optimisation be started as early as possible, i.e., at the design stage. Such an approach allows the consideration of all options and makes it possible to take a comprehensive look at the regeneration process and its investment. An example of such an optimisation at this stage can include the size and shape of the design solutions, type of material and technology used or proposed structural solutions. Such an early and comprehensive analysis is also rational from the point of view of the efficiency of spending. It should be emphasised that

as the scope and size of investments increase, the complexity of their analysis increases, but at the same time the possibility for optimising solutions also increases.

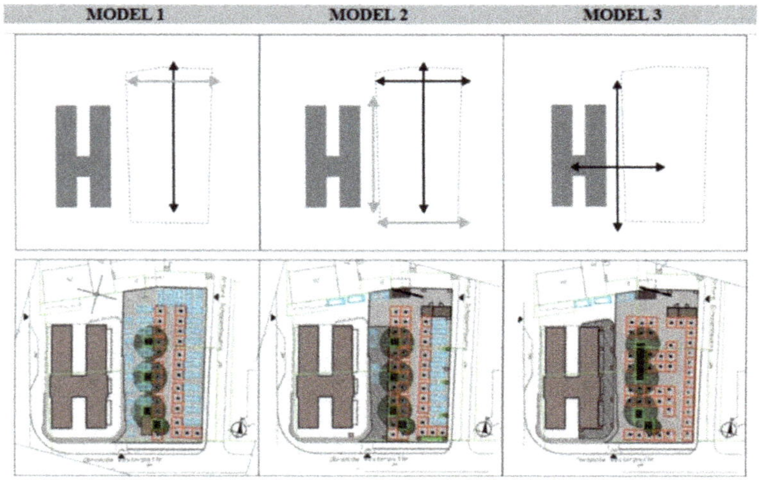

Figure 11. Three selected design models scheme drawing—three variants of the case study design proposals taken into account for the research.

Table 1. Midterm and final evaluation of the three design models—three variants and the chosen model (suggestion) of the most optimal solution to be implemented.

Criteria/MODEL	MODEL 1	MODEL 2	MODEL 3
Spatial Social	1 1	2 3	3 2
Mid Evaluation	2	5	5
additional factors needed		no final choice M2 = M3	
Time (working shifts) Cost	2 3	3 2	1 1
Final Evaluation	7	10 data	7 data

3. The regeneration process is undoubtedly complex, and must take into account social, spatial, cost, time, functionality, safety and environmental criteria. Optimisation of solutions should be carried out in many forms through workshops, public consultation, environmental interviews, on many occasions and at various stages of preparation, sometimes also during the implementation of the investment. The recommended solution (Model 2) takes into account the long-term forecast of pedestrian traffic, as well as other parameters. Consequently, this solution ensures the safety of the public space is maintained at the highest level, whilst allowing successful redevelopment of the marketplace and the whole district, and it takes into account issues such as a reduction in the degradation of the public space, and the material, environmental and economic losses of its users (sellers, buyers, inhabitants).

4. The analysis presented in the article and the proposed solution for the regeneration of the marketplace (Model 2) takes into account all of the current and future social and spatial needs, expectations of marketplace users and social and functional requirements. In a broader context, it also meets the objectives of sustainable development and aspects of creating places, with a particular focus on their flexibility, and urban development

impact. Implementation of the optimal design variant is crucial in the context of successful regeneration and ensuring a so-called "domino effect" of the project. In this context, the future of the marketplace after implementation of the chosen variant-model seems to be very promising for the further development of the entire district.

Moreover, it is important to note that the tested implementation model is currently at the implementation stage in the broader context of the project. Subsequent implementation could be the basis for discussions with other cities, where the problem of maintaining markets has clearly emerged. We are gratified to see that our study is perceived as relevant and can be potentially useful to other cities and communities around the country. This is an urgent problem that needs to be addressed, particularly in view of the generational change facing these areas.

Author Contributions: Conceptualization, J.B., W.M. and M.S.; Data curation, D.N.; Formal analysis, W.M. and M.S.; Funding acquisition, J.B. and P.C.; Investigation, J.B., W.M. and M.S.; Methodology, J.B., W.M. and M.S.; Project administration, J.B.; Resources, J.B. and D.N.; Software, W.M., D.N. and M.S.; Supervision, J.B., P.C., G.G., W.M. and M.S.; Validation, J.B., W.M. and M.S.; Visualization, J.B., W.M. and M.S.; Writing—original draft, J.B., W.M. and M.S.; Writing—review and editing, J.B., W.M., P.C. and G.G. All authors have read and agreed to the published version of the manuscript.

Funding: This research was funded by the Polish National Centre for Research and Development under the a research and development project carried out under the grant of the strategic program of scientific research and development work "Social and economic development of Poland in the conditions of globalizing markets" GOSPOSTRATEG—contract for the implementation and financing of the project number Gospostrateg1/392278/6/NCBR/2018. The research team of Gospostrateg project consists of three units represented successively by: the Stowarzyszenie Inicjatywa Miasto—Project Leader Association, the Gdańsk University of Technology and the Academy of Fine Arts. The issue/subject of the article is one of the tasks of the case study implementation of the project, which the Gdańsk University Technology team was responsible for.

Institutional Review Board Statement: Not applicable.

Informed Consent Statement: Not applicable.

Data Availability Statement: The data cited in this study are openly available in Open Research Data Catalog on the Bridge of Knowledge Portal (Most Wiedzy) at https://mostwiedzy.pl/en/open-research-data/catalog accessed on 26 June 2021 at https://doi.org/10.34808/yn27-gb21 accessed on 26 June 2021; https://doi.org/10.34808/58s9-fg37 accessed on 26 June 2021; https://doi.org/10.34808/qarn-c096 accessed on 26 June 2021; https://doi.org/10.34808/s5b1-9e72 accessed on 26 June 2021.

Acknowledgments: It should be emphasised that the analysis presented in the article should be treated as preliminary research on the diagnosed problem of the pilot project. Note: The study is strictly linked to and supported by the research project entitled: Public market revitalisation strategy usingthe social catalyst entrepreneurship method, brand repositioning and placemaking as a tool for local development policy (in Polish, the title reads as follows: "Strategia rewitalizacji obiektów handlu targowego z wykorzystaniem metody społecznego katalizatora przedsiębiorczości, repozycjonowania marki oraz placemakingu jako narzędzie polityki lokalnej"), project number: Gospostrateg1/392278/6/NCBR/2018. It is conducted within the framework of the GOSPOSTRATEG program, financed by the National Centre for Research and Development in Poland. According to the authors' assumptions and expectations, this research will be continued within the above-mentioned project in order to prepare a strategy of regeneration of marketplaces in Poland and elsewhere within eastern Europe.

Conflicts of Interest: The authors declare no conflict of interest.

Appendix A

A clear explanation of the factors actually measured during the testing phase has been added in the Appendix A, as an attribute summary with details and data supplemental to space syntax analysis of the models.

Attribute Summary

	Attribute	Minimum	Average	Maximum
1	Connectivity	2	10.9333	33
2	Line Length	5.82553	25.9758	110.069
3	Choice	1	106.187	1104
4	Choice [Norm]	0.000370233	0.0393138	0.408738
5	Entropy	1.14773	1.66832	2.01792
6	Integration [HH]	1.57464	2.76622	5.03885
7	Integration [P-value]	1.57464	2.76622	5.03885
8	Integration [Tekl]	0.69574	0.77798	0.897682
9	Intensity	0.803583	1.24223	2.02169
10	Harmonic Mean Depth	5.10811	15.4297	39.0863
11	Mean Depth	1.74324	2.43495	3.37838
12	Node Count	75	75	75
13	Relativised Entropy	1.2489	1.74067	2.23086

OK

Attributionsummary with details and data supplemental to space syntax analysis of MODEL 1

Attribute Summary

	Attribute	Minimum	Average	Maximum
1	Connectivity	4	185.273	466
2	Line Length	0.753007	31.7609	110.069
3	Choice	0	1275.64	13480
4	Choice [Norm]	0	0.00209895	0.0221801
5	Entropy	0.985697	1.38848	1.69289
6	Integration [HH]	2.8904	6.91974	12.1312
7	Integration [P-value]	2.8904	6.91974	12.1312
8	Integration [Tekl]	0.792552	0.887443	0.966661
9	Intensity	0.549531	1.26574	1.9506
10	Harmonic Mean Depth	5.85033	89.6503	532.719
11	Mean Depth	1.62103	2.15652	3.60653
12	Node Count	1104	1104	1104
13	Relativised Entropy	1.33678	1.7507	2.78389

OK

Attribution summary with details and data supplemental to space syntax analysis of MODEL 2

Attribute Summary

	Attribute	Minimum	Average	Maximum
1	Connectivity	3	108.532	212
2	Line Length	7.40154	27.2773	110.069
3	Choice	0	203.079	1654
4	Choice [Norm]	0	0.00531259	0.043269
5	Entropy	0.721513	1.22828	1.60695
6	Integration [HH]	3.04807	9.12793	23.0197
7	Integration [P-value]	3.04807	9.12793	23.0197
8	Integration [Tekl]	0.791331	0.949678	1.17185
9	Intensity	0.506842	1.95006	3.51604
10	Harmonic Mean Depth	4.74399	61.3619	120.592
11	Mean Depth	1.24188	1.73314	2.82671
12	Node Count	278	278	278
13	Relativised Entropy	1.1093	1.42681	2.73285

OK

Attribution summary with details and data supplemental to space syntax analysis of MODEL 3

References

1. Janssens, F.; Sezer, C. Marketplaces as an Urban Development Strategy. *Built Environ.* **2013**, *39*, 169–171. [CrossRef]
2. Caramaschi, S. Public markets: Rediscovering the centrality of markets in cities and their relevance to urban sustainable development. In *The Sustainable City IX*; Marchettini, N., Brebbia, C.A., Pulselli, R., Bastianoni, S., Eds.; WIT Press: Southampton, UK, 2014; pp. 1187–1197.
3. Schappo, P.; Van Melik, R. Meeting on the marketplace: On the integrative potential of The Hague Market. *J. Urban. Int. Res. Placemaking Urban Sustain.* **2017**, *10*, 318–332. [CrossRef]
4. Black, R. Porta Palazzo: The Anthropology of an Italian Market. Pennsylvania: University of Pennsylvania Press 2012. Available online: www.jstor.org/stable/j.ctt3fhzsr (accessed on 26 June 2021).
5. Ercan, M.A. *Regeneration, Heritage and Sustainable Communities in Turkey: Challenges, Complexities and Potentials*; Routledge: London, UK; New York, NY, USA, 2020.
6. Özgür, E.F. Urban design projects and the planning process: The Kadıköy Old Market Area Revitalization Project and the Kartal Industrial Area Regeneration Project. *Cities* **2013**, *31*, 208–219. [CrossRef]
7. Janssens, F. Through Istanbul's marketplaces: The materiality of the market. In Proceedings of the RC21 International Conference on The Ideal City: Between Myth and Reality. Representations, Policies, Contradictions and Challenges for Tomorrow's Urban Life, Urbino, Italy, 27–29 August 2015; pp. 27–29.
8. Greater London Authority. *The Local Enterprise Partnership for London, Understanding London's Markets*; Report; Greater London Authority City Hall: London, UK, 2017.
9. Whyte, W.H. *The Social Life of Small Urban Spaces*; Project for Public Spaces: New York, NY, USA; The Conservation Foundation: Washington, DC, USA, 1980.
10. Bennett, R.; Savani, S. The Rebranding of City Places: An International Comparative Investigation. *Int. Public Manag. Rev.* **2003**, *4*, 70–87.
11. Bieszk-Stolorz, B.; Felsztyńska, I. Analysis of the attractiveness of large cities with respect to the development of marketplace trade, (Polish title: Analiza atrakcyjności dużych miast pod względem rozwoju handle targowiskowego). *Studia I Pract. WNEiS US* **2018**, *54*, 407–419. (In Polish) [CrossRef]
12. Czyż, P.; Hanzel, P. Przestrzenie handlu tradycyjnego w kontekście procesu rewitalizacji—Szkic problematyki. *Studia Kom. Przestrz. Zagospod. Kraj. PAN* **2018**, *192*, 315–325. (In Polish)
13. Borucka, J.; Czyż, P.; Mazurkiewicz, W.; Pancewicz, Ł.; Perzyna, I. Improving social competencies of architecture students through participatory design of marketplace regeneration. *World Trans. Eng. Technol. Educ.* **2021**, *19*, 71–78.
14. City Initiative Association (In Polish: Inicjatywa Miasto Stowarzyszenie). *SocialCatalyst of Entrepreneurship Project (In Polish: Projekt—Społeczny Katalizator Przedsiębiorczości)*; 2018. Available online: http://inicjatywamiasto.pl/portfolio/spoleczny-katalizator-przedsiebiorczosci/ (accessed on 7 July 2020). (In Polish).
15. National Centre for Research and Development (In Polish: Narodowe Centrum Badań i Rozwoju). *Strategic Program "Social and Economic Development of Poland in the Conditions of Globalising Markets -GOSPOSTRATEG" from National Centre for Research and Development in Poland (In Polish: GOSPOSTRATEG—Strategiczny Program Badań Naukowych i Prac Rozwojowych "Społeczny i Gospodarczy Rozwój POLSKI w Warunkach Globalizujących się Rynków")*; 2018. Available online: https://www.ncbr.gov.pl/programy/programy-strategiczne/gospostrateg/ (accessed on 7 July 2020). (In Polish)
16. Chang, C.H.; Lin, J.J.; Lin, J.H.; Chiang, M.C. Domestic open-end equity mutual fund performance evaluation using extended TOPSIS method with different distance approaches. *Expert Syst. Appl.* **2010**, *37*, 4642–4649. [CrossRef]
17. Roszkowska, E. Teoria Podejmowania Decyzji a Teorie Zarządzania. Optim. *Studia Ekon.* **2010**, *44*, 97–119. (In Polish)
18. Berčič, T.; Bohanec, M.; AžmanMomirski, L. Role of decision models in the evaluation of spatial design solutions. *Ann.-Anal. Za IstrskeMediter. Studije-Ser. Hist. Et Sociol.* **2018**, *28*, 621–636.
19. Hillier, B.; Leaman, A.; Stansall, P.; Bedford, M. Space Syntax. *Environ. Plan. B* **1976**, *3*, 147–285. Available online: http://en.wikipedia.org/wiki/Space_syntax (accessed on 10 June 2021). [CrossRef]
20. Hillier, B.; Hanson, J. *The Social Logic of Space*; Cambridge University Press: Cambridge, UK, 1984.
21. Hillier, B. *Space Is the Machine: A Configurational Theory of Architecture*; Space Syntax: London, UK, 1996.
22. Hajrasouliha, A.; Li, Y. The Impact of Street Network Connectivity on Pedestrian Volume. *Urban Stud.* **2015**, *52*, 2483–2497. [CrossRef]
23. Rokem, J.; Vaughan, L. Segregation, Mobility and Encounters in Jerusalem: The Role of Public Transport Infrastructure in Connecting the 'Divided City'. *Urban Stud.* **2018**, *55*, 3454–3473. [CrossRef]
24. Dettlaff, W. Space Syntax Analysis—Methodology of Understanding Space. *PhD Interdiscip. J.* **2014**, *1*, 283–291.
25. Seamon, D. Understanding place holistically: Cities, synergistic relationality, and space syntax. *J. Space Syntax* **2015**, *6*, 19–33. Available online: http://Joss.bartlett.ucl.ac.uk/journal/index.php/joss/article/view/246 (accessed on 18 March 2021).
26. Fridgeirsson, T.; Rosłon, J. *Optymalizacja Procesów Budowlanych*; ERASMUS+ proj. nr. 2015-1-PL01-KA202-016454; Biblioteka Menedżerów Budowlanych: Islandia, Poland, 2017; Volume 12. (In Polish)
27. Zhou, J.; Love, P.E.; Wang, X.; Teo, K.L.; Irani, Z. A review of methods and algorithms for optimising construction scheduling. *J. Oper. Res. Soc.* **2013**, *64*, 1091–1105. [CrossRef]

28. Hejducki, Z.; Podolski, M. *Harmonogramowanie Przedsięwzięć Budowlanych z Zastosowaniem Algorytmów Metaheurystycznych*; ERASMUS+ proj. nr. 2015-1-PL01-KA202-016454 63; Biblioteka Menedżerów Budowlanych: Wrocław, Poland, 2012; pp. 68–79. (In Polish)
29. Anysz, H. The profit as in-company evaluation of the construction site effectiveness. In Proceedings of the MATEC Web of Conferences. EDP Sciences, Seoul, Korea, 24 July 2017; Theoretical Foundation of Civil Engineering: Warsaw, Poland, 2017; pp. 1–6. [CrossRef]
30. Nguyen, A.T.; Sigrid, R.; Philippe, R. A review on simulation-based optimization methods applied to building performance analysis. *Appl. Energy* **2014**, *113*, 1043–1058. [CrossRef]
31. Jaśkowski, P.; Sobotka, A. Scheduling construction projects using evolutionary algorithm. *J. Constr. Eng. Manag.* **2016**, *132*, 861–870. [CrossRef]
32. Książek, M.V.; Nowak, P.O.; Kivrak, S.; Rosłon, J.H.; Ustinovichius, L. Computer-aided decision-making in construction project development. *J. Civ. Eng. Manag.* **2015**, *21*, 248–259. [CrossRef]
33. Ucar, A.; Inalli, M. Exergoeconomic analysis and optimization of a solar-assisted heating system for residential buildings. *Build. Environ.* **2016**, *41*, 1551–1556. [CrossRef]
34. Wright, J.A.; Loosemore, H.A.; Farmani, R. Optimization of building thermal design and control by multi-criterion genetic algorithm. *Energy Build.* **2002**, *34*, 959–972. [CrossRef]
35. Liao, T.W.; Egbelu, P.J.; Sarker, B.R.; Leu, S.S. Metaheuristics for project and construction management—A state-of-the-art review. *Autom. Constr.* **2011**, *20*, 491–505. [CrossRef]
36. Belniak, S.; Leśniak, A.; Plebankiewicz, E.; Zima, K. The influence of the building shape on the costs of its construction. *J. Financ. Manag. Prop. Constr.* **2013**, *18*, 90–102. [CrossRef]
37. Ramzi, O.; Krarti, M. Building Shape Optimization Using Neural Network and Genetic Algorithm Approach. *Ashrae Trans.* **2016**, *112*, 484–491.
38. Hillier, B.; Turner, A.; Yang, T. Metric and topo-geometric properties of urban street networks: Some convergences, divergences and new results. *J. Space Syntax* **2010**, *1*, 258–279.
39. Pujadas, P.; Pardo-Bosch, F.; Aguado-Renter, A.; Aguado, A. MIVES multi-criteria approach for the evaluation, prioritization, and selection of public investment projects. A case study in the city of Barcelona. *Land Use Policy* **2017**, *64*, 29–37. [CrossRef]
40. Cichocka, J.; Browne, W.N. Multicriteria optimization in architectural design Goal-oriented methods and computational morphogenesis. In *Shapes of Logic Everything What Surround Us Can Be Described*; Oficyna Wydawnicza Politechniki Wrocławskiej: Wrocław, Poland, 2016.
41. Plebankiewicz, E.; Meszek, W.; Zima, K.; Wieczorek, D. Probabilistic and Fuzzy Approaches for Estimating the Life Cycle Costs of Buildings under Conditions of Exposure to Risk. *Sustainability* **2020**, *12*, 226. [CrossRef]
42. Gobis, A.; Jamroz, K.; Jeliński, Ł. Zastosowanie metody szacowania kosztów cyklu życia w zarządzaniu infrastrukturą transportową. *Pract. Nauk. Politech. Warsz. Transp.* **2019**, *126*, 5–14. (In Polish)
43. Szczepański, M.; Grzyl, B. Technical and Economic Analysis of the Implementation of Selected Variants of Road Investment. *Buildings* **2020**, *10*, 97. [CrossRef]
44. Kiss, B.; Silvestre, J.D.; Andrade Santos, R.; Szalay, Z. Environmental and Economic Optimisation of Buildings in Portugal and Hungary. *Sustainability* **2021**, *13*, 13531. [CrossRef]
45. Borucka, J. Regeneration Project of Market Places GOSPOSTRATEG—"Polanki" Market in Gdańsk-Oliwa Pilot Project Monitoring Dataset. In *Sharing Research Data Across Disciplines*; Wałek, A., Ed.; Politechnika Gdańska: Gdańsk, Poland, 2022; pp. 240–246. [CrossRef]
46. Borucka, J.; Mazurkiewicz, W.M.; Nałęcz, D. *Monitoring of Activities Carried Out as Part of Prototyping and Implementation of the Pilot Project in the Area of the "Polanki" Market and Its Direct Neighbourhood, in the Gdańsk-Oliwa District, Step1; Stage from July 2020 Year*; Gdańsk University of Technology Open Research Data Catalog on the Bridge of Knowledge Portal (Most Wiedzy). 2020. Available online: https://mostwiedzy.pl/en/open-research-data/catalog (accessed on 26 June 2021). [CrossRef]
47. Borucka, J.; Mazurkiewicz, W.M.; Nałęcz, D. *Monitoring of Activities Carried Out as Part of Prototyping and Implementation of the Pilot Project in the Area of the "Polanki" Market and Its and Its Direct Neighbourhood, in the Gdańsk-Oliwa District, Step1; Stage from August 2020 Year*; Gdańsk University of Technology Open Research Data Catalog on the Bridge of Knowledge Portal (Most Wiedzy). 2020. Available online: https://mostwiedzy.pl/en/open-research-data/catalog (accessed on 26 June 2021). [CrossRef]
48. Borucka, J.; Mazurkiewicz, W.M.; Nałęcz, D. *Monitoring of Activities Carried Out as Part of Prototyping and Implementation of the Pilot Project in the Area of the "Polanki" Market and Its and Its Direct Neighbourhood, in the Gdańsk-Oliwa District, Step1; Stage from September 2020 Year*; Gdańsk University of Technology Open Research Data Catalog on the Bridge of Knowledge Portal (Most Wiedzy). 2020. Available online: https://mostwiedzy.pl/en/open-research-data/catalog (accessed on 26 June 2021). [CrossRef]
49. Borucka, J.; Mazurkiewicz, W.M.; Nałęcz, D. *Monitoring of Activities Carried Out as Part of Prototyping and Implementation of the Pilot Project in the Area of the "Polanki" Market and Its and Its Direct Neighbourhood, in the Gdańsk-Oliwa District, Step1; Stage from October 2020 Year*; Gdańsk University of Technology Open Research Data Catalog on the Bridge of Knowledge Portal (Most Wiedzy). 2020. Available online: https://mostwiedzy.pl/en/open-research-data/catalog (accessed on 26 June 2021). [CrossRef]

50. Hillier, B.; Burdett, R.; Peponis, J.; Penn, A. Creating Life: Or, Does Architecture Determine Anything? *Archit. Compon./Archit. Behav.* **1987**, *3*, 233–250.
51. Szczepańska, J. Democracy by Design? Using Space Syntax Theory to Understand the Co-Presence of Various Users in Public Space. (Polish Title: Demokracja Przez Projekt?—Wykorzystanie Teorii Space Syntax do Zrozumienia Współobecności Różnych Użytkowników w Przestrzeni Publicznej). MSc Thesis, University of Warsaw, Warsaw, Poland, 2011. (In Polish)

sustainability MDPI

Article

Environmental Performances of a Cubic Modular Steel Structure: A Solution for a Sustainable Development in the Construction Sector

Sebastian George Maxineasa *, Dorina Nicolina Isopescu *, Ioana-Roxana Baciu and Marius Lucian Lupu

Department of Civil and Industrial Engineering, Faculty of Civil Engineering and Building Services, "Gheorghe Asachi" Technical University of Iasi, 700050 Iasi, Romania; ioanaroxana.baciu@yahoo.com (I.-R.B.); mariuslucianlupu@gmail.com (M.L.L.)

* Correspondence: sebastian.maxineasa@tuiasi.ro (S.G.M.); dorina_isopescu@yahoo.co.uk (D.N.I.)

Abstract: The production of building materials is a significant component of the impact the construction sector has on the natural environment. Steel is among the most utilized materials, having various applications specific to the built environment. Therefore, understanding the impact of this structural material represents an important step in achieving global sustainable development. The paper aims to analyze the effects of different steel structural elements on the Earth's ecosystem with respect to concerns over sustainability. In order to reach this goal, the authors have analyzed a dwelling steel structure based on cubic modules with high structural modularity. In addition, the study looks at the influence of an over the floor reinforced concrete slab in order to gain an overall view regarding environmental performances. The impact on the natural environment has been analyzed by considering the cradle-to gate with options Life Cycle Assessment study. The paper provides up-to-date knowledge on the environmental performances of the analyzed structure, presenting encouraging conclusions for construction sector specialists with respect to the use of steel as a material that can represent a solution in the current global effort to minimize the environmental burdens imposed by the construction sector.

Keywords: natural environment; construction sector; steel; sustainable development; cradle-to-gate with options; Life Cycle Assessment

Citation: Maxineasa, S.G.; Isopescu, D.N.; Baciu, I.-R.; Lupu, M.L. Environmental Performances of a Cubic Modular Steel Structure: A Solution for a Sustainable Development in the Construction Sector. *Sustainability* **2021**, *13*, 12062. https://doi.org/10.3390/su132112062

Academic Editor: Oleg Kapliński

Received: 8 September 2021
Accepted: 29 October 2021
Published: 1 November 2021

Publisher's Note: MDPI stays neutral with regard to jurisdictional claims in published maps and institutional affiliations.

1. Introduction

It is well-known that the impact on the natural environment resulting from people's daily activities is constantly increasing every year and that this impact threatens to limit the abilities of future generations to ensure proper development. At the global scale, this negative influence is characterised by an alarming increase in the amount of greenhouse gas emissions into the atmosphere, as well as the quantity of non-renewable energy consumed in order to satisfy living standards. In order to have a complete view of the present negative situation regarding the natural environment, we need to take into consideration, too, the distressing rates of raw material consumption at the global scale. The latest report of the Global Footprint Network shows that globally we are consuming approximately 70% more raw materials than Earth has the capacity to naturally renew [1]. Therefore, understanding and taking steps to achieve sustainability is the most significant challenge that must be faced in the near future by all global industrial sectors, with the goal of drastically reducing their environmental burdens.

It is common knowledge that the building sector is responsible for one of the most substantial impacts on the natural environment. The construction industry is responsible for the consumption of approximately 60% of the entire volume of natural resources consumed at the global scale. Furthermore, the construction industry is responsible for consuming nearly 40% of the global amount of energy produced, generating at the same time approximately 25% of the total global waste and over 40% of the overall volume of greenhouse gases emitted into the atmosphere. Another critical environmental effect of the sector is represented by the amount of building materials consumed, which is more than half of the volume of materials used globally [2–15].

Even though the last decades have seen the adoption of different rules and regulations (e.g., the enforcement of the Nearly Zero Energy Buildings regulation in the European Union) intended to reduce non-renewable energy consumption in building work, as well as in minimizing carbon dioxide emissions, the construction sector remains one of the biggest polluters globally. It must be clearly understood that an important part of the total impact of the construction sector is influenced in a substantial manner by the amount and type of materials employed for maintaining the state of the existing built environment and, most importantly, for creating new buildings.

Thus, taking into account the fact that the world's population is significantly increasing, it is expected that in the near future the consumption of construction materials will increase due to the need for enlarging the existing built environment. Taking this into account, it can with justification be said that besides improving energy efficiency standards in the construction sector, we also need to improve the types and amounts of materials consumed in this sector. The negative effects resulting from the use of massive volumes of materials can be reduced by finding different solutions and/or materials that can be used in building design (e.g., by considering structural systems which can be disassembled, or by using various highly recyclable structural materials). Therefore, fully evaluating and understanding the impact of various construction materials and applying that knowledge to the choice of products used in this sector will have a major influence on the global sustainable development of the built environment [4–7,16–25].

Steel is considered to be one of the principal materials used in the construction sector. Whether it is used as bars in reinforced concrete elements, as connectors or fixing elements for timber structures, or as structural steel sections, this material is consumed in huge amounts in construction. From this perspective, it must be mentioned that the material's specific manufacturing processes are responsible for about 9% of the overall volume of carbon dioxide (CO_2) emitted worldwide [5,26].

At the global scale, in the last four decades, the quantity of steel manufactured has increased by approximately 2.6 times, reaching 1600 million tons in 2013 [5,27]. The global consumption of this material is rapidly increasing year after year; the value registered for steel consumption in 2017 is approximately 0.2 billion tons higher than the one reported in 2013 [28,29]. Taking into account that approximately half of the total amount of steel fabricated worldwide is consumed in processes specific to the built environment, consideration of the ecological effects of this material in the construction sector is highly consequential [5,26].

Part of the impact on the natural environment associated with steel-based products is due to the methods employed for manufacturing the component material. Most commonly, steel is produced using the electric arc furnace (EAF) method or the basic oxygen furnace (BOF) method [5,30–33]. The EAF technique consumes electricity, while the BOF production method uses substantial amounts of natural gas and coal. It should be noted that in the former method, large quantities of scrap materials are consumed; it can therefore be said that EAF products have a smaller negative ecological impact [5,34].

Compared with other traditional building materials, steel has a unique ecological characteristic. It is a material that can be fully recycled numerous times without diminishment of its mechanical properties [5,31,32]. Steel producers have taken serious steps to minimize the ecological impact of their products. One visible result of these environmental policies has been observed in the United States of America, where the carbon footprint of steel production is 47% lower than what it was in 1990 [5,34].

Considering the above, and also taking into account that the ecological effects of the built environment are expected to increase, civil engineering specialists must seek to put into effect measures that will significantly reduce the environmental burdens specific to the construction industry. Furthermore, as previously mentioned, this sector exerts a considerable influence over the ecological impact of the steel manufacturing industry; therefore, the use of this material in the built environment should involve an understanding of the effects on the Earth's ecosystem and efforts to diminish them.

Seeing that, in recent years, different norms and European directives have been promoted with the declared goal of significantly reducing energy consumption while creating optimal interior living conditions during the usage phase of a construction, the authors believe that at the present moment it is of paramount importance to fully understand and improve the environmental performance of the structural materials used in building construction. Therefore, the goal of the present paper is to determine the environmental consequences of using steel as a structural material. This aim is achieved by analysing a dwelling structure using cubic modules made of steel square hollow sections (SHS). The environmental burdens this structure imposes were determined and interpreted by employing the Life Cycle Assessment (LCA) methodology.

2. Case Studies

The objective of the study was achieved by using the international standards ISO 14040:2006 and ISO 14044:2006, which define LCA as the "compilation and evaluation of the inputs, outputs and the potential environmental impacts of a product system throughout its life cycle" [35,36]. In the present study, the authors have considered the cradle-to-gate with options LCA study in order to assess the impact on the environment of the analyzed structure. In order to enable a better understanding of the boundaries of the study, Table 1 presents the life cycle stages that have been used. These modules have been characterized using the European standards EN 15978:2011 [37] and EN 15804+A1:2013 [38].

Table 1. Life cycle phases considered.

Life Cycle Phase	Life Cycle Module
Extraction of raw materials	A1
Processing of raw materials and production of construction materials	A3
De-construction/Demolition	C1
Waste processing	C3
Reuse/Recycling of materials	D
Transportation phases	A2, A4, C2

The authors have considered the above-mentioned type of LCA study due to the fact that no type of maintenance work is required during the operation stage (e.g., reapplying a protective coating), seeing as the steel elements will be extremely well-protected against moisture in a highly energy-efficient building. This protection is a direct result of the different construction details, materials and technologies (e.g., vapour barrier foil, a thick layer of thermal insulating material, indoor heat recovery ventilation system) that have to be used in order to achieve a building with a low level of energy consumption that is able to create and maintain optimal indoor climate conditions. Therefore, the present study has some limitations regarding the life cycle modules considered in the assessment. The authors haven't considered all the modules for several reasons. For one, module A5 has not been taken into account due to the fact that the existing databases are not clear on the impact resulting from different technologies and machines used in the installation process. Therefore, in order to avoid influencing the final results and deriving invalid conclusions, the authors decided to not consider this module. As stated before, according to the EU regulations on energy consumption for heating and cooling, the levels of energy used are going to be significantly reduced, and therefore we think that the main environmental problem for the construction sector will be the negative impact resulting from the use of different materials. This is the reason we have not considered module B6. Additionally, taking into account the nZEB regulations, of the final amount of operational energy consumed, a substantial part should come from renewable sources, and the values for heating and cooling will be comparable in many EU countries. Thus, in the future, in many EU countries the impact resulting from energy use during the usage stage of a building should be at least similar, depending on the regulations adopted by each country. As stated before, the authors are only interested in evaluating the ecological influence of the materials that are used for realizing the analyzed structural system.

The analyzed single floor dwelling is formed by using nine cubic modules, as presented in Figure 1. A single steel unit has a length and a height equal to 3.6 m. The vertical and horizontal structural elements of the module are made from square hollow section (SHS) profiles, 180 mm in depth and width, and with a specified thickness of 12 mm (Figure 2). The structural behaviour of the cubic modules has been analyzed for different structural configurations in a study conducted at the Faculty of Civil Engineering and Building Services of Iasi [39]. The thermal performances of a structure made of steel cubic modules with similar characteristics are presented in Isopescu et al. [40].

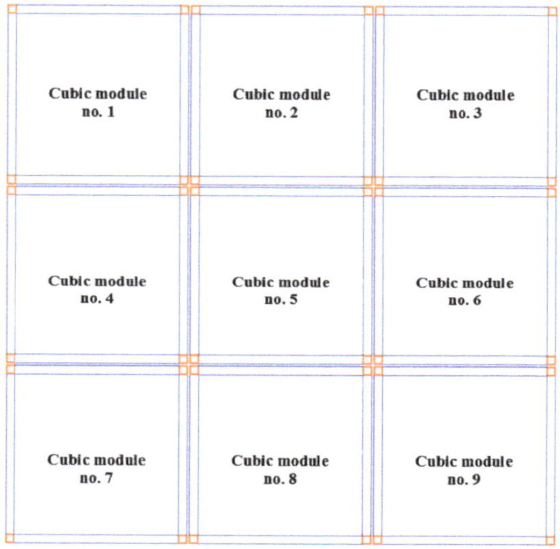

Figure 1. The in-plan configuration of the analyzed structures.

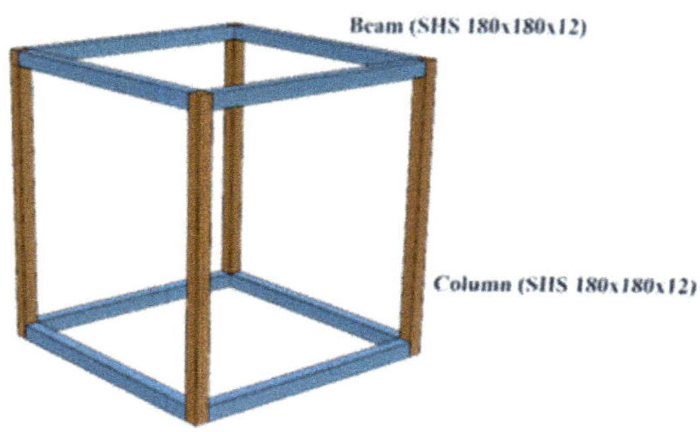

Figure 2. Steel cubic module.

Taking into account that the aim of the study is to determine the environmental implications of using steel as a structural material, the performed analysis only considers the beams, columns, and connection components. The authors have also evaluated the ecological burdens of an over the floor reinforced concrete slab, which has a height of

17 cm. The composite floor is made of C 20/25 concrete and a steel deck with a thickness of 1 mm. All structural elements have been designed by considering the European codes and national standards. As a final step of the assessment, the environmental performances of the steel structure were compared with those of the reinforced concrete slab. In order to provide a clear view regarding the cubic module and the quantities of components used for assembling the structure, Figure 3 shows the two types of junctions between the linear steel elements.

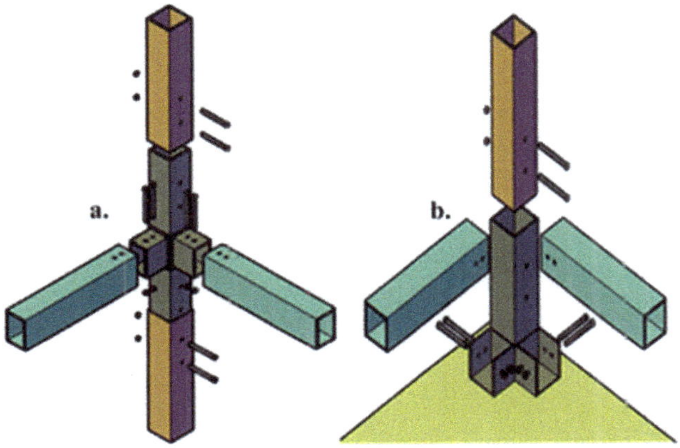

Figure 3. (**a**) Joint between columns and beams. (**b**) Joint between columns and ground floor. Source: [39,40].

So as to include the influence of the transportation phases along with the environmental influence of the analyzed elements, a diesel truck with a Euro 6 engine and a payload capacity of 3.3 tons has been considered during the study. Table 2 displays the transport distances that have been used in the assessment. All transport distances have been considered, as they were identified on a real case at the local scale.

Table 2. Transportation distances.

Material	Distances (km)	From → To
SHS steel profiles	10	Steel mill → construction site
Steel screws	10	Steel mill → construction site
Steel deck	10	Steel mill → construction site
Steel reinforcement	10	Steel mill → construction site
Fine aggregate	30	Quarry → concrete mixing plant
Coarse aggregate	30	Quarry → concrete mixing plant
Cement CEM I 32.5	165	Quarry → concrete mixing plant
Concrete	25	Concrete mixing plant → construction site
Scrap steel	10	Construction site → recycling unit
Recycled concrete	30	Construction site → concrete mixing plant

Table 3 shows the impact categories that have been considered for assessing the effects of the considered products over the natural environment. The environmental impact indicators used for performing the analysis were selected by considering the recommendations of the European Commission's Joint Research Centre—Institute for Environment and Sustainability [41]. Additionally, the assessment incorporates the act L124 2013/179/EU [42] and the European norm EN 15804+A1:2013 [38]. In order to determine the ecological effects of the assessed products, the midpoint approach and the GaBi ts software have been utilised, using the software's database.

Table 3. Environmental impact indicators considered.

Impact Category	Parameter	Unit
Global Warming (Climate Change) (including biogenic carbon)	Radiative forcing global warming potential (GWP)	kg CO_2-eq.
Human Toxicity (cancer effects)	Human toxicity potential, cancer effects (HTPc)	CTUh
Ozone Depletion	Depletion potential of the stratospheric ozone layer/Ozone depletion potential (ODP)	kg CFC-11 eq.

3. Analysing the Environmental Impact from the Cradle-to-Gate Perspective

The first phase of the analysis consists in determining the ecological burdens of the pre-operation phase of the steel cubic modules that form the considered dwelling structure. Therefore, the A1, A2, A3, and A4 life cycle modules have been considered at this stage of the study. Table 4 displays the amount of component materials that were considered in determining the impact of the construction products. Even if the columns and beams are made from SHS profiles with the same characteristics, the authors decided to evaluate the impact of these linear structural elements separately (i.e., the impact of beams and the impact of columns), with the goal of gaining a better comprehension regarding the obtained results. The pre-operation ecological impact of the reinforced concrete slab has been determined by considering the amounts of component materials presented in Table 5.

Table 4. The quantities of materials used in the steel structure analyzed.

Component Material	Quantity (kg)
SHS steel profiles—Columns	8048.16
SHS steel profiles—Beams	16,096.32
Steel screws	601.34

Table 5. The quantities of component materials used in the analysis of the reinforced concrete slab.

Component Material	Quantity (kg)
Fine aggregate	18,495
Coarse aggregate	18,495
Cement (CEM I 32.5)	7471
Water	3736
Steel reinforcement	2376.20
Steel deck (steel sheet)	2376.20

The values describing the implications for the natural environment resulting from the use of steel cubic modules for the construction of the analyzed structure are presented in Figure 4. These results show that the beams have the greatest negative influence, while the SHS steel profiles used for columns have a carbon footprint that is almost 2.5 times larger than that of the fixing elements. Compared to all the analyzed steel products, the amount of diesel and the transportation phase have an insignificant impact on the GWP parameter.

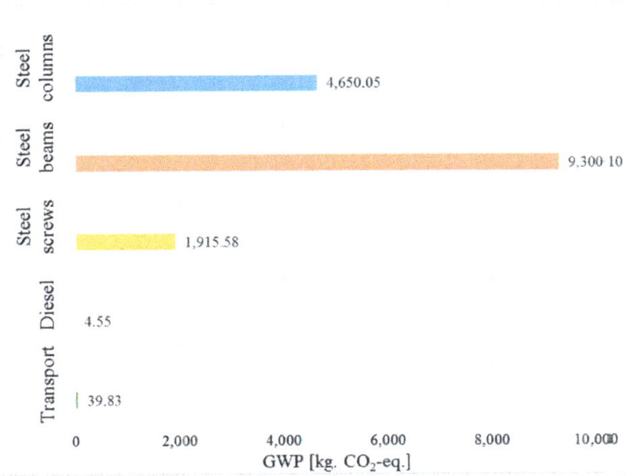

Figure 4. Cradle-to-gate impact of the steel structure for the Global Warming indicator.

In the case of the HTPc environmental parameter (Figure 5), the steel beams have the most notable impact. The steel columns have the second greatest negative effect, followed by the impact exerted by the steel screws, the amount of diesel used, and, finally, the transportation phase. The results for the Ozone Depletion environmental indicator are presented in Figure 6. As in the case of the two previously analyzed environmental parameters, the beams have the most important impact over the stratospheric ozone layer, followed by the negative effects of the columns, screws, and the amount of diesel consumed in the transportation phase.

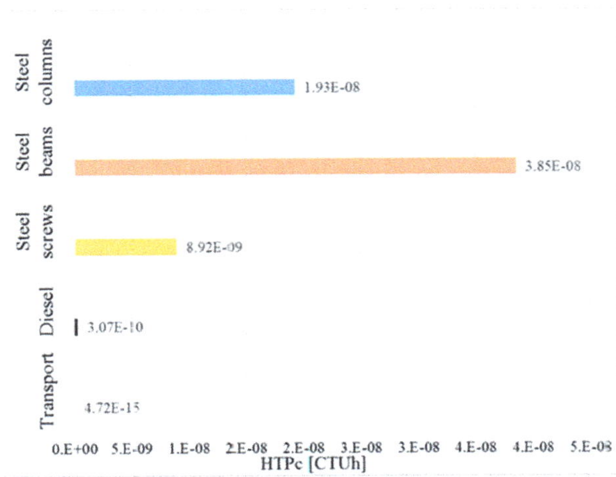

Figure 5. Cradle-to-gate impact of the steel structure for the Human Toxicity indicator.

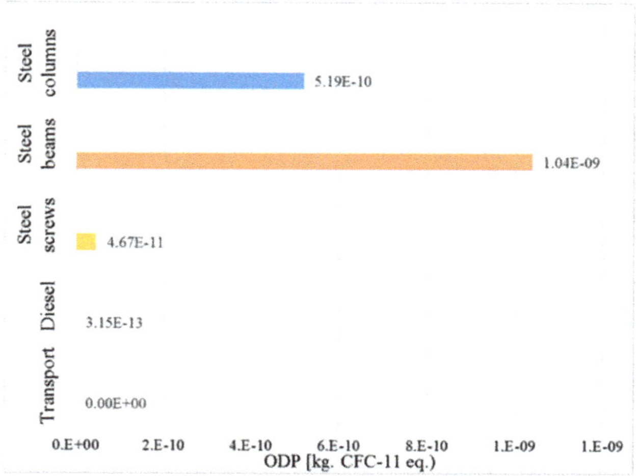

Figure 6. Cradle-to-gate impact of the steel structure for the Ozone Depletion indicator.

The values that describe the impact of the product under analysis over the pre-operation stage are presented in Figures 7–9. The impact of the reinforced concrete slab for the GWP parameter is described in Figure 7. It can be noticed that the negative effect of cement represents approximately 84% of the combined impact of the steel deck and steel reinforcements. Additionally, the three above mentioned component materials are responsible for nearly the entire carbon footprint of the reinforced concrete slab.

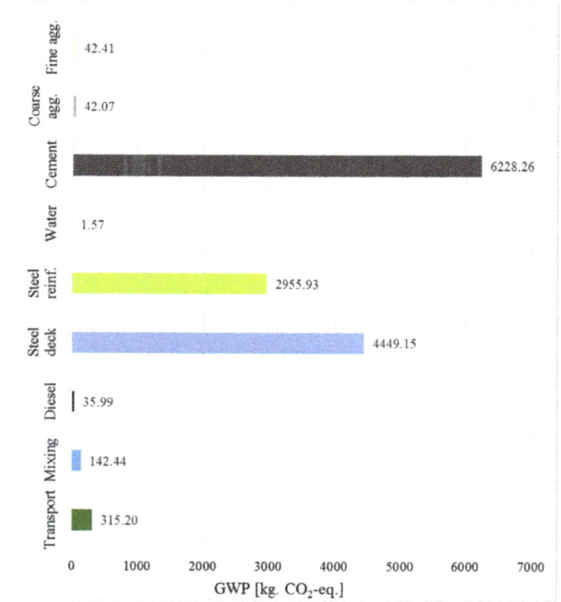

Figure 7. Cradle-to-gate impact of the concrete slab for the Global Warming indicator.

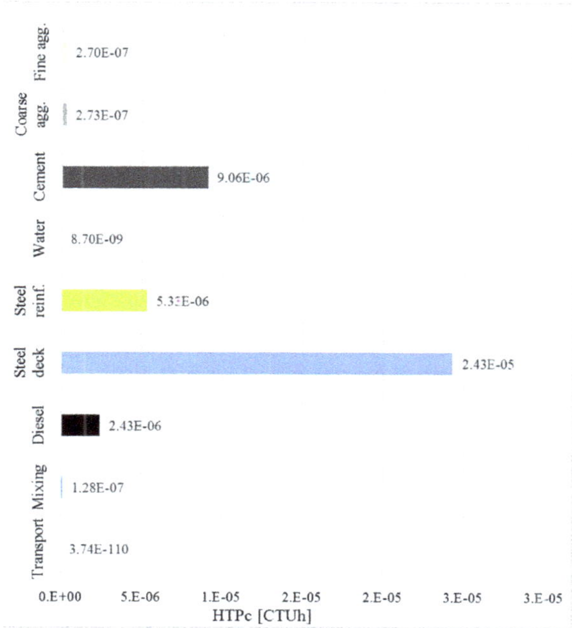

Figure 8. Cradle-to-gate impact of the concrete slab for the Human Toxicity indicator.

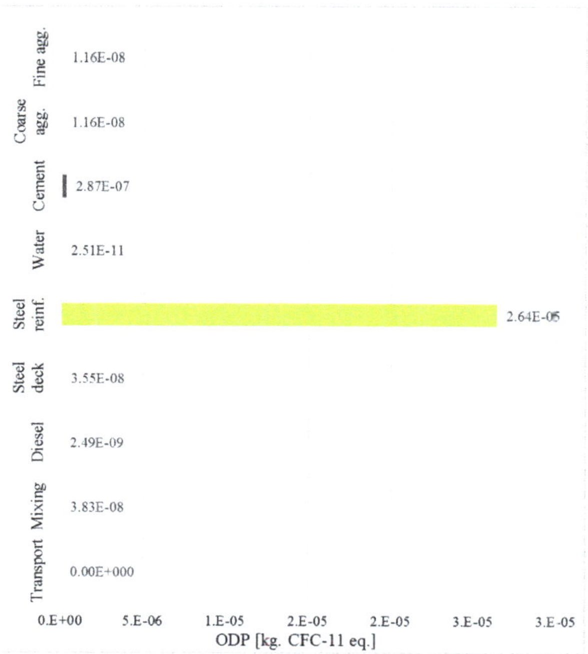

Figure 9. Cradle-to-gate impact of the concrete slab for the Ozone Depletion indicator.

In Figure 8, it can be seen that the overall value for the HTPc parameter is highly influenced by the steel deck, this product being responsible for more than 50% of the total cancerous effects. As represented by the results, the cement, steel reinforcement, and diesel have an important impact on the overall result as well. Figure 9 presents the environmental effects of the analyzed construction product in the case of the ODP impact indicator. In this case, the amount of steel reinforcement makes up for about 99% of the overall negative impact on the stratospheric ozone layer.

Comparing the values for the cradle-to-gate evaluation (Table 6), it can be ascertained that the steel cubic modules have the most important effect on the total environmental footprint for the two considered environmental parameters. The steel structural elements account for about 53% of the overall carbon footprint and for approximately 62% of the total impact on human health. At the same time, the RC slab exerts 94% of the overall negative influence on the environment in the ODP category.

Table 6. Cradle-to-Gate impact.

Environmental Parameter	Total Impact	Steel Structure Impact	RC Slab Impact
Global warming potential (kg CO_2-eq.)	30,123.13	15,910.11	14,213.02
Human toxicity potential, cancer effects (CTUh)	1.09×10^{-4}	6.70×10^{-5}	4.18×10^{-5}
Ozone depletion potential (kg CFC-11 eq.)	2.84×10^{-5}	1.61×10^{-6}	2.68×10^{-5}

4. End-of-Life Assessment

In this final part of the study, the authors have determined and interpreted the ecological benefits of the considered structure over the post-operation life cycle phase by taking into account the C1, C2, C3, and D life cycle modules. The analysis considered a mechanized process for the demolition of the reinforced concrete slab using a hydraulic breaker, a compact excavator, an on-site concrete crusher, and also a vibrating screen. In order to obtain a higher volume of scrap material that can be used for producing a new batch of steel-based products, the demolition of the steel module structure was considered to be completed by workers using different hand tools.

For the analyzed end-of-life (EoL) scenario, the recovery percentages of the materials consumed to build the construction under analysis are assumed to be 80% in the case of the steel cubic structure and 70% in the case of the RC slab component materials. It is also considered that in the EoL assessment the scrap steel is used for manufacturing new steel products, while the recovered concrete is used as crushed aggregates, 50% as fine aggregate and 50% as coarse aggregate (Table 7).

Table 7. Quantities of materials considered in the end-of-life assessment.

Material	Quantity (kg)
SHS steel profiles	19,315.58
Steel screws	481.07
Concrete	33,737.90
Steel reinforcement	1663.34
Steel deck (steel sheet)	1413.84

Table 8 shows the environmental values for the end-of-life phase of the considered structure. As can be observed, the EoL scenario has a positive impact (negative values) only in the case of the GWP parameter. By comparing the global warming potential of the steel structure with that of the RC slab, it can be concluded that the recycling of the cubic modules has a negative carbon footprint, almost seven times higher than that resulting from the recycling of all the RC slab component materials. The reinforced concrete element has a lower environmental impact in terms of the HTPc and ODP parameters.

Table 8. End-of-life environmental impact.

Environmental Parameter	Total EoL Impact	Recycling of the Steel Structure	Recycling of the RC Slab
Global warming potential (kg CO_2-eq.)	−34,388.59	−29,904.34	−4484.25
Human toxicity potential, cancer effects (CTUh)	2.13×10^{-5}	1.26×10^{-5}	8.17×10^{-5}
Ozone depletion potential (kg CFC-11 eq.)	1.11×10^{-3}	9.57×10^{-4}	1.49×10^{-4}

5. Discussion

The goal of the research was to evaluate, from a cradle-to-gate with options viewpoint, the environmental impact of a structure made using a steel cubic module. The study has been completed by taking into account the following three stages of a Life Cycle Assessment analysis: the pre-operation stage, the post-operation stage, and an analysis of the overall ecological effects.

Table 9 shows the values that offer a clear description of the environmental impact of the entire structure that has been analyzed over the considered life cycle. The assessed cubic assembly has a positive effect in the case of the Global Warming Potential environmental impact parameter and a negative environmental impact in the other two considered impact categories (i.e., Human Toxicity, cancer effects, and Ozone Depletion).

Table 9. Overall impact of the considered structure.

Environmental Parameter	Total Impact	Steel Structure Impact	RC Slab Impact
Global warming potential (kg CO_2-eq.)	−4265.46	−13,994.23	9728.77
Human toxicity potential, cancer effects (CTUh)	1.30×10^{-4}	7.96×10^{-5}	5.05×10^{-5}
Ozone depletion potential (kg CFC-11 eq.)	1.14×10^{-3}	9.59×10^{-4}	1.76×10^{-4}

As observed in Figure 10 and Table 9, only the steel linear structural components of the analyzed assembly have a negative value regarding the amount of carbon dioxide emitted into the atmosphere. Therefore, it can be stated that by using the assessed products, the global phenomenon of climate change is massively influenced in a positive manner. In addition, by analysing the results it can be noticed that the steel elements of the structure are responsible for approximately 61% of the total impact over human health, expressed by using the HTPc indicator. At the same time, these structural components account for almost 85% of the overall impact on the stratospheric ozone layer, evaluated according to the ODP parameter. Therefore, the analysis shows that the reinforced concrete slab registers a significantly lower negative influence in the case of the last two considered environmental impact indicators.

Analysing the entire set of results, it can be concluded that the steel structural system can represent a viable solution for supporting present efforts that are being made at the global scale regarding the implementation of sustainable development measures in the construction sector—this despite the fact that the linear steel products have a negative effect for two of the considered environmental impact categories. This idea is supported by the massive positive influence in the case of the Global Warming indicator, which without doubt represents the most important environmental parameter, seeing that the emissions of equivalent carbon dioxide into the atmosphere represent the main reason for the temperature anomalies that are registered yearly. Another ecological benefit of the steel cubic modules is represented by the high level of recyclability of its component material, which can be translated into a significant reduction of the amount of raw materials globally extracted, therefore reducing the negative pressure on the natural ecosystem with respect to the Earth's capacity for renewing the global stock of natural resources.

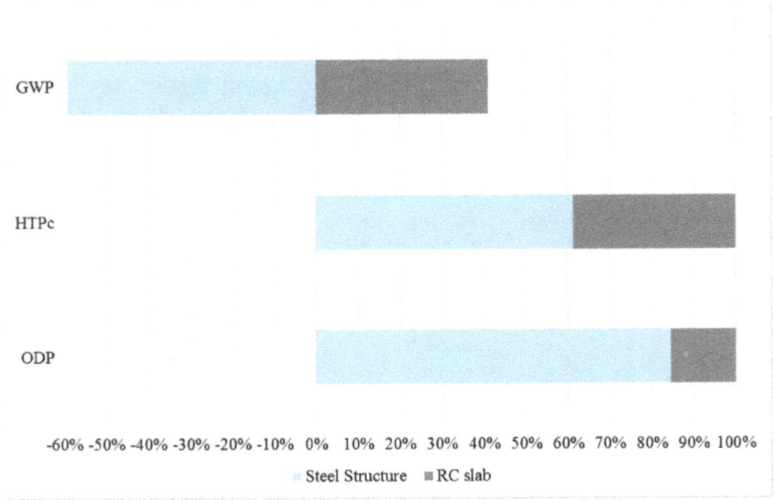

Figure 10. Total environmental influence of the considered structure.

6. Conclusions

As argued, the construction sector has one of the most substantial negative ecological influences at the global scale, which makes it a crucial area when it comes to achieving sustainability goals at the global level (i.e., the environmental dimension). The rapidly increasing global population phenomenon can be translated into a high demand for continuously expanding the built environment, which further leads to a greater pressure on the construction industry's global efforts to minimize its environmental impact. Considering the above, all the natural resources consumed by the activities that are specific to this sector must be handled in a more rational way. At the same time, civil engineering specialists must significantly increase the search for new innovative solutions that target both structural and environmental problems.

This study undoubtedly shows that the steel cubic modules analyzed could represent a solution for minimizing the level of environmental burdens resulting from the activities of the construction sector, especially with regard to the carbon footprint associated with the built environment. Moreover, the high degree of recyclability of steel represents another important environmental aspect of this traditional construction material that can transform this material into a more suitable solution within the current context of global sustainable development. Furthermore, the modularity of the analyzed structure can lead to a lower cost for the final users, thus also positively influencing the economic dimension of the sustainability concept. In conclusion, it can be stated that using steel as a structural material can represent a sustainable solution for the built environment. In addition, if in the near future the steel production industry will supplement its current efforts, steel could have a neutral environmental footprint.

Author Contributions: Conceptualization, S.G.M. and D.N.I.; methodology, S.G.M. and D.N.I.; software, S.G.M. and I.-R.B.; validation, D.N.I., I.-R.B., and M.L.L.; formal analysis, S.G.M. and I.-R.B.; investigation, S.G.M., D.N.I., I.-R.B., and M.L.L.; resources, S.G.M.; data curation, S.G.M., I.-R.B., and M.L.L.; writing—original draft preparation, S.G.M. and D.N.I.; writing—review and editing, I.-R.B.; visualization, I.-R.B. and M.L.L.; supervision, S.G.M.; project administration, D.N.I.; funding acquisition, S.G.M. All authors have read and agreed to the published version of the manuscript.

Funding: This work is funded by TUIASI Internal Grants Program (GI_P9/2021), financed by the Romanian Government.

Institutional Review Board Statement: Not applicable.

Informed Consent Statement: Not applicable.

Data Availability Statement: Not applicable.

Acknowledgments: This paper was produced with the support of the TUIASI Internal Grants Program (GI_P9/2021), financed by the Romanian Government.

Conflicts of Interest: The authors declare no conflict of interest.

References

1. Global Footprint Network. Available online: https://www.overshootday.org/ (accessed on 1 September 2020).
2. Bribian, I.Z.; Capilla, A.V.; Usón, A.A. Life cycle assessment of building materials: Comparative analysis of energy and environmental impacts and evaluation of the eco-efficiency improvement potential. *Build. Environ.* **2011**, *46*, 1133–1140. [CrossRef]
3. Ding, G. Life cycle assessment (LCA) of sustainable building materials: An overview. *Eco-Effic. Constr. Build. Mater.* **2014**, 38–62. [CrossRef]
4. Maxineasa, S.G.; Taranu, N. Environmental impact of fibre-reinforced polymer strengthening solutions of reinforced con-crete columns. *Ann. Acad. Rom. Sci.* **2015**, *7*, 41–52.
5. Maxineasa, S.G.; Taranu, N. Life cycle analysis of strengthening concrete beams with FRP. *Woodhead Publ.* **2018**, 673–721. [CrossRef]
6. Maxineasa, S.G.; Taranu, N.; Bejan, L.; Isopescu, D.; Banu, O.M. Environmental impact of carbon fibre-reinforced polymer flexural strengthening solutions of reinforced concrete beams. *Int. J. Life Cycle Assess.* **2015**, *20*, 1343–1358. [CrossRef]
7. Maxineasa, S.G.; Entuc, I.-S.; Taranu, N.; Florenta, I.; Secu, A. Environmental performances of different timber structures for pitched roofs. *J. Clean. Prod.* **2018**, *175*, 164–175. [CrossRef]
8. Messari-Becker, L.; Bollinger, K.; Grohmann, M. Life-cycle assessment as a planning tool for sustainable buildings. In *Life-Cycle and Sustainability of Civil Infrastructure Systems; Proceedings of the Third International Symposium on Life-Cycle Civil Engineering (IALCCE 2012), Vienna, Austria, 3–6 October 2012*; Strauss, A., Frangopol, D.M., Bergmeister, K., Eds.; CRC Press: London, UK; pp. 1558–1562.
9. Miller, A.; Ip, K. Sustainable construction materials. In *Design and Management of Sustainable Built Environments*; Yao, R., Ed.; Springer: London, UK, 2013; pp. 341–358.
10. Mokhlesian, S.; Holmén, M. Business model changes and green construction processes. *Constr. Manag. Econ.* **2012**, *30*, 761–775. [CrossRef]
11. Pacheco-Torgal, F.; Labrincha, J. The future of construction materials research and the seventh UN Millennium Development Goal: A few insights. *Constr. Build. Mater.* **2013**, *40*, 729–737. [CrossRef]
12. Ramesh, T.; Prakash, R.; Shukla, K. Life cycle energy analysis of buildings: An overview. *Energy Build.* **2010**, *42*, 1592–1600. [CrossRef]
13. Romano, E.; Cascini, L.; D'Aniello, M.; Portioli, F.; Landolfo, R. A simplified multi-performance approach to life-cycle assessment of steel structures. *Structures* **2020**, *27*, 371–382. [CrossRef]
14. Vitale, P.; Spagnuolo, A.; Lubritto, C.; Arena, U. Environmental performances of residential buildings with a structure in cold formed steel or reinforced concrete. *J. Clean. Prod* **2018**, *189*, 839–852. [CrossRef]
15. Blankendaal, T.; Schuur, P.; Voordijk, H. Reducing the environmental impact of concrete and asphalt: A scenario approach. *J. Clean. Prod.* **2014**, *66*, 27–36. [CrossRef]
16. Juan, I.A.; Müller, F.; Hack, N.; Wangler, T.; Habert, G. Potential benefits of digital fabrication for complex structures: Environ-mental assessment of a robotically fabricated concrete wall. *J. Clean. Prod.* **2017**, *154*, 330–340. [CrossRef]
17. Brejnrod, K.N.; Kalbar, P.; Petersen, S.; Birkved, M. The absolute environmental performance of buildings. *Build. Environ.* **2017**, *119*, 87–98. [CrossRef]
18. Iuorio, O.; Napolano, L.; Fiorino, L.; Landolfo, R. The environmental impacts of an innovative modular lightweight steel system: The Elissa case. *J. Clean. Prod.* **2019**, *238*. [CrossRef]
19. Kamali, M.; Hewage, K.; Sadiq, R. Conventional versus modular construction methods: A comparative cradle-to-gate LCA for residential buildings. *Energy Build.* **2019**, *204*, 109479. [CrossRef]
20. Lu, Y.; Le, V.H.; Song, X. Beyond Boundaries: A Global Use of Life Cycle Inventories for Construction Materials. *J. Clean. Prod.* **2017**, *156*, 876–887. [CrossRef]
21. Pacheco-Torres, R.; Roldán, J.; Gago, E.; Ordóñez, J. Assessing the relationship between urban planning options and carbon emissions at the use stage of new urbanized areas: A case study in a warm climate location. *Energy Build.* **2017**, *136*, 73–85. [CrossRef]

22. Sathre, R.; González-García, S. Life cycle assessment (LCA) of wood-based building materials. *Eco-Effic. Constr. Build. Mater.* **2014**, 311–337. [CrossRef]
23. Vacek, P.; Struhala, K.; Matějka, L. Life-cycle study on semi intensive green roofs. *J. Clean. Prod.* **2017**, *154*, 203–213. [CrossRef]
24. Yao, R. Sustainability in the built environment. In *Design and Management of Sustainable Built Environments*; Yao, R., Ed.; Springer: London, UK, 2013; pp. 1–22.
25. Zhao, D.; McCoy, A.P.; Du, J.; Agee, P.; Lu, Y. Interaction effects of building technology and resident behavior on energy consumption in residential buildings. *Energy Build.* **2017**, *134*, 223–233. [CrossRef]
26. Moynihan, M.C.; Allwood, J.M. The flow of steel into the construction sector. *Resour. Conserv. Recycl.* **2012**, *68*, 88–95. [CrossRef]
27. WSA. World Steel in Figures, World Steel Association, Brussels. 2013. Available online: https://www.worldsteel.org/steel-by-topic/statistics/World-Steel-in-Figures.html (accessed on 27 October 2021).
28. Wang, Y.; Wen, Z.; Cao, X.; Zheng, Z.; Xu, J. Environmental efficiency evaluation of China's iron and steel industry: A process-level data envelopment analysis. *Sci. Total. Environ.* **2019**, *707*, 135903. [CrossRef] [PubMed]
29. WSA. Steel Statistical Yearbook 2019, World Steel Association, Brussels. 2019. Available online: https://www.worldsteel.org/media-centre/press-releases/2019/2019-Steel-Statistical-Yearbook-published.html (accessed on 27 October 2021).
30. Bjorn, A.; Owsianiak, M.; Laurent, A.; Olsen, S.I.; Corona, A.; Hauschild, M.Z. Scope Definition. In *Life Cycle Assessment. Theory and Practice*; Hauschild, M.Z., Rosen-baum, R.K., Olsen, S.I., Eds.; Springer International Publishing AG: Cham, UK, 2018; pp. 75–116.
31. Estrada, H.; Borja, D.H.; Lee, L. Sustainability in infrastructure design. In *Fiber Reinforced Polymer (FRP) Composites for Infrastructure Applications*; Jain, R., Lee, L., Eds.; Springer Science+Business Media B.V.: Dordrecht, UK, 2012; pp. 23–52.
32. Maxineasa, S.G.; Taranu, N. Traditional building materials and fibre reinforced polymer composites. A sustainability aproach in construction sector. *Bull. Polytech. Inst. Iasi Constr. Archit. Sect.* **2013**, *LIX(LXIII)*, 55–68.
33. Nidheesh, P.; Kumar, M.S. An overview of environmental sustainability in cement and steel production. *J. Clean. Prod.* **2019**, *231*, 856–871. [CrossRef]
34. AISC. Structural steel: The premier green construction material. In *Sustainable Manufacturing Process*; American Institute of Steel Construction: Chicago, IL, USA, 2016.
35. ISO. *Environmental Management—Life Cycle Assessment—Principles and Framework (ISO 14040:2006)*; The International Or-ganization for Standardization (ISO): Geneva, Switzerland, 2006.
36. ISO. *Environmental Management—Life Cycle Assessment—Requirements and Guidelines (ISO 14044:2006)*; The International Organization for Standardization (ISO): Geneva, Switzerland, 2006.
37. EN. *Sustainability of Construction Works—Assessment of Environmental Performance of Buildings—Calculation Method (EN 15978:2011)*; European Committee for Standardization: Brussels, Belgium, 2011.
38. EN. *Sustainability of Construction Works—Environmental Product Declarations—Core Rules for the Product Category of Construction Products (EN 15804:2012+A1:2013)*; European Committee for Standardization: Brussels, Belgium, 2013.
39. Isopescu, D.; Neculai, O.; Maxineasa, S.G. *Report on Steel Cubic Module Structure Used for Prefabricated Constructions (In Romanian)*; "Gheorghe Asachi" Technical University of Iasi, Faculty of Civil Engineering and Building Services: Iasi, Rumania, 2016.
40. Isopescu, D.N.; Maxineasa, S.G.; Neculai, O. Thermal analysis of a structural solution for sustainable, modular and prefabri-cated buildings. *Iop Conf. Ser. Mater. Sci. Eng.* **2017**, 209. [CrossRef]
41. European Commission–Joint Research Centre–Institute for Environment and Sustainability. *International Reference Life Cycle Data System (ILCD) Handbook—Recommendations for Life Cycle Impact Assessment in the European context, EUR 24571 EN*; Publications Office of the European Unio: Luxemburg, Luxemburg, 2011.
42. *L 124. Commission Recommendation of 9 April 2013 on the use of common methods to measure and communicate the life cycle environmental performance of products and organisations (2013/179/EU)*; Official Journal of the European Union, Publica-tions Office of the European Union: Luxemburg, Luxemburg, 2013.

sustainability

Article

A Comparative Study for Forced Ventilation Systems in Industrial Buildings to Improve the Workers' Thermal Comfort

Mohamed I. Elhadary [1], Abdullah Mossa Y. Alzahrani [2], Reda M. H. Aly [3] and Bahaa Elboshy [4,*]

1 Department of Mechanical Engineering, Faculty of Engineering, Tanta University, Tanta 31511, Egypt; mohamed.elhadary@f-eng.tanta.edu.eg
2 Department of Civil Engineering, Faculty of Engineering, Taif University, Taif 21944, Saudi Arabia; amyalzahrani@tu.edu.sa
3 Department of Architectural Engineering, Faculty of Engineering, Al Azhar University, Cairo 11765, Egypt; redda.ali70@azhar.edu.eg
4 Department of Architectural Engineering, Faculty of Engineering, Tanta University, Tanta 31511, Egypt
* Correspondence: bahaa.elboshi@f-eng.tanta.edu.eg

Abstract: The appropriate ventilation for factory spaces with regard to volume flow rate and air velocity inside the factory is one of the most important factors in the improvement of the thermal comfort of workers and in the reduction of the percentage of pollution they are exposed to, which in turn helps to improve the work environment and increase productivity. It also could improve the performance of machines. Hence, overheating can cause various problems and malfunctions. In this study, three types of mechanical ventilation systems are compared: the wall fan extract ventilation system, the roof fan extract ventilation system, and the spot cooling system. The Ansys software has been used to conduct the computational fluid dynamics (CFD) simulations for the different cases and the ventilation effectiveness factor (VEF) has been used to compare the performances of the three systems. The ventilation factor notably relies on the temperature distribution produced through the modeling and the results show that the most optimal system that can be used for similar factory spaces is the forced ventilation system. Finally, it is also the best in terms of energy consumption, despite the increase in the initial cost of its installation.

Keywords: CFD; mechanical ventilation; industrial buildings; thermal comfort; air quality

Citation: Elhadary, M.I.; Alzahrani, A.M.Y.; Aly, R.M.H.; Elboshy, B. A Comparative Study for Forced Ventilation Systems in Industrial Buildings to Improve the Workers' Thermal Comfort. *Sustainability* **2021**, *13*, 10267. https://doi.org/10.3390/su131810267

Academic Editors: Oleg Kapliński, Lili Dong, Agata Bonenberg and Wojciech Bonenberg

Received: 15 July 2021
Accepted: 10 September 2021
Published: 14 September 2021

1. Introduction

The Egyptian state's plan for urban development is mainly accompanied by industrial development, which helps to expand development axes and attract inhabitants to new cities [1]. Notably, one of the most critical problems facing urban and industrial expansion is the variety in the climate between different areas, which is hoped to be expanded particularly in the southern areas characterized by high temperatures [2,3]. Nowadays, there is great emphasis on reducing the pollution resulting from industrial activities and rationalizing energy consumption within the industrial environment, and indeed these issues are the most important elements when it comes to creating a sustainable design for industrial buildings [4]. Indeed, industrial spaces are the most exposed to high temperatures, which have a harmful effect on workers, causing many diseases [5] and damage to the machines in the factory [6]. With the above in mind, it becomes clear that studying industrial buildings and finding appropriate solutions that create a comfortable work environment, as well as reduce energy consumption, helps to attract industrial investments.

Hence, ventilation is one of the most important elements concerning the improvement of thermal conditions inside factories and it could reduce pollution [7]. Thus, it plays a significant role in improving the quality of the industrial environment as a whole and contributes to avoiding many damages to workers and machines [8]. Notably, mechanical ventilation methods are the most commonly used systems in factories [9], despite there being several types of mechanical ventilation systems that differ in terms of air distribution. Indeed, selecting the appropriate method for the type of industry and nature of the climate is one of the most important factors for obtaining effective ventilation.

Several types of forced ventilation systems are used to improve the air quality and temperature in industrial buildings, and this research study focuses on two main types of ventilation: extract ventilation and spot cooling. While the former system is based on a natural supply via trickle vents or mechanical extraction, the latter can reduce exposure to pollutants in building spaces [10]. Notably, extract ventilation systems work by depressurizing a structure, namely the system exhausting air from the space, which in turn causes a change in pressure that pulls in the make-up from the outside via leaks in the building shell and intentional, passive vents. Indeed, exhaust ventilation systems are relatively simple and inexpensive to install [5], with a typical exhaust ventilation system consisting of fans connected to a centrally located exhaust point [11]. Furthermore, adjustable, passive vents through windows or walls can be installed in other rooms to introduce fresh air, thus leaks in the building's envelope do not have to be relied upon. However, passive vents may require larger pressure differences than those induced by the ventilation fan to work properly. Notably, one concern of exhaust ventilation systems is that, along with fresh air, they may draw in pollutants, which can include radon and molds from crawlspaces, dust from the attic, fumes from the attached garage, and flue gases from the fireplace, fossil-fuel-fired water heater, or furnace [12].

This research study aims at comparing some of the commonly used mechanical ventilation systems in Egyptian factories to reveal the most effective system in terms of providing thermal comfort and reducing energy consumption. The first two types that have been investigated in this research study are different types of extract ventilation: wall fan extract systems and roof fan extract systems, as shown in Figure 1. The third type of ventilation system that has been investigated in this research study is the spot cooling system, the concept of which focuses on cooling the occupants in a room rather than randomly dispersing the cooling air everywhere to bring down the entire room temperature, as shown in Figure 2. Notably, spot cooling is intended to help reduce energy consumption without sacrificing consumer comfortability [13,14].

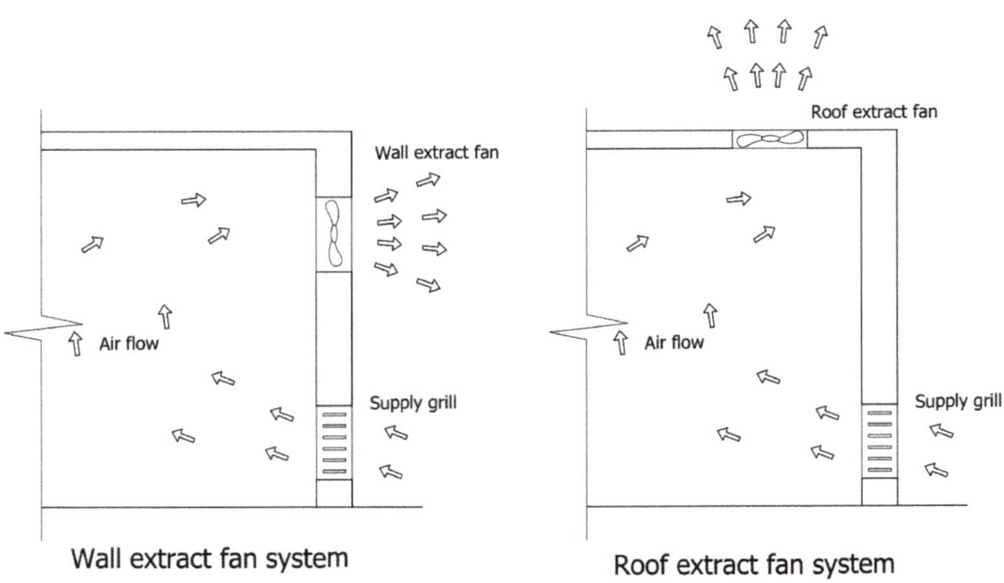

Figure 1. Exhaust ventilation system.

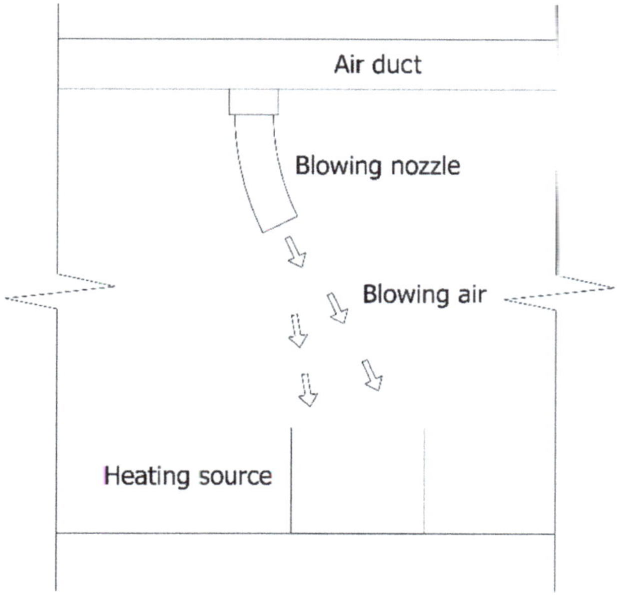

Figure 2. Spot cooling system (https://www.taikisha-group.com/service/spot.htm, accessed at 15 July 2021).

2. Methodology and Simulation Method

To study and analyze the ventilation systems' performances in large building scales such as factories, there is a need for an inexpensive method for the purpose of comparing the different systems [15]; on this note, CFD is a powerful tool for studying ventilation performance in factories, as they are inexpensive and able handle most of the thermo-fluid boundary conditions encountered in real-life scenarios [16]. Particularly in cases whereby the airflow is represented by strong streamline curvature due to wind forces, CFD modeling is considered to be the most suitable tool for reliable airflow simulation [17].

With this in mind, a number of researchers have used the CFD in studying and evaluating the building's indoor mechanical and natural ventilation systems, and while CFD uses approximations, a number of previous research studies have validated the simulated results with experimental data, which approve the efficiency and reliability of CFD results [16]. Further to this, some research studies have used the CFD in simulating the natural ventilation in different building types [15,17–21]. Additionally, mechanical ventilation has been analyzed using the CDF method in different research studies for residential and office buildings [9,22,23]. Finally, a number of researchers have investigated the mechanical ventilation systems in factories using CFD and validated the results, in turn revealing the high efficiency and reliability of this method in studying such buildings [16,24].

This research study aims to test and evaluate the performance of different forced ventilation systems in industrial buildings by comparing three popular systems used in factories: the wall fan extract ventilation system, roof fan extract ventilation system, and spot cooling system. Notably, an oil factory production space has been selected for conducting such a comparison between the different systems.

Initially, the estimation and design for each system were done for the purpose of achieving the optimum benefit in every case, but after that, CDF simulations were done using ANSYS software, considering the different effective factors and the surrounding environment. Notably, the simulation result has been compared in consideration of the different thermal comfort items and the different systems have been compared in consideration of both the initial cost and energy consumption for each one. Consequently, multi-criteria feasibility has been conducted to help choose the appropriate system for the factories' ventilation.

3. Simulation Work

3.1. Case Study

This research study discusses a comparison between the ventilation efficiency of an oil plant consisting of a number of machines operating at a capacity of 100 watts per cubic meter and located in Kharga Oases in the Western Desert of Egypt. The area's climate is arid, with sporadic rainfall and hot temperatures during the summer months [25]. The temperature occasionally approaches 50 °C and the mean monthly relative humidity ranges from 30% during the summer to 56% during the winter [26]. Kharga Oases is included in the Egyptian state development plan to increase the inhabited and agricultural area, as shown in Figure 3.

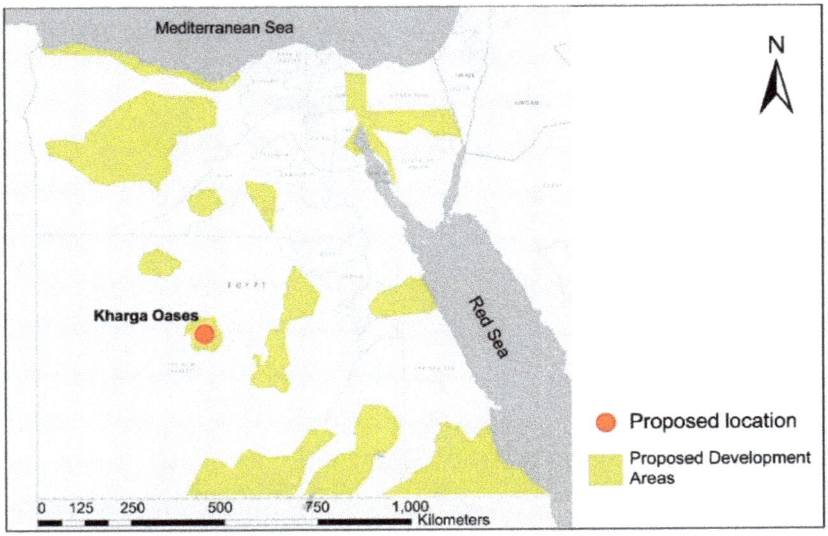

Figure 3. The case study location.

The case study factory building occupies a built area of 5000 m^2 and includes an industrial space with different workspaces and storage areas. In addition, there is a loading area and an administration building attached to the factory, as shown in Figure 4.

The design of the three systems has been done according to the ASHRAE code (Smacna STANDARDS) and is compatible with the ventilation systems used in such buildings through experience in this field. Furthermore, the three designed systems aim to provide the appropriate indoor ventilation to achieve thermal comfort for the factory laboratories. According to ASHRAE, the aimed ventilation rate is 20 times per hour, whereas the arrangement of the fans and supply grills has been chosen according to experience in this field. The specification for each ventilation system has been reached, as indicated in Table 1.

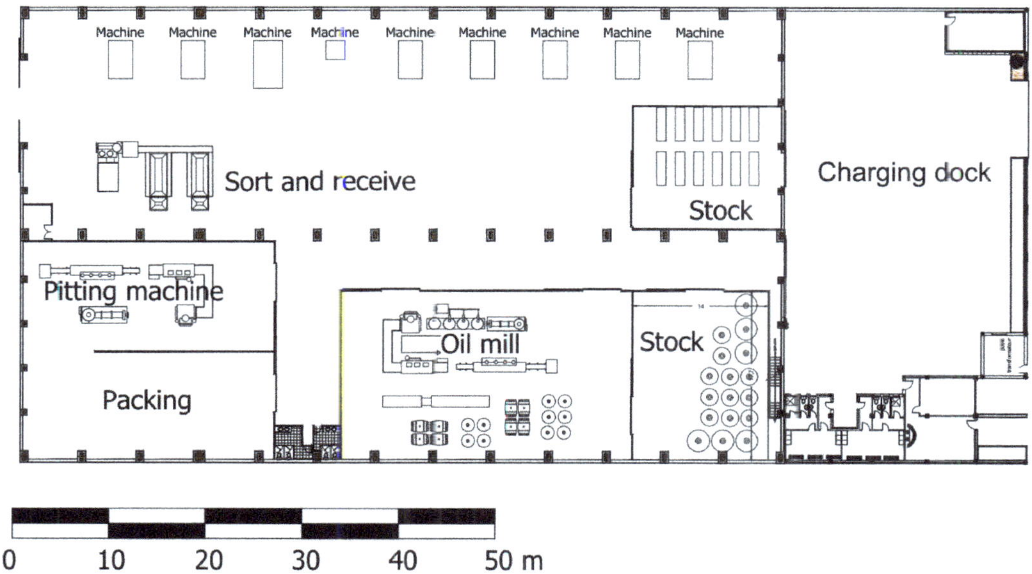

Figure 4. Factory plan.

Table 1. Specification for each ventilation system.

Components	Number	Specifications	Initial Cost (L.E.)	Power Consumption Per Year (KW)
		Basic Components		
Factory solid dimensions (ground/roof /walls)	–	Size: (x = 105 m, y = 12 m, z = 49.9 m)	–	–
Machines	13	Each machine produced = 100 w/m^3	–	–
		System A		
Wall exhaust fan	10	v = 6000 rpm, gauge pressure = 2 pa	260,000	35
Grill	10	Air velocity = 2.5 m/s		
Input doors	2	V = 1 m/s, T = 45 °C		
		System B		
Roof exhaust fan	11	v = 6000 rpm, gauge pressure = 2 pa	320,000	45
Grill	10	Air velocity = 2.5 m/s		
Input doors	2	V = 1 m/s, T = 45 °C		
		System C		
Duct	2	Q = 7500 cfm, velocity = 750 fpm, friction loss in inches of water per 100 feet = 0.1, cubic feet of air per minute = 1000.	550,000	25

The first system (System A: wall fan extract ventilation) relies on the installation of exhaust wall fans at the height of three meters from the floor with a reciprocal distribution for the fans on both sides of the factory, which helps to improve the distribution of the ventilation space. Furthermore, natural air supply grills are additionally installed at the height of 0.5 m from the factory floor, with the sizes and numbers shown in Table 1. Notably, Figure 5 shows the installation method for the ventilation elements and illustrates the expected ventilation distribution through the adopted system: the air enters from the grills at a speed of 2.5 m/s and the velocity wall exhaust fans reach 6000 rpm.

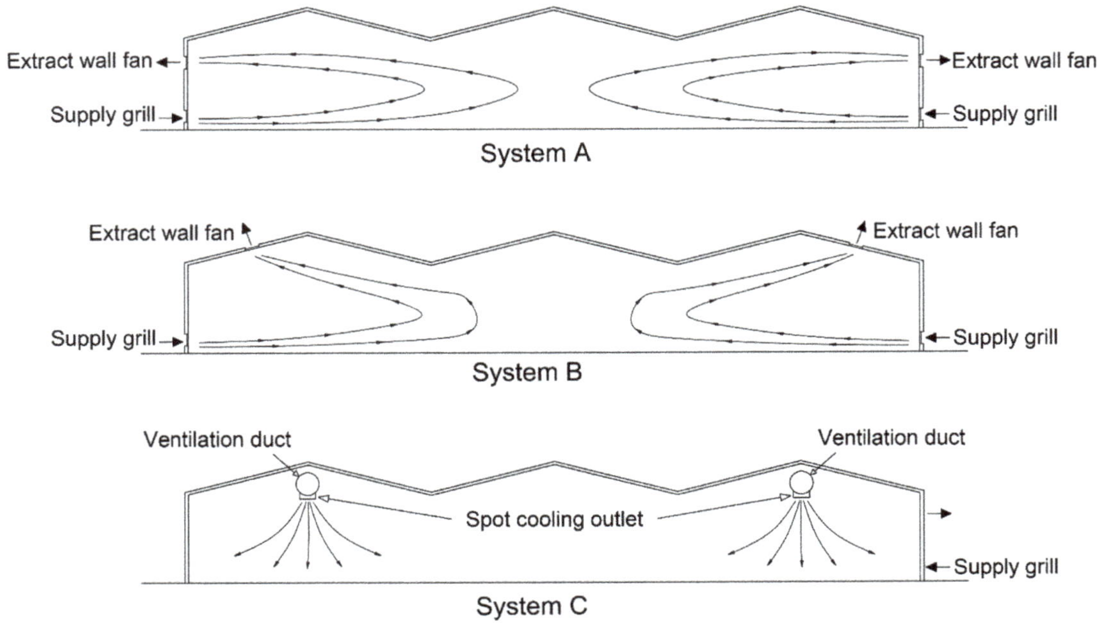

Figure 5. Conceptual air flow for different ventilation systems.

Meanwhile, the second system (System B: roof fan extract ventilation) relies on the installation of extraction fans in the ceiling, which can improve the distribution of the air inside the space by simulating natural ventilation. Furthermore, natural air supply grills are installed at the height of 0.5 m from the factory floor, with the sizes and numbers shown in Table 1 and Figure 1 indicating the installation method for the ventilation elements, as well as the expected ventilation distribution through the adopted system. The specification of the grills and fans are the same as the first system.

The third system (System C: spot cooling) is based on the use of the fan sections connected to the ducks that extend the factory space's length, with every duct being equipped with output nozzles with an outer diameter of 10 ft, which focus on the machine's areas as the source of the heating to transmit a temperature of about 20 °C, considering that the air did not enter from the entrance doors. Notably, the fans and ducks were installed (sizes and numbers shown in Table 1, with Figure 4 showing the installation method for the ventilation elements and illustrating the expected ventilation distribution through the adopted system).

3.2. Simulation Method and Weather Conditions

The numerical simulation in this study has been conducted using a commercial CFD program (ANSYS Fluent) to utilize the finite volume method under steady-state conditions. Furthermore, the investigation used the Reynolds-averaged Navier Stokes equations with the renormalization group RNG k–ε model for modeling turbulent flow and the equations being discretized with the second-order upwind scheme. The pressure and velocity were coupled by the SIMPLE algorithm. See the data in Table 2 for the weather conditions. Most of the time, the model reached a quasi-stable state after 4000–5000 iterations.

Table 2. Simulation weather conditions.

Item	Condition
Longitude (deg)	27.17
Latitude (deg)	24.24
Solar irradiation method	Fair weather condition
Sunshine factor	1
Humidity	14%
Outdoor temperature	45 °C
Outdoor air velocity	23 km/hr
Simulation time	Date 21/6, Hour 13:00

4. Simulation Results

According to the CFD simulations for the three systems, the temperature disruption and airflow velocity have been conducted to compare these systems and identify the most suitable system for the provision of a suitable work environment and for the reduction of energy consumption. Notably, the temperature disruption in the space for different systems has been conducted using the CFD model, as shown in Figures 6 and 7, with the temperature volume rendered in the former revealing that for systems A and B, there was a variation in the temperature between the parts of the working area. Although some areas had a low temperature, there were other areas that recorded a high temperature. According to System C, the temperature volume rendering the homogeneity in the temperature had fairly good records over the entire factory area. Figure 7 shows the temperature contour plan for the levels 0.4 m and 2 m, ensuring the observation in temperature volume rendering.

Furthermore, Figures 8 and 9 show the airflow velocity distribution in the factory space, with that in the former showing the air velocity distribution for Systems A and B as possessing a wide range variety, in turn causing a high velocity in some areas and low velocity in others. Meanwhile, System C shows a uniform distribution of air velocity over the entire factory space. These observations are also ensured in the ventilation velocity cross-section in Figure 9.

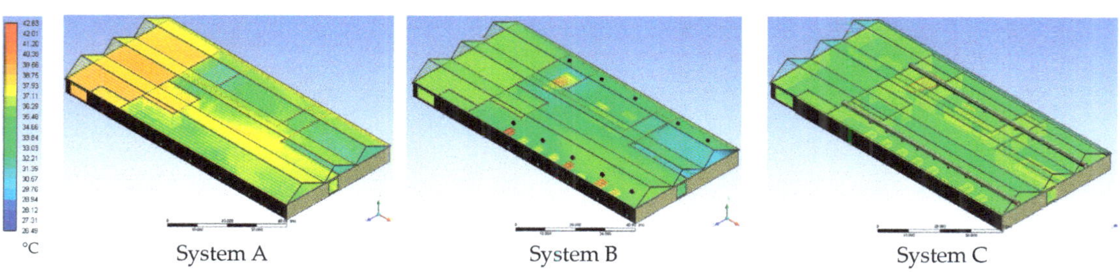

Figure 6. Temperature volume rendering.

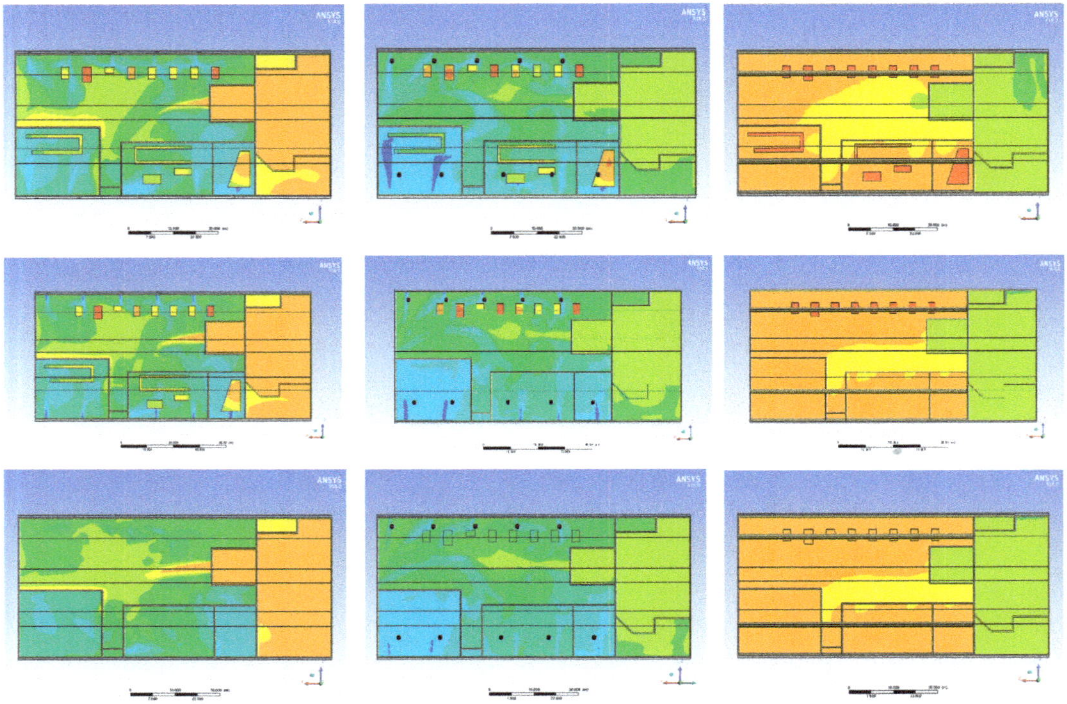

Y = 0.4 m

Y = 1.5 m

Y = 2 m

System A System B System C

Figure 7. Temperature contour-plane.

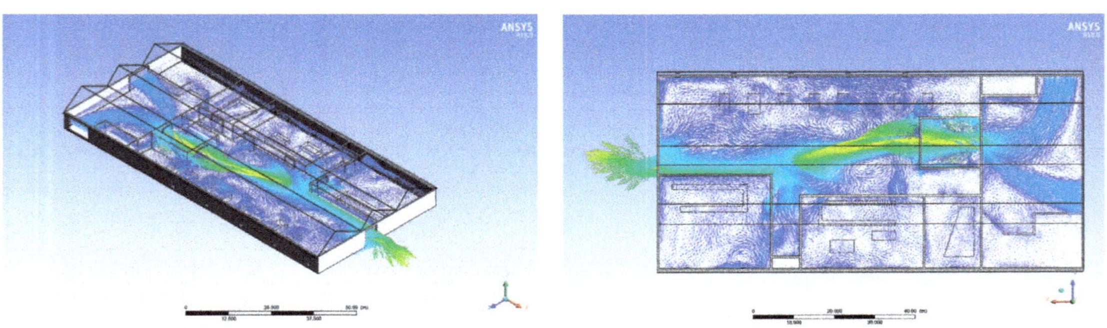

System A

Figure 8. *Cont.*

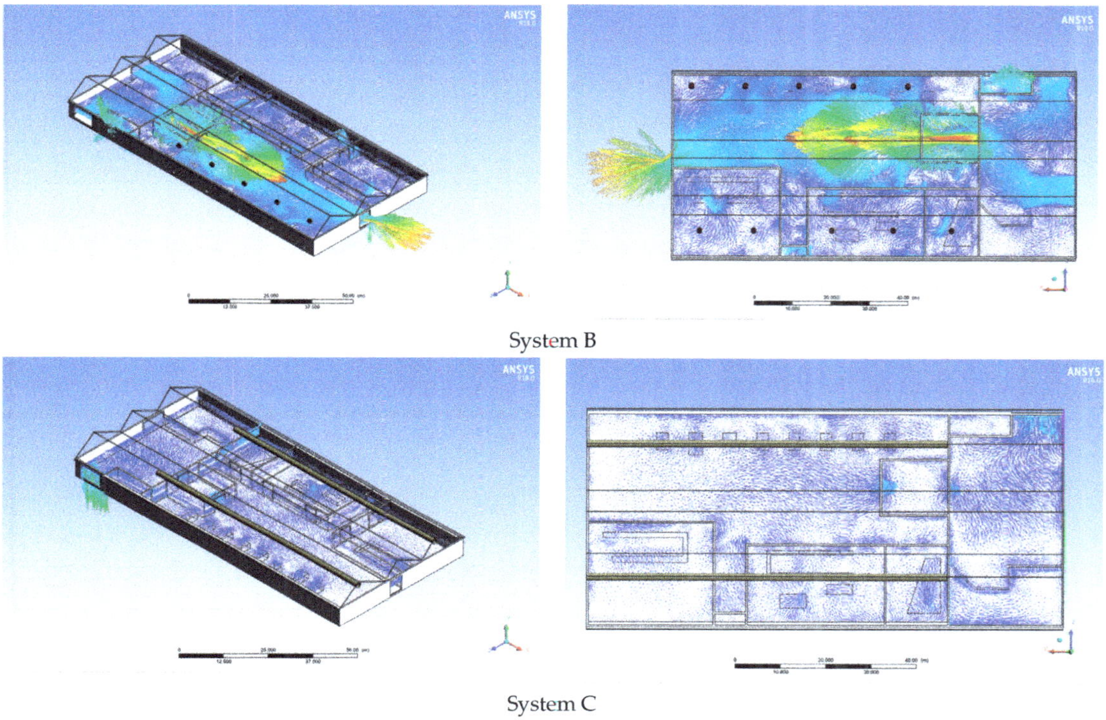

Figure 8. Ventilation velocity volume rendering.

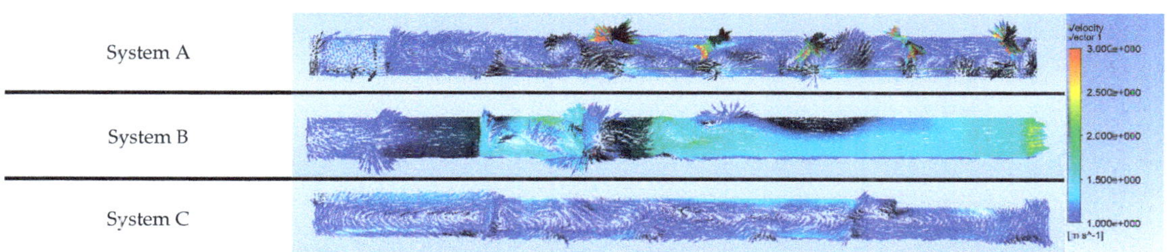

Figure 9. Ventilation velocity cross-section.

5. Discussion

After obtaining the simulation result, different investigations have been adopted to compare the different ventilation systems. For investigating the thermal comfort for the different systems, the Fagner equation has been used by considering the ambient air temperature, mean radiant temperature, and relative air velocity for different cases. In addition to assuming that the metabolic energy production was 2.2 met, the relative humidity was 50% and the basic clothing insulation was 1 clo. The predicted mean vote (PMV) and predicted percentage dissatisfied (PPD) have been calculated as shown in Table 3.

According to Fagner's values of the thermal sensation in Table 4, the PMV values for the first two cases lay between slightly warm and warm sensations, whereas in the case of system C, the value lies between cool and slightly cool sensations.

Table 3. PVM and PPD for different cases.

	PMV	PPD (%)
System A	1.54	52.9
System B	1.31	40.5
System C	−1.50	50.5

Table 4. Values of the thermal sensation [27].

Thermal Sensation	Cold	Cool	Slightly Cool	Neutral	Slightly Warm	Warm	Hot
Value	−3	−2	−1	0	1	2	3

In addition to thermal comfort investigation, the ventilation systems' performance in the factory has been investigated using the ventilation effectiveness factor (VEF), as shown in the following equations.

Ventilation factor:

$$\text{VEF.1} = \frac{T_o - T_i}{T_{avg} - T_i} = \frac{42 - 45}{34.5 - 45} = 0.285$$

$$\text{VEF.2} = \frac{T_o - T_i}{T_{avg} - T_i} = \frac{39 - 45}{33 - 45} = 0.5$$

$$\text{VEF.3} = \frac{T_o - T_i}{T_{avg} - T_i} = \frac{26 - 45}{23 - 45} = 0.863$$

where T_o is the output temperature outside the space, T_i is the input temperature inside the space, and T_{avg} is the average temperature in the space.

The ventilation effectiveness factor comparison for the three systems (Figure 10) reveals that the most effective ventilation system is System C, which represents the spot cooling system. Furthermore, an economic comparison has been conducted between the three systems, as shown in Figures 11 and 12, with the former illustrating the initial cost for the different systems. This reveals that the most expensive system is C, while the cheapest is System A, which explains the preference of most investors to use System A in factories. Meanwhile, Figure 12 shows the annual energy consumption costs for the different systems, revealing that the most energy-saving system is System C. The energy consumption cost was conducted by estimating that the kWh costs EGP 1.15 for business use according to [28].

Accordingly, this study found that using System C (the spot cooling system) could be the most effective in terms of its ventilation effectiveness, thermal comfort for workers, and energy consumption. Although the initial cost for System C is higher than the others, this cost could be compensated for by the higher productivity of the workers due to the provided thermal comfort and reduction of energy consumption. According to the literature, the spot cooling system is one of the most efficient ventilation systems due to the concentration of the air nozzles on the heating source, establishing it the most energy-efficient ventilation system [27], which complies with the result in this research study.

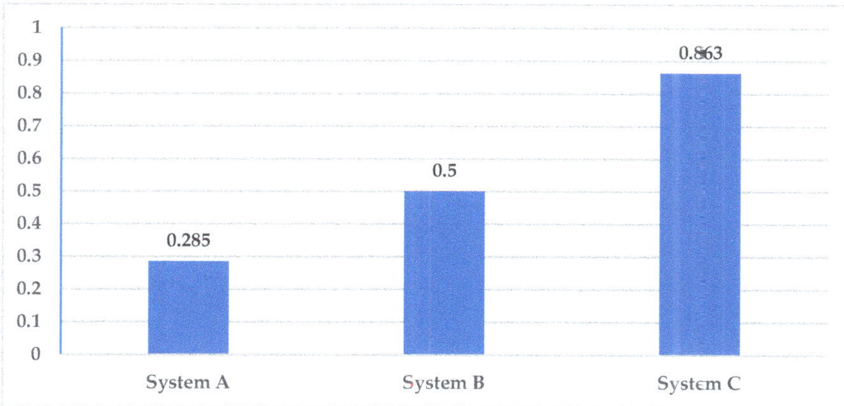

Figure 10. Ventilation effectiveness factor.

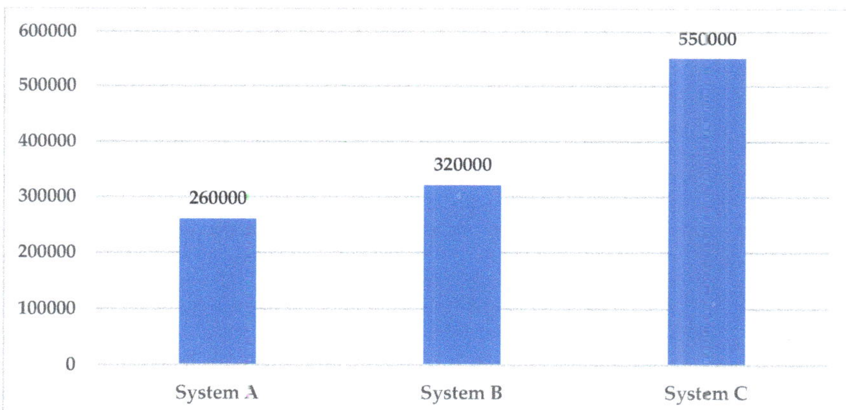

Figure 11. Ventilation systems' initial costs (EGP).

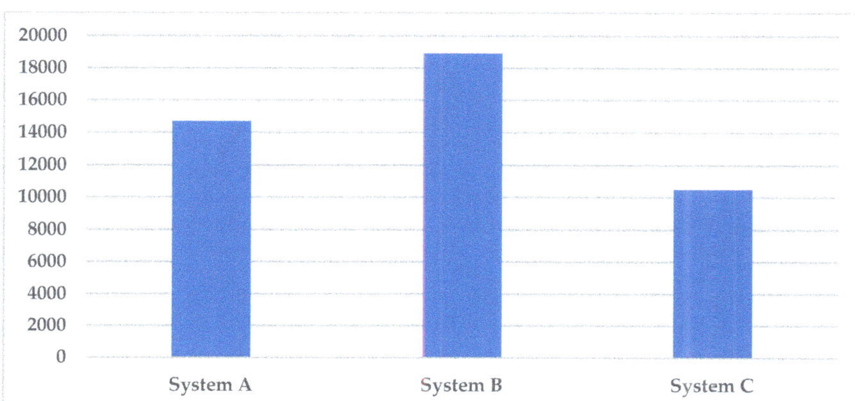

Figure 12. Ventilation systems' annual energy cost (EGP).

6. Conclusions

The feasibility of the study of industrial building ventilation systems is a crucial step for ensuring the use of the most effective system in terms of providing a suitable work environment and reducing energy consumption. Indeed, traditional ventilation systems are the most commonly used in Egyptian industrial buildings due to their low initial costs and ease of installation. However, according to the type of industry and weather conditions, the chosen ventilation system should provide the most effective ventilation and ensure lower energy consumption.

In this research study, three mechanical ventilation systems have been simulated using the CFD models, considering this is the most effective and inexpensive method used and validated in several research studies to investigate indoor ventilation. The systems included in this study were the wall fan extract ventilation system, the roof fan extract ventilation system, and the spot cooling system. The three chosen systems are distinguished by their convergent costs and ease of implementation, and the comparison between the three system results reveal that the spot cooling system is the most effective system according to the ventilation factor. In addition, System B (roof extraction fans) shows an increase in ventilation efficiency compared to System A (wall extraction fans), although there is no significant difference in the cost. With this in mind, this research study recommends using the spot cooling ventilation method for its high efficiency in industrial spaces, while roof extraction fans fall in second place. Due to the high cost required to validate the results in a real factory building, it is recommended to conduct an experimental investigation for these different systems in future research. However, testing the efficiency of these systems on different industrial building sizes and different types of industries is planned to be included in future research work. In addition, studying the effect of these systems on the concentration of pollutants resulting from some types of industries can provide a comprehensive vision for choosing the best systems that save energy, provide thermal comfort, and a clean environment within such buildings.

Author Contributions: Conceptualization, M.I.E. and B.E.; methodology, M.I.E. and B.E.; software, M.I.E.; validation, B.E. and R.M.H.A.; formal analysis, B.E.; investigation, M.I.E.; resources, A.M.Y.A.; writing—original draft preparation, B.E.; writing—review and editing, B.E. and R.M.H.A.; visualization, B.E.; supervision, R.M.H.A. and A.M.Y.A.; project administration, A.M.Y.A. and R.M.H.A. All authors have read and agreed to the published version of the manuscript.

Funding: This research received no external funding.

Institutional Review Board Statement: Not applicable.

Informed Consent Statement: Not applicable.

Conflicts of Interest: The authors declare no conflict of interest.

References

1. Hegazy, I.R.; Moustafa, W.S. Toward revitalization of new towns in Egypt case study: Sixth of October. *Int. J. Sustain. Built Environ.* **2013**, *2*, 10–18. [CrossRef]
2. Agrawala, S.; Moehner, A.; Raey, M.E.; Conway, D.; Van Aalst, M.; Hagenstad, M.; Smith, J. Development and climate change in Egypt: Focus on coastal resources and the Nile. *Organ. Econ. Co-Oper. Dev.* **2004**, *1*, 1–68.
3. *The Strategic Plan for Urban Growth and Development and Related Development Areas*; The General Organization of Physical Planning (GOPP): Cairo, Egypt, 2014. Available online: http://gopp.gov.eg/plans/ (accessed on 15 July 2021).
4. Cao, Z.; Zhai, C.; Wang, Y.; Zhao, T.; Wang, H. Flow characteristics and pollutant removal effectiveness of multi-vortex ventilation in high pollution emission industrial plant with large aspect ratio. *Sustain. Cities Soc.* **2020**, *54*, 101990. [CrossRef]
5. Meng, X.; Wang, Y.; Xing, X.; Xu, Y. Experimental study on the performance of hybrid buoyancy-driven natural ventilation with a mechanical exhaust system in an industrial building. *Energy Build.* **2020**, *208*, 109674. [CrossRef]
6. Kang, J.H.; Lee, S.J. Improvement of natural ventilation in a large factory building using a louver ventilator. *Build. Environ.* **2008**, *43*, 2132–2141. [CrossRef]
7. Murga, A.; Long, Z.; Yoo, S.-J.; Sumiyoshi, E.; Ito, K. Decreasing inhaled contaminant dose of a factory worker through a hybrid Emergency Ventilation System: Performance evaluation in worst-case scenario. *Energy Built Environ.* **2020**, *1*, 319–326. [CrossRef]
8. Fletcher, B.; Johnson, A.E. Ventilation of small factory units. *J. Wind Eng. Ind. Aerodyn.* **1992**, *40*, 293–305. [CrossRef]

9. Pakari, A.; Ghani, S. Comparison of different mechanical ventilation systems for dairy cow barns: CFD simulations and field measurements. *Comput. Electron. Agric.* **2021**, *186*, 106207. [CrossRef]
10. Pollet, I.; Laverge, J.; Vens, A.; Losfeld, F.; Reeves, M.; Janssens, A. Performance of automated demand controlled mechanical extract ventilation systems for dwellings. *J. Sustain. Eng. Des.* **2013**, *1*, 1–8.
11. Russell, M.; Sherman, M.; Rudd, A. Review of residential ventilation technologies. *HVAC R Res.* **2007**, *13*, 325–348. [CrossRef]
12. U.S. Department of Energy Whole-House Ventilation | Department of Energy. Available online: https://www.energy.gov/energysaver/weatherize/ventilation/whole-house-ventilation (accessed on 21 May 2021).
13. Miqdad, A.; Ali, A.; Kadir, K.; Ahmed, S.F.; Malik, M.A.A. Development of system to control air conditioner's airflow for spot cooling. In Proceedings of the 2017 International Conference on Engineering Technology and Technopreneurship (ICE2T), Kuala Lumpur, Malaysia, 18–20 September 2017; Volume 2017, pp. 1–4. [CrossRef]
14. Kabeel, A.E.; Sultan, G.I.; Zyada, Z.A.; El-Hadary, M.I. Performance study of spot cooling of tractor cabinet. *Energy* **2010**, *35*, 1679–1687. [CrossRef]
15. Bangalee, M.Z.I.; Lin, S.Y.; Miau, J.J. Wind driven natural ventilation through multiple windows of a building: A computational approach. *Energy Build.* **2012**, *45*, 317–325. [CrossRef]
16. Zhang, J.; Long, Z.; Liu, W.; Chen, Q. Strategy for studying ventilation performance in factories. *Aerosol Air Qual. Res.* **2016**, *16*, 442–452. [CrossRef]
17. Stavrakakis, G.M.; Koukou, M.K.; Vrachopoulos, M.G.; Markatos, N.C. Natural cross-ventilation in buildings: Building-scale experiments, numerical simulation and thermal comfort evaluation. *Energy Build.* **2008**, *40*, 1666–1681. [CrossRef]
18. Gaczoł, T. Living quarters. A natural balanced ventilation system. Simulations part 1. *E3S Web Conf.* **2018**, *49*, 00025. [CrossRef]
19. Cheung, J.O.P.; Liu, C.H. CFD simulations of natural ventilation behaviour in high-rise buildings in regular and staggered arrangements at various spacings. *Energy Build.* **2011**, *43*, 1149–1158. [CrossRef]
20. Ayad, S.S. Computational study of natural ventilation. *J. Wind Eng. Ind. Aerodyn.* **1999**, *82*, 49–68. [CrossRef]
21. Pérez, M.M.; Patiño, G.L.; Jiménez, P.A.L. International journal of comparison between natural and forced air flow. *Energy Environ.* **2013**, *4*, 357–368.
22. Karimipanah, T.; Awbi, H.B. Theoretical and experimental investigation of impinging jet ventilation and comparison with wall displacement ventilation. *Build. Environ.* **2002**, *37*, 1329–1342. [CrossRef]
23. Kobayashi, N.; Chen, Q. Floor-supply displacement ventilation in a small office. *Indoor Built Environ.* **2003**, *12*, 281–291. [CrossRef]
24. Tian, G.; Fan, Y.; Wang, H.; Peng, K.; Zhang, X.; Zheng, H. Studies on the thermal environment and natural ventilation in the industrial building spaces enclosed by fabric membranes: A case study. *J. Build. Eng.* **2020**, *32*, 101651. [CrossRef]
25. Mahmod, W.E.; Watanabe, K.; Zahr-Eldeen, A.A. Analyse du débit de nappe souterraine dans une zone aride avec des données hydrogéologiques limitées utilisant le Modèle de Grey: Étude de cas du Grès Nubien, oasis de Kharga, Egypte. *Hydrogeol. J.* **2013**, *21*, 1021–1034. [CrossRef]
26. Lamoreaux, P.E.; Memon, B.A.; Idris, H. Groundwater development, Kharga Oases, Western Desert of Egypt: A long-term environmental concern. *Environ. Geol. Water Sci.* **1985**, *7*, 129–149. [CrossRef]
27. Pinto, N.d.M.; Xavier, A.A.d.P.; Hatakeyama, K. Thermal Comfort in Industrial Environment: Conditions and Parameters. *Procedia Manuf.* **2015**, *3*, 4999–5006. [CrossRef]
28. Global Petrol Prices Egypt Electricity Prices. Available online: https://www.globalpetrolprices.com/Egypt/electricity_prices/ (accessed on 15 July 2021).

sustainability

MDPI

Article

Computer-Aided Automated Greenery Design—Towards a Green BIM

Dominik Sędzicki [1,*], Jan Cudzik [1], Wojciech Bonenberg [2] and Lucyna Nyka [1]

1 Department of Urban Architecture and Waterscapes, Faculty of Architecture, Gdańsk University of Technology, 11/12 Narutowicza Street, 80-233 Gdańsk, Poland; jan.cudzik@pg.edu.pl (J.C.); lucyna.nyka@pg.edu.pl (L.N.)

2 Institute of Architecture and Spatial Planning, Faculty of Architecture, Poznań University of Technology, ul. Jacka Rychlewskiego 2, 61-131 Poznań, Poland; wojciech.bonenberg@put.poznan.pl

* Correspondence: dominik.sedzicki@pg.edu.pl

Abstract: Contemporary climate challenges are changing the architect's awareness, which results in a broader spectrum of interest. The available software enables the design of vegetation, but it is often very limited and requires specialist knowledge. The available software allows the creation of individual solutions based on visual algorithms or writing scripts; however, they are still not common methods used in architecture and urban planning. The study proposes a new complex digital method of selection and design of greenery based on a new parameter spreadsheet. The proposition is supported by the review and investigation of the software used by designers identifying a range of tools for the design of greenery. The study proposes a theoretical model for automated plant selection and variations of possible greenery scenarios that could be integrated into the design process at the early stages of concept development.

Keywords: greenery; automated design; plant selector; sustainability; algorithmic design; landscape design; BIM; green BIM

Citation: Sędzicki, D.; Cudzik, J.; Bonenberg, W.; Nyka, L. Computer-Aided Automated Greenery Design—Towards a Green BIM. *Sustainability* **2022**, *14*, 8927. https://doi.org/10.3390/su14148927

Academic Editor: Yoshiki Shimomura

Received: 30 June 2022
Accepted: 19 July 2022
Published: 21 July 2022

1. Introduction

We are heading for a scenario in which technology will greatly help us to maximize the roles and skills of architectural professionals, making room to plan, design, build and manage buildings and infrastructures in a much more economical and sustainable way [1,2]. Technology brings numerous advantages and challenges that are known to professionals, who currently, due to the complexity of new procedures, often rely on basic analog and simple digital tools [3,4]. Considering the cultural and processual obstacles, there is a need for design tools and methods to adjust to the production of architectural and urban structures and spaces with the best fit for climatic challenges [5–7]. The design process is shifting towards data analytics and automatization which is based on accessible data sources [8–12]. New technologies are already changing the way buildings and cities are being designed and managed, but there is still a huge potential in that field [13,14].

Contemporary climate challenges impact the awareness of architects [15]. Designers acknowledge the importance of a multidisciplinary approach in terms of building environments both in architecture and urban design [16,17]. However, still, landscape design is often considered just an additional process to the core of architectural activities and thus often neglected. Due to the rise in ecological awareness among professionals involved in the discipline, this situation is evolving [18,19]. As stated in the European Commission document entitled Green Infrastructure (GI)-Enhancing Europe's Natural Capital and European Commission Directive (EU) 2018/844, energy efficiency and scenarios for greenery are important factors of global policy [20,21]. The change in this matter procures the need for the inclusion of greenery design and the creation of tools for architects enabling its selection at an early stage of design.

At the same time, the landscape design field faces barriers in the implementation of the system because most of the available software does not include specific tools to meet the needs of this sector [22]. That makes the challenge greater and more complex since it is necessary to develop an action plan that addresses not only the procedures related to the use of the software but also a methodology that smartly adapts the tools offered by the industry [23]. Landscaping often refers to the practice of landscape design and gardening, which traditionally concerns designing sites with vegetation for aesthetic, cultural, social, and other purposes [24,25]. Landscape architecture and landscape engineering, on the other hand, are interdisciplinary professions that integrate technical considerations such as environmental engineering and the means to reduce the carbon footprint of built structures [26,27].

In the era when automation processes are entering the discipline of architecture, the question should be posed, whether there are any automation processes for greenery design already incorporated into the software used by architects and landscape planners?

If not, the first general question appears, namely, whether the new digital methods related to automation of processes can be applied in greenery design. If so, what kind of environmental and vegetation parameters should be considered in the development of an automation model for greenery?

The study aims to propose a new complex digital method of selection and design of greenery as a useful tool for architects and other professionals for landscape design. The secondary goal is to review and investigate the software used by designers identifying a range of tools for the design of greenery. The scope of the study is based on available software used in the field of architecture and landscape planning. The third goal is to create a parameters spreadsheet of greenery that could be used as a data set for the automatic selection and design process. The ultimate goal of a study is to propose a theoretical model for automated plant selection and variations of possible greenery scenarios that could be integrated into the design process at the early stages of concept development. The model of automated greenery design proposed in this paper will not only contribute to practical implications but also advance the theoretical discourse on the integration of technological advancements with emerging new thematic areas [28–31]. Moreover, a new comprehensive approach toward integration of architectural, structural and greenery projects in early-stage design processes would create a new framework for further research.

Existing standards and workflows for landscape design, at present, rely more on analog methodologies. However, researchers, scholars, and some practitioners suggest that diverse possibilities, thanks to BIM (Building Information Modelling) processes and available technology, allow for better integration of vegetation into the spatiality of design projects, as well as the opportunity for acquiring information using various spatial analysis tools and visualization approaches [32,33]. Each plant object in the BIM environment model has precise data related to its actual nature, and this allows for quantitative planning, bill of materials, model coordination, automatization of the process of creating documentation, and many more [34]. Therefore, the advantages of using more advanced software solutions are obvious. However, visions and predictions of adopting new information modeling technologies in landscape architecture vary significantly.

As Jillian Walliss and Heike Rahman suggest, BIM-related software packages currently fail to provide a versatile and easy-to-use platform for landscape designers and other professionals. They identify two main reasons for this particular problem [35]. Firstly, creating complex 3D geometries and topography are usually beyond landscape designers' skills in terms of software handling. Secondly, planting design in BIM software is currently the most underdeveloped element of landscape architectural work. The authors also argue that more recently software developers have conceived specialized software add-ons or plug-ins that are supposed to support the integration of landscape works into the BIM environment, but they proven to have limited success in compatibility with the original BIM environment. In some cases, difficulties are the result of a lack of interoperability, whereas in other cases, software packages have limited tools, which in turn remove their parametric capability. Authors also suggest that other avenues for research should include exploration into programming data for plant growth prediction and plant selection.

Some authors provide evidence that designing individual tools in the form of coding and programming is the way forward in developing solutions for custom-made interventions using greenery for landscape [36,37]. Creating a toolbox for a specific situation using general-purpose programming languages (C++, Python, JavaScript, etc.) requires a lot of effort and is less flexible; however, it may determine a successful and more accurate outcome in the real setting. Creating a code is less common than using off-shelf software packages, but is widening the spectrum of possibilities for finding a perfect fit for a landscape design problem. Although using a common procedural programming language proves to be more open for data collection from analysis and simulations, in turn, it is not as accessible due to the lack of a user-friendly interface.

Some various experimental urban initiatives and interventions use digital tools to design experimental constructions that support the growth of plants. A prominent example is Urban Microclimate Canopy developed at the Technical University of Munich [38]. The authors present a digitally infused workflow that led to the development of lightweight elements that can accommodate climbing plants. The process of creation allowed the structure to adapt to different spatial requirements and local conditions but also provided all the benefits that come from the integration of greenery with the design, for instance, improved outdoor comfort. The use of new technologies involved innovative parametric design, robotic manufacturing tasks, calculating and visualizing load distribution within the structure, creating climatic and sun path analysis, and simulations of growth patterns. However, despite all the digitalization in the design process, the plant species selection for the experiment was carried out without the use of computers.

Rüdiger Clausen in his report on BIM in landscape architecture argues that information-sharing technology is likely to become the standard not only for the architectural and landscape design process but also for planning in the field of urban and vegetation design [39]. This is especially important since landscape architects are in pursuit of the right planning approach in coordination with the essential technological improvements for plant selection in their designs. It is to be expected that through the development of ever more efficient planning tools and the increasing distribution of BIM as a planning method in architecture over the next few years, the additional efforts and expenditures for model-based planning will decrease in the field of landscape architecture as well. This will have implications in the whole spectrum from early design stages to on-site maintenance of vegetation. However, at present, when looking from a practical point of view, gardening and landscaping companies are unable at this time to process the data information models and it often comes down to working with 2D plans because designers need years to become proficient in BIM software packages. The author also argues that since landscape architecture always focuses on implementing something new into an environment, the software interfaces may require many individual adaptations.

As demand for BIM for landscape rises, some research teams, mandated by governmental institutions, undertake the task of standardization of Building Information Modeling databases for landscape design. Knut Hallgeir describes such development by taking a Norwegian standardization project as an example [40]. To achieve a more holistic design approach for architects, combining indoor and outdoor schemes, the unification of databases is required. Since 2014, Norwegian BIM for landscape initiative has been trying to develop its system of creating standards for the profession by basing its work on British Landscape Institute Product Data Templates, but also by identifying different plant parameters needed for different design stages.

The purpose of PDTs (Product Data Templates) is comprehensively explained in the book "BIM for Landscape" released by the British Landscape Institute [41]. PDTs are being developed by several professional institutions to create easy-to-use digital catalogs providing greater access to information on species of vegetation. Such spreadsheets provide great exchangeability between different parties involved in the landscape design. Most importantly, databases such as this specify various plant parameters which are crucial for vegetation livability.

2. Materials and Methods

2.1. Methods

The work consists of two stages, starting with the review and investigation and recognition of software used in greenery design and characteristic parameters of greenery (Figure 1). The review of software functional capabilities in greenery design including plant selection, plant visual representation (two-dimensional and three-dimensional), characteristic parameters involvement, growth patterns, and more.

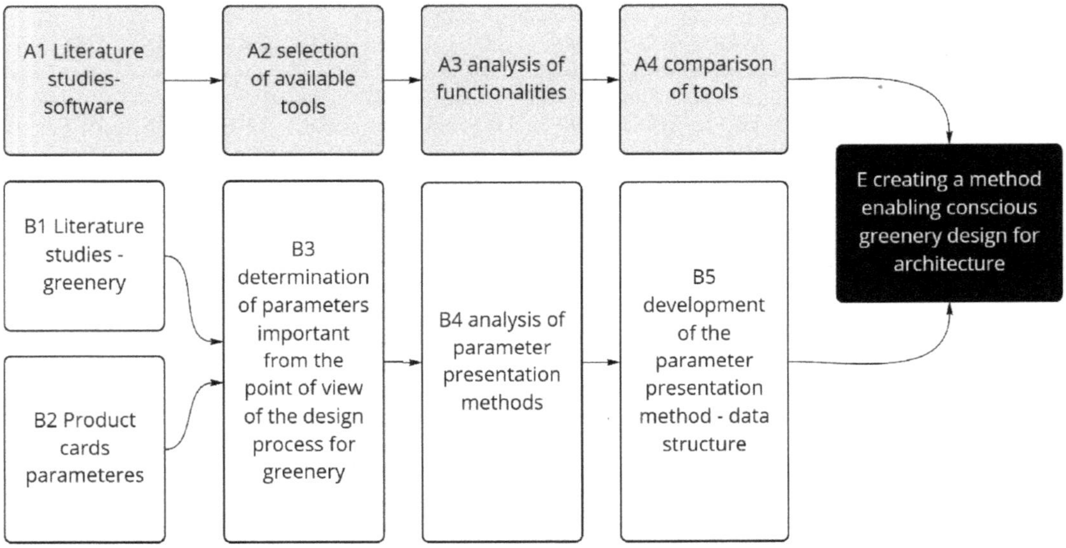

Figure 1. Research method.

The investigation of greenery consists of a selection of criteria for the categorization of plants for creating data structures from which designers could benefit or that could become a database for semi-automated and automated systems. The research process is summarized with a new proposition for a complex digital method in the process of designing greenery that would involve all the necessary parameters involved in the greenery design process, that are not always fully considered, such as humidity or soil characteristics.

The software study is performed as a review of tools available for architects. The tools are rated in 10 categories crucial for the process of landscape design; in each one, a range of marks from 0 to 10 (integers) points is applied. The total number of points possible to achieve is 100 points. The classification should not be considered a direct comparison but rather an inside process of landscape design within them.

The proposed categories in this study involve manual landscape design, visual representation of greenery (two-dimensional and three-dimensional), the possibility of introducing plants to the library (introduction of the product card and its specified requirements), climatic data from external open sources (including EPW (EnergyPlus Weather Format), CLM (ESP-r weather format), WEA (Daysim weather format), and more), climatic analysis of the macro site (location, world directions, height, climate zone), sun hours analysis, airing analysis, simulation of growth and patterns–plants, selection of species (selection of species based on parameters) and automation of plant selection based on the selection of parameters.

The greenery literature study was performed with the reference to the recent publications that address the greenery selection process. This led to the determination of parameters important in the design process. Then, the analysis of possible greenery types and their parameters presentation methods were conducted. The study concluded with the development of possible parameters, presentation method, and data structure. The classification led to highlighting the most useful features in the matter of study in available software and a new proposition for a complex digital method in the process of designing greenery.

2.2. Review: Available Software Applied in Landscape Design

Digital tools have been used in the design industry for decades. They can complement, and, in some cases, completely replace traditional design procedures such as drawing, drafting, and modeling. Digital technologies increase the efficiency, speed, precision, and complexity of expressions in design. In addition to the basic functions provided by computer techniques, such as 2D drawing and 3D modeling, the programs offer many intelligent and complex functions that can improve the building and landscape design, building climate systems, and project material specifications to allow for better customization.

This section aims to assess the state of functionality of the software in the field of landscape design (Table 1). To determine the versatility of such programs, a set of functions was defined and tested for each program. The usefulness of programs to architects and landscape designs was measured by counting available program features. There are 4 main types of software in landscape design. As follows they are 2D graphic design software, CAD (Computer-aided design), BIM (Building Information Modeling), and visual algorithm platforms. The first group and the most basic one is 2D graphic design programs which are generally used for representation purposes. Designers often choose to develop their designs within this form of the digital environment, due to the ease of use. Such programs cannot properly test out schemes in 3D settings. The second type is CAD software that enables two-dimensional vector layout design and three-dimensional modeling. The third type is BIM-based software which is more complex and responsive. What distinguishes BIM from CAD is the information data included in the model. This allows for quantitative planning, bill of materials, and model coordination, and automates the process of creating documentation and visualization. The fourth group is visual algorithm software enabling an algorithmic approach to the process of design, which makes the design more flexible. It is based on a method of linking parameters and components that can be parametrized.

- **Adobe Photoshop** for landscape design, is useful for graphic processing of visualizations and the creation of plans. This program does not have a specific function for this field, its tools are universal, which allows them to be used for many purposes. An example of such a solution could be the use of different brush shapes as symbols of plant species on the set.
- **Adobe Illustrator**, similar to Photoshop, is a universal program in many design fields. It is useful for creating various types of graphics in vector format. It does not have landscape design features. In this field, a user can create, for example, land plans, and graphic symbols. It allows showing the project in an individual graphical style.
- **AutoCAD** (and similar ZWCAD, GSTARCAD) is CAAD software aimed at architects, engineers, and construction professionals. It allows the creation of 2D drawings—from sketchy to precise design documentation. The program also enables 3D modeling. For landscaping, it offers a range of graphic symbols of trees and plants to be placed on the plan.
- **SketchUp** is a 3D modeling program that is very intuitive and versatile; additionally, it has many useful features and add-ons for landscape designers. It gives the possibility of using geo-location, for example, to obtain a terrain model or study the sun setting. Objects of different kinds, for example, specific plant species, can be modeled from scratch or downloaded from the online library.

- **Rhinoceros** is a CAAD program designed for precise 3D modeling. It gives the possibility to create any shape from small to large scale with the accuracy needed for design, prototyping, construction, analysis, and production. Additional collaborative software and plugins for design, drafting, CAM, engineering, analysis, rendering, animation, and illustration provide many possibilities in effects and project implementation. The Rhino software itself does not propose specific solutions for landscape designers.
- **Grasshopper** was initially developed as a plug-in for Rhino, which is currently a part of Rhinoceros software. It is mainly used to create algorithms for parametric modeling of 3D geometry in conjunction with data analysis. The software allows for developing individual solutions for different fields of design such as production lighting performance analysis or building energy consumption. The visual language of the software provides an intuitive way to explore projects without having to learn a scripting language.
- **Landsdesign** is strictly dedicated to landscape design. It enables 2D drafting, 3D modeling, and creating a visualization. Additionally, it allows the analysis and evaluation of landscape features. After modifying the site in the project, the program will show operations related to earthworks for execution purposes. To include vegetation in the project, you can use the database with over 1800 species that are represented in 2D and 3D elevations, which are conceptual, detailed, and realistic. Placed vegetation and other design elements detect the topography and are automatically placed on the ground. To find the appropriate plant species, you can use a filter that will adjust the search results based on the given characteristics.
- **3ds Max** offers a rich and flexible set of tools for creating projects. It gives the possibility of modeling interiors and 3D objects or even game characters and creating high-quality visualizations or animations. When focusing on the field of landscape design, proxy objects are a useful feature.
- **Revit** is an example of a BIM program used in many design areas, mainly intended for architecture and engineering but also used in landscape design. The program allows the creation of a project throughout its cycle from conceptual design, visualization, and analysis to production and construction. It gives us the possibility of making 2D drawings, 3D models, documentation, and using data on objects. This allows for the automation of routine and repetitive tasks as well as precision in execution.
- **ArchiCAD**, similar to Revit, is professional BIM software that offers an intuitive design environment, accurate information management, open collaboration, and automated documentation. It is used for the design and implementation of architectural projects using data analysis. It enables 3D modeling, documentation, and rendering of realistic visualizations.
- **LandFX** is a BIM plug-in for landscape design targeted at software such as AutoCAD, SketchUp, and Rhino. The program allows the design of details, vegetation, and irrigation, creation of a 2D design by selecting symbols, plantings, and labels, and the actions are automated.
- **Vectorworks Landmark** is a BIM program aimed at the professional landscape designer, offering project support from start to finish or at any stage. In total, 2D and 3D functions, visualizations, and project documentation are proposed. The program enables creative-free 3D modeling. There are many functions available in the subject of vegetation. Some of them, for example, have access to the species database in the default version, or it is possible to select a catalog from online resources. The designer can specify his expectations regarding the features of the species he wants to include in the project and the program will filter the database proposing a specific range. Plant objects are displayed in 2D and 3D and use parameters assigned by the user. The 3D representations can be schematic or photorealistic forms.
- **MicroStation** is mainly intended for architectural and construction design. It is a 2D and 3D CAD program with the possibility of using BIM functions and parametric design. The available tools are universal for many design processes: from creating

concepts, 2D documentation, and detailed BIM models to creating visualizations with the possibility of using geo-location and lighting analysis.

- **Dynascape** is a CAD program created for landscape design, based mainly on the basic representative pillars with an emphasis on individual and effective presentation of the project: 2D drawings and documentation, graphic representation of the project, and 3D model and visualizations.
- **Lumion** allows work between developing the detailed renders of your landscape architecture projects, and working in and updating your models in CAD or 3D modeling software. LiveSync works with all leading CAD programs including SketchUp, Revit, ArchiCAD, Rhino, Vectorworks, and AutoCAD. Going beyond the usual renders that intend to only convey some form of plant life, Lumion's content library offers a large variety of vegetation, allowing you to create a scene that is richly detailed with specifically chosen nature and vegetation.

Table 1. Comparison of available software applied in landscape design.

Name	*Manual Landscape Design*	*Visual Representation of Greenery*	*Plant Library*	*Climatic Data from External Sources*	*Climatic Analysis*	*Sunhours Analysis*	*Airing Analysis*	*Simulation of Growth*	*Selection of Species*	*Automatic Plant Selection*	*Result*
Adobe Photoshop	+	+	–	–	–	–	–	–	–	–	2/10
Adobe Illustrator	+	+	–	–	–	–	–	–	–	–	2/10
AutoCAD	+	+	–	–	–	–	–	–	–	–	2/10
SketchUp	+	+	–	+	–	–	–	–	–	–	3/10
Rhinoceros	+	+	–	–	–	–	–	–	–	–	2/10
Grasshopper	+	+	+	+	+	+	+	+	+	–	9/10
Landsdesign	+	+	+	+	+	+	–	–	+	–	7/10
3ds Max	+	+	+	–	–	–	–	–	–	–	3/10
Revit	+	+	+	+	–	+	–	–	+	–	6/10
ArchiCAD	+	+	+	+	–	–	–	–	–	–	4/10
LandFX	+	+	+	+	–	–	–	–	+	–	5/10
Vectorworks Landmark	+	+	+	+	+	–	–	–	+	–	6/10
MicroStation	+	+	+	–	–	–	–	–	–	–	3/10
Dynascape	+	+	+	–	–	–	–	–	–	–	3/10
Lumion	+	+	+	+	–	–	–	–	–	–	4/10

2.3. Review: Parameters of Greenery

The section aims to select the parameters that can be used in the model for the automated greenery design (AGD). The first step in the process is a general review of available plant databases applied in architectural software. The second step is evaluating applicability in the automatic selection and design process. The last step is an attempt to create a model of a primary parameter table-selected greenery parameters (SGP) that could become a base for a full greenery parameters table (FGP).

The programs used in the field of architecture contain databases with plant species for use in the project. From several hundred or several thousand species, you can choose the right plants based on your preferences. Each species in the database is described in more detail in some programs, less in others. Various characteristics are determined, such as appearance, height, spread, climatic zone, soil type, and others. In this chapter, the product sheets available in Landsdesign and Vectorworks Landmark will be analyzed. A proposal of what such a product card could look like is also shown in the BIM for Landscape study, which will also be analyzed in terms of its practicality and the possibility of the operation of such a card.

Landsdesign provides a plant database of over 1800 species. You can select a genre that fits your project by filtering the list according to various criteria, depending on your needs and expectations. To search for suitable species, you can select different characteristics in this topic: plant type, leaf type, shape, flowering, fructification, water needs, and climatic zones. In addition to these features, it is possible to select specific characteristics: fragrant, attracts birds, suitable for narrow streets, suitable for poor soil, attractive fruit, attractive bark, suitable for screen, autumn color change, edible/medicinal fruit, resistant to vandalism, suitable for interiors, requires pruning. It is also possible to determine the type of soil on which a given species is to grow: acid, clayey, sandy, basic, fertile, fresh, humic, moist, neutral, permeable, deep, siliceous, or dry. After selecting the desired traits, the list of species in the database narrows down to show only the ones that match. The product card of the selected species includes detailed photos, Latin and common names, information about the origin and family, and information about the requirements and application of the plant. The characteristics of the species define those features that could be marked in the search process: fragrance attracts birds, etc., shown as markers with some of these characteristics. The same applies to the type of soil. Additionally, information is presented (along with sample answers): pollution—high, marine environment—low, waterlogging—low, wind—low, plagues—high, sun—high. To represent the design, genres have different kinds of graphics, namely, 2D drawings for projection, simplified 3D conceptual shapes, detailed 3D shapes, and realistic 3D shapes for visualization purposes. Visual features such as height, span, density, and many others can be freely modified to suit your project. The main filters are plant type, leaf type, shape, flowering, water needs, soil type and individual needs (Figure 2).

Figure 2. Landsdesign plant selection.

Vectorworks Landmark has its own species database as well as the ability to use catalogs from online resources. It is possible to search for suitable plants based on the given characteristics required by the project. Searchability is mainly based on category, persistence, and floral color. After selecting a specific species, a product card appears. It contains detailed photos of the plant and the following information: Latin name, common name, category, class code, landscape use, growth habit, mature height, mature spread, flower characteristics, floral color, blooms begin, foliage characteristics, foliage color, fall colors, fruit characteristics, fruit color, persistence, tolerances, water range, soil range, ph range, light range, plant zone, comments. Despite many detailed issues, the list does not contain all the information, for example, the required soil or irrigation. The main filters include category, persistence and floral color (Figure 3).

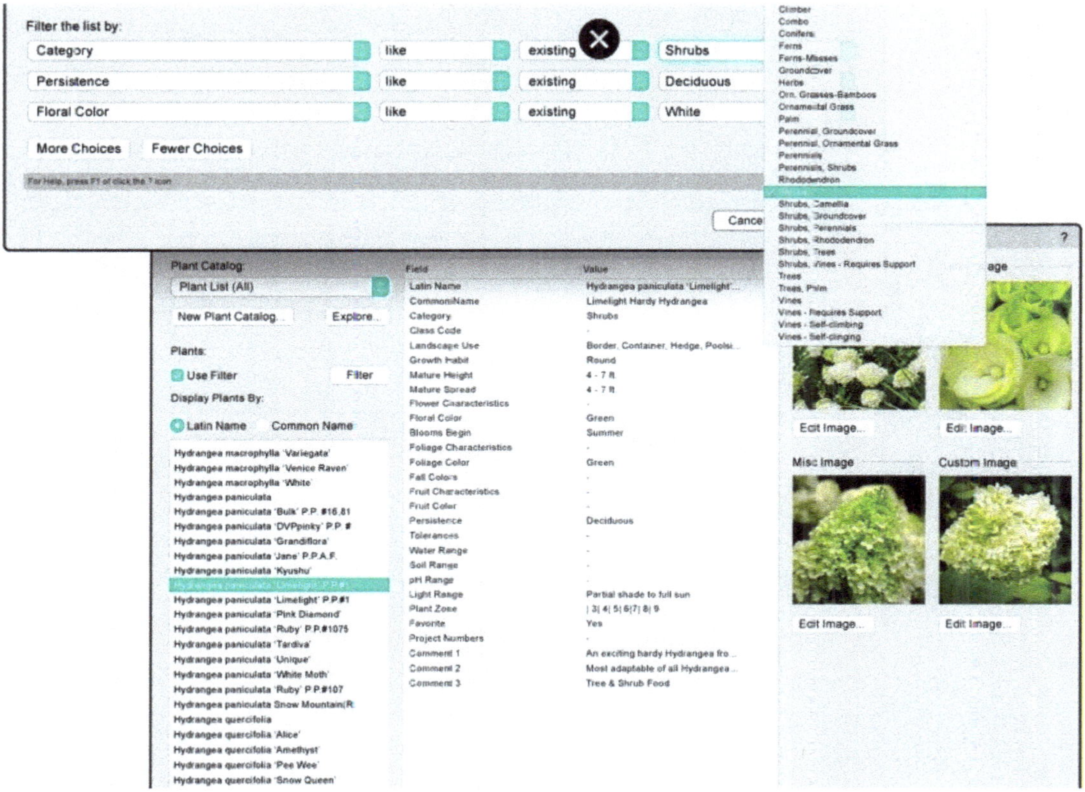

Figure 3. Vectorworks Landmark plant selection.

BIM for Landscape. The BIM for Landscape study presents a product sheet that could be a model for a detailed description of a given product. It contains several dozen terms that fall into several categories: main information, manufacturer data, naming data, nursery stock data. planting requirements, planting selection data, performance data, sustainability, operations and maintenance. Each of these categories includes several specific points. The data allow selection through manufacturer data, naming, nursery, planting requirements and selection data and maintenance (Figure 4).

The analysis of available greenery tables showed the necessity of creating a universal and simplified table that could become an open data source for the development of a system responsible for plant selection and a method of automated design. In the research process, the criteria and data content were established based on futures examined in the process of traditional plant selection. The outcome of the study is a selected greenery parameters (SGP) (Table 2) table consisting of two groups of data: basic information (Category, Gener, 3d Representation, 2d representation, Photo) and greenery properties (Height, Spread, Solar exposure, Moisture, Soil, Maintenance, Root volume, Temperature range, Life expectancy, Planting distance). The criteria selected in this part are useful for automated selection and automated design purposes.

Landscape Institute
Inspiring great places

Flora
Product Data Template

Template Category	Flora			
Template Version	v6.1			
Category Description	Plant species grown for the purpose of planting out in a landscape.			
Classification System				
Classification	Value			
Suitability for Use				
Template Custodian	Landscape Institute			
Information Category	**Parameter Name**	**Value**	**Units**	**Notes**
	Manufacturer Data			
Specifications	Supplier		Text	
Specifications	Supplier Website		URL	
Specifications	Product Range		Text	
Specifications	Product Model Number		Text	Or Code
Specifications	CE Approval		Text	Number, Yes, No
Specifications	Product Literature Webpage		URL	
Specifications	Product Features		Text	Free text to describe product
	Naming Data			
Specifications	Product Code		Text	
Specifications	Botanical Name		Text	
Specifications	Alternative Botanical Name		Text	Or Names
Specifications	Common Name		Text	Or Names
Specifications	Category or Class		List	Or Type. Select from list
Specifications	Sub-Category or Sub-Class		List	Select from list or type to define new value
	Nursery Stock Data (taken from BS 3639 and NPS)			
Specifications	Height		cm	Range of values or Minimum value
Specifications	Spread		cm	Range of values or Minimum value
Specifications	Girth		cm	Range of values or Minimum value
Specifications	Clear Stem Height		cm	Range of values or Minimum value
Specifications	Number of Breaks or Buds		Nr	Minimum number of breaks or buds
Specifications	Form Specified		List	Select from list or type to define new value
Specifications	Age and Condition		List	Select from list or type to define new value
Specifications	Root Condition and Protection		List	Select from list or type to define new value
Specifications	Cell or Container Size		List	Select from list or type to define new value
Specifications	Planting Medium		List	Select from list or type to define new value
Specifications	Fertiliser		List	Select from list or type to define new value
Specifications	Origin and Provenance		Text	
Specifications	Country or Place Grown		Text	

Figure 4. BIM for Landscape product data template.

Table 2. Selected greenery parameters (SGP).

Number	Name	Nursery	Grown	Type of Value	Unit
1.	*Category*			*text*	-
2.	*Name*			*text*	-
3.	*3D Representation*			*brep*	-
4.	*2D representation*			*spline/point*	-
5.	*Photo*			*image*	-
6.	*Height*			*number domain*	*cm/inch*
7.	*Spread*			*number domain*	*cm/inch*
8.	*Solar exposure*			*number domain*	*Sun hours*
9.	*Humidity*			*number*	*%*
10.	*Soil*			*number domain*	*ph*
11.	*Maintenance*	*1–10 low-hight*	*1–10 low-hight*	*number*	*integer*
12.	*Root volume*			*number*	*Cubic meters*
13.	*Temperature range*			*number domain*	*C/F*
14.	*Life expectancy*			*number domain*	*integer*
15.	*Planting distance*			*number*	*cm/inch*
16.	*Architecture proximity*			*number domain*	*cm/inch*

3. Results

3.1. Theoretical Model of Computer-Aided Greenery Design

We can categorize the requirements related to the design of greenery, an example of which is the proposal of a universal tabular presentation of the requirements related to a given species, which is the result of the analysis of greenery parameters. The creation of a universal tabular form will constitute a database for the proposed computer system; however, is only the first step in the entire process. It is necessary to create a comprehensive catalog of plants that responds to the given issues and can be the subject of later computer-automated analyses (Figure 5).

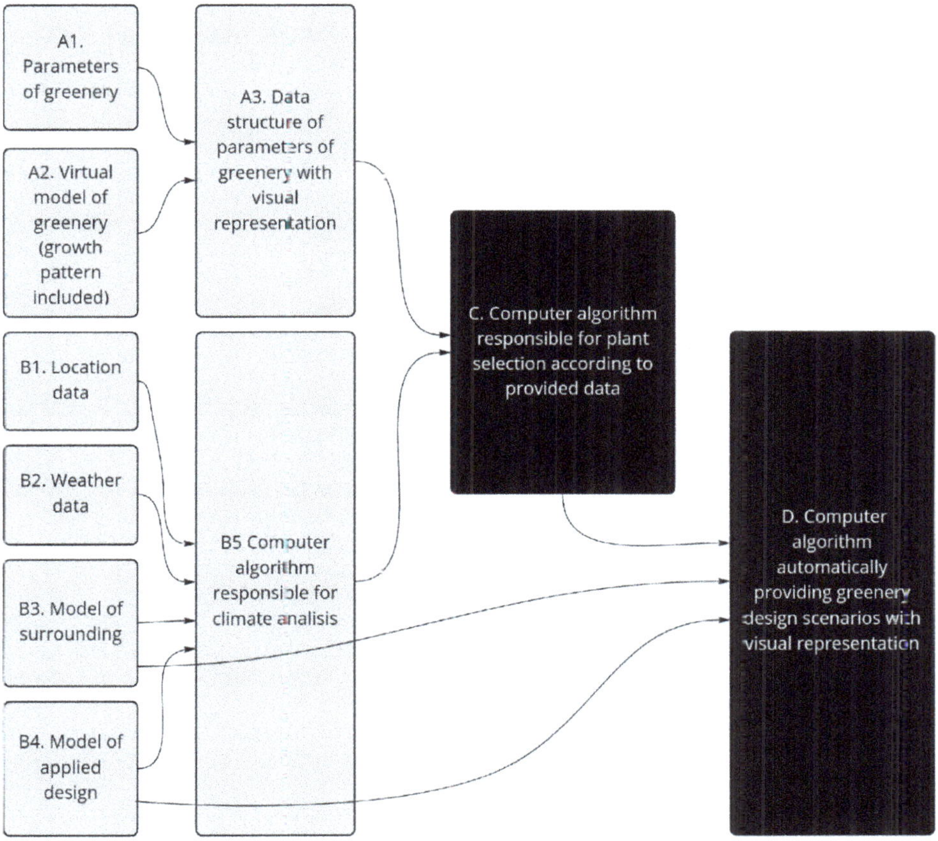

Figure 5. Theoretical model of computer-aided greenery design-automated greenery design (AGD).

The theoretical model is based on two sets of parameters—the first one consists of greenery characteristics which include parameters of greenery and virtual models of greenery (growth pattern included), resulting in the creation of data structure with the parameters of greenery and their visual representation. The second set considers design characteristic parameters and results in the creation of a computer algorithm responsible for climate analysis based on location data, weather data, model of surroundings, and a model of applied design. The two features enable the creation of computer algorithms responsible for plant selection according to provided data. The final step of the theoretical model is the computer algorithm automatically providing visual representation for greenery design scenarios.

The data necessary to conduct the analysis are divided into three groups. The first is the parameters and characteristics of plants, which will be developed in the simplified tabular form (SGP—selected greenery parameters) proposed by the authors, allowing for the assessment and verification of the applicability of a given species and the differences resulting from the growing process. The elements supporting the assessment and enabling the visual presentation of the species are also three-dimensional models and two-dimensional views.

The second group includes parameters and data related to the location and climatic data such as sun path, humidity, amount of precipitation, and climate. Appropriate data input and selection allows for accurate and necessary analyzes in the plant selection process to determine the requirements that must be met by species in a given place. However, the analysis cannot be carried out without the third data group consisting of the environment model and the designed object with elements of land development, which has a direct impact on, for example, the size of the possible root ball, humidity, or shading resulting from specific elements of land development.

In the next stage, the collected data are compiled and compared, which allows you to create a simulation related to the selection of the possibility of using given plant species in terms of the requirements related to the location of the project and its architectural dimension. The last, and at the same time the most advanced element of the proposed method, is a system proposing variants of the selection of vegetation for a given project, which may be the basis for further work.

3.2. Test Model for Single Scenario

Checking the test model is an important stage that allows for the verification of the programming environment and the possibility of data presentation and the selection of basic criteria for generating spatial solutions. A test was carried out on a limited number of parameters and species to verify the possibility of creating an automated greenery designer (AGD). For this purpose, selected greenery parameters (SGP) were created for three selected species of vines: *hedera helix, clematis, and fallopia* (Tables 3–5), which constituted the basis for further research. The tables were made based on the available greenery parameters, which were unified and reduced to a given format. Then, on their basis, data lists were created directly in Rhinoceros using the Grasshopper (Figure 6). In parallel, a test spatial model limited to a simple environment was produced (Figure 7)— plane with dimensions of 500 × 500 cm and a cuboid with dimensions of 250 × 20 × 200 cm. Gdańsk in Poland was selected as the location, and the source of the weather data was the EnergyPlus Weather File (EPW) for the Northern Port in Gdańsk. The plane became the test area (Figure 8) for selected species.

The algorithm, in line with the assumptions made in the theoretical model, has been divided into two modules—the first one that allows determining the possibility of using a given plant and the second one that suggests its specific location. First, an algorithm module was programmed to verify the possibility of using a given species based on selected parameters such as soil type (pH), and annual temperature range. In the next step, a module was created that allowed for spatial analysis such as the number of sun hours (Figure 9) produced with the Ladybug plugin and the distance from the architectural object, based on which the algorithm selects and proposes the location of the species.

The conducted research indicated the possibility of using all 3 species in a given location. The automated greenery designer model suggested the optimal location for each of them in terms of the analyzed parameters. (Figure 10) The test model shows that the process can be automated and therefore based on mathematical criteria. The final solution should be evaluated and developed to take into account additional criteria that may be important for different situations and species. All of them can be described and tested with more complex solutions. The created model could become a useful tool for professionals allowing more control and better design decisions in the early design stages.

Table 3. Selected greenery parameters (SGP) for hedera helix.

Number	Name	Nursery	Grown	Type of Value	Unit
1.	Category	climber	climber	text	-
2.	Name	Hedera helix	Hedera helix	text	-
3.	3d Representation	Violet dot	Violet dot	brep	-
4.	2d representation	Violet dot	Violet dot	spline/point	-
5.	Photo			image	-
6.	Height	80 cm–120 cm	<30,000 cm	number domain	cm/inch
7.	Spread	10 cm–30 cm	<10,000 cm	number domain	cm/inch
8.	Solar exposure	0–1400	0–1400	number domain	Sun hours
9.	Humidity	85%	85%	number domain	%
10.	Soil	<pH 8.0	<pH 8.0	number domain	ph
11.	Maintenance	2	1	number	integer
12.	Root volume	0.3 m^3	6 m^3	number	Cubic meters
13.	Temperature range	(−25)–(+35)	(−25)–(+35)	number domain	C/F
14.	Life expectancy	500 years	500 years	number domain	integer
15.	Planting distance	200 cm	500 cm	number	cm/inch
16.	Architecture proximity	0 cm–75 cm	0 cm–500 cm	number domain	cm/inch

Table 4. Selected greenery parameters (SGP) for clematis.

Number	Name	Nursery	Grown	Type of Value	Unit
1.	Category	climber	climber	text	-
2.	Name	Clematis	Clematis	text	-
3.	3d Representation	Red dot	Red dot	brep	-
4.	2d representation	Red dot	Red dot	spline/point	-
5.	Photo			image	-
6.	Height	80 cm–120 cm	<10,000 cm	number domain	cm/inch
7.	Spread	10 cm–30 cm	<6000 cm	number domain	cm/inch
8.	Solar exposure	1400–3100	1400–3100	number domain	Sun hours
9.	Humidity	75%	75%	number domain	%
10.	Soil	pH 6.5–pH 7.0	pH 6.5–pH 7.0	number domain	ph
11.	Maintenance	4	3	number	integer
12.	Root volume	0.3 m^3	2 m^3	numer	Cubic meters
13.	Temperature range	(−20)–(+35)	(−20)–(+35)	number domain	C/F
14.	Life expectancy	<100 years	<100 years	number domain	integer
15.	Planting distance	150 cm	300 cm	number	cm/inch
16.	Architecture proximity	0 cm–40 cm	0 cm–75 cm	number domain	cm/inch

Table 5. Selected greenery parameters (SGP) for clematis.

Number	Name	Nursery	Grown	Type of Value	Unit
1.	*Category*	*climber*	*climber*	*text*	-
2.	*Name*	*Fallopia*	*Fallopia*	*text*	-
3.	*3d Representation*	*Orange dot*	*Orange dot*	*brep*	-
4.	*2d representation*	*Orange dot*	*Orange dot*	*spline/point*	-
5.	*Photo*			*image*	-
6.	*Height*	*80 cm–120 cm*	*<12,000 cm*	*number domain*	*cm/inch*
7.	*Spread*	*10 cm–30 cm*	*<8000 cm*	*number domain*	*cm/inch*
8.	*Solar exposure*	*3100–4500*	*3100–4500*	*number domain*	*Sun hours*
9.	*Humidity*	*70%*	*70%*	*number domain*	*%*
10.	*Soil*	*pH 6.0–pH 7.0*	*pH 6.0–pH 7.0*	*number domain*	*ph*
11.	*Maintenance*	*4*	*3*	*number*	*integer*
12.	*Root volume*	*0.3* m^3	*4* m^3	*numer*	*Cubic meters*
13.	*Temperature range*	*(−20)–(+35)*	*(−20)–(+35)*	*number domain*	*C/F*
14.	*Life expectancy*	*<100 years*	*<100 years*	*number domain*	*integer*
15.	*Planting distance*	*200 cm*	*400 cm*	*number*	*cm/inch*
16.	*Architecture proximity*	*0 cm–75 cm*	*0 cm–200 cm*	*number domain*	*cm/inch*

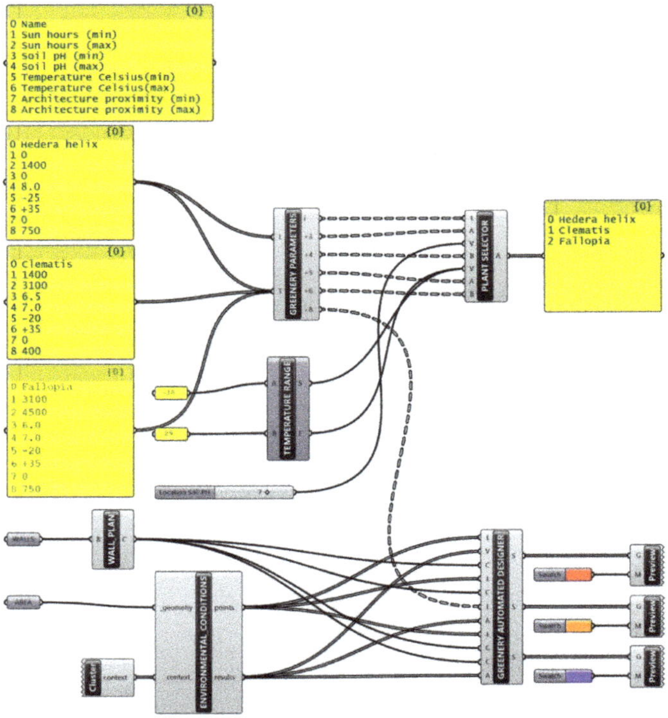

Figure 6. Automated greenery design (AGD) test model created with Grasshopper for Rhinoceros.

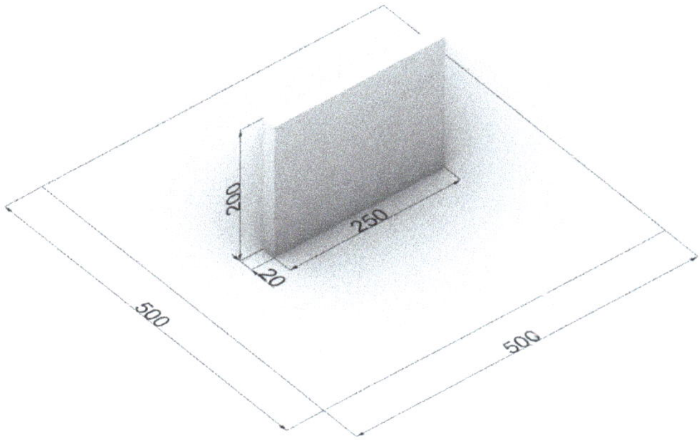

Figure 7. Test spatial model.

Figure 8. Test area.

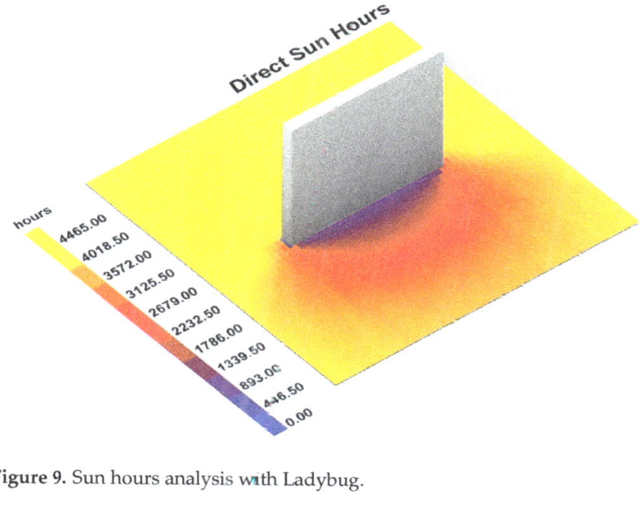

Figure 9. Sun hours analysis with Ladybug.

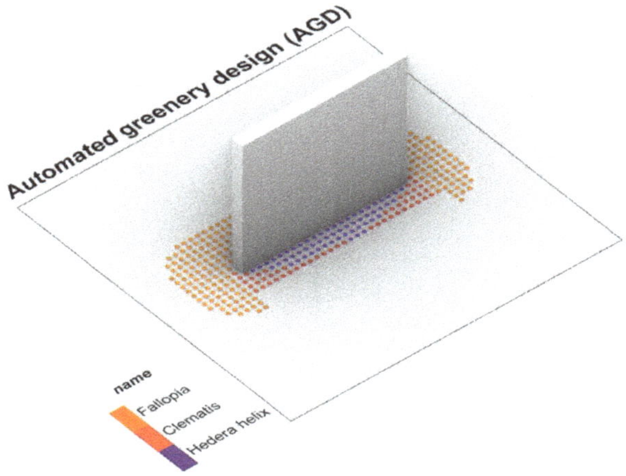

Figure 10. Sun hours analysis with Ladybug.

4. Discussion

Contemporary climate challenges are changing the architect's awareness, which not only deals with the shell but also has to deal with greenery, which is the natural direction of expanding competencies. The available software enables the design of vegetation, but it is often very limited and requires specialist knowledge that goes beyond the competencies of the architect [42]. Most of the software available on the market allows the creation of proprietary solutions based on the creation of algorithms or writing scripts, but these are not currently common methods used in architecture and urban planning due to their high complexity [43,44]. Changing the method and introducing digital data-based systems will make architects keener to include green solutions at an early stage of projects and their solutions will be better adapted to the climate and requirements related to a given location [45,46].

One interesting issue also takes into account aspects related to fauna and their requirements, such as maintaining green corridors or maintaining biodiversity, which in turn strongly influence the development of greenery as a complete structure. Providing an appropriate environment for birds or insects in the city has a positive effect on the development of plant tissue and increases its importance in the entire ecosystem [47].

The currently proposed model is limited to selected parameters but in future research, the approach can be broadenedby the consideration of various environmental issues. For example, the impact of vegetation design parameters on air quality should be taken into account [48–51]. In particular, the aerodynamic and deposition effects of vegetation should be considered [52–54]. This can pertain to the selection of vegetation species most efficient in pollution deposition but also to selecting vegetation design characteristics so that it does not obstruct airflow, especially in densely built-up urban areas and street canyons [54–56]. In many cities, the scientific knowledge on these processes is already disseminated in the form of practical guidelines [57,58]. However, it is still not widespread among planners and designers and therefore not commonly applied in practice [48,59]. Therefore, design-support tools which facilitate the consideration of the impact of vegetation on local air quality, especially alongside other issues, are required.

The selection of plants should also consider such factors as the lifecycle of plants, the waste that is produced as well as the contribution of a specific type of greenery to the local CO2 balance. The design should also consider the three R's concept (reduce, reuse, recycle) applied to the greenery as an important component of the architectural and urban project [60]. It is also crucial to develop the system for lifecycle management with a focus on environmental performance. In this matter, redevelopment of existing structures and strategies for heritage buildings should be considered with the availability to improve the analysis of existing buildings and their maintenance [28]. The new experience coming from the recent pandemic shows the need to include in future automated models for greenery design, factors related to the prevention of widespread diseases, peoples' anxiety, social distancing intentions, cultural differences [61,62], healing properties of plants and their impact on peoples' wellbeing.

It should be noted that BIM is a relatively new dynamically developing tool, about which there are not yet clearly defined boundaries of application. These boundaries are constantly being expanded to include more and more relationships between the built environment and the natural environment [63]. Our proposal makes it possible to improve the efficiency of greenery design by dynamically controlling important parameters that determine the environmental balance. It is a proposal for new applications of BIM with respect to Greenery design in landscape contexts.

The introduction of new tools and design methods should be supported by an increase in architects' awareness of the importance of greenery and its impact on the comfort of use. In this context, it is also important to assess and analyze the impact of the entire project on the surroundings and the environment. This gives a chance to change the organization of the design process and make informed decisions not only in terms of the choice of greenery but the entire project.

The proposed computer method would allow for a more precise and knowledge-based design of greenery at the early stages of architectural design. The solution will require supplementation and consultation with a landscape architect, but it would constitute a good basis, supported by hard data, for the development of the concept at later stages of design. The proposed method could be a response to the often-complicated interface or redundant data provided at various stages of the work. The plant cards created for the system could consist of selected greenery parameters (SGP), taken into account during computer analyzes, but it could also include an additional full greenery parameters table (FGP) allowing for additional verification of special cases. In the authors' opinion, creating cards and their variants is the greatest challenge, and an attempt to unify individual requirements may be a great research challenge.

5. Conclusions

Despite the continuous development of software in the field of architecture and urban planning, software in the field of greenery design does not propose advanced solutions for architects. The research indicates the necessity to propose a solution enabling conscious green design in a form available to architects. The proposed solution assumes the possibility of designing greenery at the early stages of an architectural design process with the use of automated plant selection and enabling the development of multiple scenario variations. Thanks to this, we can see the possibility of enabling the architect in the full design process, it has enabled comprehensive thinking about the location where architecture meets greenery. This offers a chance to organize the entire design process and make better decisions not only in terms of the choice of greenery but also the entire created structure.

In this context, Green BIM makes it possible to coordinate the most important components of an architectural project. These diagnostic factors—thanks to BIM technology—become available directly to designers, planners, investors and users. They make it possible to make design decisions supported by objectified measures from the early stages of design concepts. An important advantage of the proposed approach is the possibility of using simulation and evaluation based on BIM technology in greenery design.

Author Contributions: Conceptualization, D.S., J.C., W.B. and L.N.; methodology, D.S., J.C. and L.N.; software, D.S. and J.C.; validation, D.S. and J.C.; formal analysis, D.S., J.C., W.B. and L.N.; investigation, D.S., J.C., W.B. and L.N.; resources, D.S. and J.C.; data curation, D.S. and J.C.; writing—original draft preparation, D.S. and J.C.; writing—review and editing, D.S., J.C., W.B. and L.N.; visualization, D.S. and J.C.; supervision, J.C. and L.N.; funding acquisition, L.N. and W.B. All authors have read and agreed to the published version of the manuscript.

Funding: This research was funded by the European Union's Horizon 2020 research and innovation programme under the Maria Skłodowska-Curie grant number 823901—sosclimatewaterfront—H2020-MSCA-RISE-2018. The APC was funded by Poznań University of Technology, Future City Lab 2022 programme under grant agreement No. 0111/SBAD/2209.

Institutional Review Board Statement: Not applicable.

Informed Consent Statement: Not applicable.

Data Availability Statement: Not applicable.

Conflicts of Interest: The authors declare no conflict of interest.

References

1. Lynn, G.; Gage, M.F.; Nielson, S.; Rappaport, N. *Composites, Surfaces, and Software: High Performance Architecture*; Yale School of Architecture: New Haven, CT, USA, 2010; p. 104.
2. Urbanowicz, K.; Nyka, L. Interactive and Media Architecture—From Social Encounters to City Planning Strategies. *Procedia Eng.* **2016**, *161*, 1330–1337. [CrossRef]
3. Cudzik, J.; Radziszewski, K. Parametric design in architectural education. *World Trans. Eng. Technol. Educ.* **2019**, *17*, 448–453.
4. Liao, L.; Zhou, K.; Fan, C.; Ma, Y. Evaluation of Complexity Issues in Building Information Modeling Diffusion Research. *Sustainability* **2022**, *14*, 3005. [CrossRef]
5. Cameron, R.W.; Taylor, J.E.; Emmett, M.R. What's 'cool' in the world of green façades? How plant choice influences the cooling properties of green walls. *Build. Environ.* **2014**, *73*, 198–207. [CrossRef]
6. Abdel-Razek, S.A.; Marie, H.S.; Alshehri, A.; Elzeki, O.M. Energy Efficiency through the Implementation of an AI Model to Predict Room Occupancy Based on Thermal Comfort Parameters. *Sustainability* **2022**, *14*, 7734. [CrossRef]
7. Gruber, P.; Imhof, B. Patterns of Growth—Biomimetics and Architectural Design. *Buildings* **2017**, *7*, 32. [CrossRef]
8. Yi, H.; Yi, Y.K.; Chan, T. Performance Based Architectural design optimization: Automated 3D space Layout using simulated annealing. In Proceedings of the 2014 ASHRAE/IBPSA-USA Building Simulation Conference, Atlanta, GA, USA, 10–12 September 2014; pp. 292–299.
9. Wu, G.; Miao, Y.; Wang, F. Intelligent Design Model of Urban Landscape Space Based on Optimized BP Neural Network. *J. Sens.* **2022**, *2022*, 9704287. [CrossRef]
10. Li, C.Z.; Guo, Z.; Su, D.; Xiao, B.; Tam, V.W.Y. The Application of Advanced Information Technologies in Civil Infrastructure Construction and Maintenance. *Sustainability* **2022**, *14*, 7761. [CrossRef]
11. Yokoi, K.; Fukuda, T.; Yabuki, N.; Motamedi, A. Integrating BIM, CFD and AR for Thermal Assessment of Indoor Greenery. In Proceedings of the 22nd International Conference on Computer-Aided Architectural Design Research in Asia (CAADRIA 2017), Suzhou, China, 5–8 April 2017. [CrossRef]
12. Claypool, M.; Retsin, G.; Garcia, M.J.; Jaschke, C.; Saey, K. Automation and the Discrete: Exploring New Potentials for Streamlining Production in Architectural Design Research. *J. Arch. Educ.* **2021**, *75*, 108–114. [CrossRef]
13. Badidi, E. Edge AI and Blockchain for Smart Sustainable Cities: Promise and Potential. *Sustainability* **2022**, *14*, 7609. [CrossRef]
14. Weber, R.E.; Mueller, C.; Reinhart, C. Automated floorplan generation in architectural design: A review of methods and applications. *Autom. Constr.* **2022**, *140*, 104385. [CrossRef]
15. Wong, N.H.; Tan, C.L.; Kolokotsa, D.D.; Takebayashi, H. Greenery as a mitigation and adaptation strategy to urban heat. *Nat. Rev. Earth Environ.* **2021**, *2*, 166–181. [CrossRef]
16. Albers, M.; Deppisch, S. Resilience in the Light of Climate Change: Useful Approach or Empty Phrase for Spatial Planning? *Eur. Plan. Stud.* **2013**, *21*, 1598–1610. [CrossRef]
17. Russo, A.; Cirella, G.T. Modern compact cities: How much greenery do we need? *Int. J. Environ. Res. Public Health* **2018**, *15*, 2180. [CrossRef]
18. Brink, E.; Aalders, T.; Ádám, D.; Feller, R.; Henselek, Y.; Hoffmann, A.; Ibe, K.; Matthey-Doret, A.; Meyer, M.; Negrut, N.L.; et al. Cascades of green: A review of ecosystem-based adaptation in urban areas. *Glob. Environ. Change* **2016**, *36*, 111–123. [CrossRef]
19. Nyka, L. From structures to landscapes-towards re-conceptualization of the urban condition. *Archit. Res. Addressing Soc. Chall.* **2017**, *1*, 509–515. [CrossRef]
20. European Commission. *Directive (EU) 2018/844 of the European Parliament and of the Council of 30 May 2018 Amending Directive 2010/31/EU on the Energy Performance of Buildings and Directive 2021/27/EU on Energy Efficiency*; L156/75–91; European Commission (EU): Brussels, Belgium, 2018.
21. Campiotti, C.A.; Gatti, L.; Campiotti, A.; Consorti, L.; De Rossi, P.; Bibbiani, C.; Muleo, R.; Latini, A. Vertical Greenery as Natural Tool for Improving Energy Efficiency of Buildings. *Horticulturae* **2022**, *8*, 526. [CrossRef]
22. Hermansdorfer, M. BIM for Landscape. In *Grading: BIM. landscapingSMART. 3D-Machine Control Systems. Stormwater Management*; Birkhäuser: Berlin, Germany; Boston, MA, USA, 2019; pp. 213–220. [CrossRef]
23. Shashua-Bar, L.; Potchter, O.; Bitan, A.; Boltansky, D.; Yaakov, Y. Microclimate modelling of street tree species effects within the varied urban morphology in the Mediterranean city of Tel Aviv, Israel. *Int. J. Clim.* **2009**, *30*, 44–57. [CrossRef]

24. Picuno, C.A.; Godosi, Z.; Picuno, P. Implementing a Landscape Information Modelling (LIM) toolfor planning leisure facilities and landscape protection. In Proceedings of the Public Recreation and Landscape Protection—with Environment Hand in Hand: Conference Proceeding, Brno, Czech Republic, 9–10 May 2022; pp. 186–190. [CrossRef]
25. Al Hattab, M.; Hamzeh, F. Simulating the dynamics of social agents and information flows in BIM-based design. *Autom. Constr.* **2018**, *92*, 1–22. [CrossRef]
26. Bonenberg, W.; Rybicki, S.M.; Schneider-Skalska, G.; Stochel-Cyunel, J. Sustainable Water Management in a Krakow Housing Complex from the Nineteen-Seventies in Comparison with a Model Bio-Morpheme Unit. *Sustainability* **2022**, *14*, 5499. [CrossRef]
27. Besir, A.B.; Cuce, E. Green roofs and facades: A comprehensive review. *Renew. Sustain. Energy Rev.* **2018**, *82*, 915–939. [CrossRef]
28. Fernández-Mora, V.; Navarro, I.J.; Yepes, V. Integration of the structural project into the BIM paradigm: A literature review. *J. Build. Eng.* **2022**, *53*, 104318. [CrossRef]
29. Chong, H.; Lee, C.; Wang, X. A mixed review of the adoption of Building Information Modelling (BIM) for sustainability. *J. Clean. Prod.* **2017**, *142*, 4114–4126. [CrossRef]
30. Nyka, L.; Cudzik, J.; Urbanowicz, K. The CDIO model in architectural education and research by design. *World Trans. Eng. Technol. Educ.* **2020**, *18*, 85–90.
31. Wei, X.; Bonenberg, W.; Zhou, M.; Wang, J. Application of BIM Simulation and Visualization in Landscape Architecture Design. *Adv. Hum. Factors Archit. Sustain. Urban Plan. Infrastruct.* **2020**, *1214*, 215–221. [CrossRef]
32. Sanchez-Sepulveda, M.; Fonseca, D.; Franquesa, J.; Redondo, E. Virtual interactive innovations applied for digital urban transformations. Mixed approach. *Future Gener. Comput. Syst.* **2019**, *91*, 371–381. [CrossRef]
33. Kamel, E.; & Memari, A.M. Review of BIM's application in energy simulation: Tools, issues, and solutions. *Autom. Constr.* **2019**, *97*, 164–180. [CrossRef]
34. Zanni, M.A.; Soetanto, R.; Ruikar, K. Towards a BIM-enabled sustainable building design process: Roles, responsibilities, and requirements. *Arch. Eng. Des. Manag.* **2016**, *13*, 101–129. [CrossRef]
35. Walliss, J.; Rahmann, H. *Landscape Architecture and Digital Technologies: Re-Conceptualising Design and Making*, 1st ed.; Routledge: London, UK, 2016; pp. 216–217. [CrossRef]
36. Wong, Y.C.; Chin, K.-Y. Plant Parameters Influencing the Cooling Performance of Vegetated Roofs: A review. *Int. J. Sci. Res. Publ.* **2018**, 8. [CrossRef]
37. Cantrell, B.; Mekies, A. (Eds.) *Codify: Parametric and Computational Design in Landscape Architecture*, 1st ed.; Routledge: London, UK, 2018; pp. 98–99. [CrossRef]
38. Shu, Q.; Middleton, W.; Dörstelmann, M.; Santucci, D.; Ludwig, F. Urban Microclimate Canopy: Design, Manufacture, Installation, and Growth Simulation of a Living Architecture Prototype. *Sustainability* **2020**, *12*, 6004. [CrossRef]
39. Clausen, R. BIM in Landscape Architecture: A Report. *J. Digit. Landsc. Archit.* **2021**, *6*, 353–369. [CrossRef]
40. Wik, K.H.; Sekse, M.; Enebo, B.A.; Thorvaldsen, J. BIM for Landscape: A Norwegian Standardization Project. *J. Digit. Landsc. Archit.* **2018**, *3*, 241–248. [CrossRef]
41. Fenby-Taylor, H. *BIM for Landscape*, 1st ed.; Routledge: London, UK, 2016. [CrossRef]
42. Marzouk, M.; Ayman, R.; Alwan, Z.; Elshaboury, N. Green building system integration into project delivery utilising BIM. *Environ. Dev. Sustain.* **2021**, *24*, 6467–6480. [CrossRef]
43. Kolarevic, B. (Ed.) *Architecture in the Digital Age: Design and Manufacturing*, 1st ed.; Taylor & Francis: Abingdon, UK, 2003. [CrossRef]
44. Han, B.; Leite, F. Generic extended reality and integrated development for visualization applications in architecture, engineering, and construction. *Autom. Constr.* **2022**, *140*, 104329. [CrossRef]
45. Keibach, E.; Shayesteh, H. BIM for Landscape Design Improving Climate Adaptation Planning: The Evaluation of Software Tools Based on the ISO 25010 Standard. *Appl. Sci.* **2022**, *12*, 739. [CrossRef]
46. Ibrahim, Y.; Kershaw, T.; Shepherd, P. A methodology For Modelling Microclimates: A Ladybug-tools and ENVI-met verification study. In Proceedings of the 35th PLEA Conference. Planning Post Carbon Cities, A Coruña, Spain, 1–3 September 2020.
47. Arévalo, C.; Amaya-Espinel, J.D.; Henríquez, C.; Ibarra, J.T.; Bonacic, C. Urban noise and surrounding city morphology influence green space occupancy by native birds in a Mediterranean-type South American metropolis. *Sci. Rep.* **2022**, *12*, 4471. [CrossRef] [PubMed]
48. Badach, J.; Dymnicka, M.; Baranowski, A. Urban Vegetation in Air Quality Management: A Review and Policy Framework. *Sustainability* **2020**, *12*, 1258. [CrossRef]
49. Badach, J.; Voordeckers, D.; Nyka, L.; Van Acker, M. A framework for Air Quality Management Zones-Useful GIS-based tool for urban planning: Case studies in Antwerp and Gdańsk. *Build. Environ.* **2020**, *174*, 106743. [CrossRef]
50. Bottalico, F.; Chirici, G.; Giannetti, F.; De Marco, A.; Nocentini, S.; Paoletti, E.; Salbitano, F.; Sanesi, G.; Serenelli, C.; Travaglini, D. Air Pollution Removal by Green Infrastructures and Urban Forests in the City of Florence. *Agric. Agric. Sci. Procedia* **2016**, *8*, 243–251. [CrossRef]
51. Baldauf, R. Roadside vegetation design characteristics that can improve local, near-road air quality. *Transp. Res. Part D Transp. Environ.* **2017**, *52*, 354–361. [CrossRef]
52. Buccolieri, R.; Santiago, J.-L.; Rivas, E.; Sanchez, B. Review on urban tree modelling in CFD simulations: Aerodynamic, deposition and thermal effects. *Urban For. Urban Green.* **2018**, *31*, 212–220. [CrossRef]

53. Janhäll, S. Review on urban vegetation and particle air pollution—Deposition and dispersion. *Atmos. Environ.* **2015**, *105*, 130–137. [CrossRef]
54. Voordeckers, D.; Meysman, F.; Billen, P.; Tytgat, T.; Van Acker, M. The impact of street canyon morphology and traffic volume on NO2 values in the street canyons of Antwerp. *Build. Environ.* **2021**, *197*, 107825. [CrossRef]
55. Xue, F.; Li, X. The impact of roadside trees on traffic released PM 10 in urban street canyon: Aerodynamic and deposition effects. *Sustain. Cities Soc.* **2017**, *30*, 195–204. [CrossRef]
56. Buccolieri, R.; Gromke, C.; Di Sabatino, S.; Ruck, B. Aerodynamic effects of trees on pollutant concentration in street canyons. *Sci. Total Environ.* **2009**, *407*, 5247–5256. [CrossRef]
57. Baldauf, R. *Recommendations for Constructing Roadside Vegetation Barriers to Improve Near-Road Air Quality*; 600/R-16/0; United States Environmental Protection Agency: Washington, DC, USA, 2016.
58. Greater London Authority Using Green Infrastructure to Protect People from Air Pollution. Available online: https://www.london.gov.uk/WHAT-WE-DO/environment/environment-publications/using-green-infrastructure-protect-people-air-pollution (accessed on 5 August 2020).
59. Badach, J.; Dymnicka, M.; Załęcki, J.; Brosz, M.; Voordeckers, D.; Van Acker, M. Exploring the Institutional and Bottom-Up Actions for Urban Air Quality Improvement: Case Studies in Antwerp and Gdańsk. *Sustainability* **2021**, *13*, 11790. [CrossRef]
60. Johnny, W.K.W.; Zhou, J. Enchencing environmental sustainability over building life cycles through green BIM: A review. *Autom. Constr.* **2015**, *57*, 156–165.
61. Qin, Z.; Song, Y. Symbol Matters: A Sequential Mediation Model in Examining the Impact of Product Design with Buddhist Symbols on Charitable Donation Intentions. *Religions* **2022**, *13*, 151. [CrossRef]
62. Luo, J.; Zhang, Y.; Song, Y. Design for Pandemic Information: Examining the Effects of Grahps on Anxiety and Social Distancing Intention in the COVID-19. *Front. Public Health* **2022**, *10*, 800789. [CrossRef]
63. Bonenberg, W.; Wei, X. Green BIM in Sustainable Infrastructure. *Procedia Manuf.* **2015**, *3*, 1654–1659. [CrossRef]

sustainability

Article

Study of the Landscape Pattern of Shuiyu Village in Beijing, China: A Comprehensive Analysis of Adaptation to Local Microclimate

Ling Qi [1], Ranqian Liu [2,*], Yuechen Cui [3], Mo Zhou [4], Wojciech Bonenberg [4] and Zhisheng Song [5]

1 School of Architecture and Urban Planning, Beijing University of Technology, Beijing 100124, China; 63651106@bjut.edu.cn
2 School of Architecture, Tianjin University, Tianjin 300072, China
3 Department of Urban Planning and Design, Faculty of Architecture, The University of Hong Kong, Hong Kong SAR 999077, China; cuicui22332@gmail.com
4 Faculty of Architecture, Poznan University of Technology, 60-965 Poznan, Poland; mo.zhou@put.poznan.pl (M.Z.); Wojciech.Bonenberg@put.Poznan.pl (W.B.)
5 Tsinghua University Architectural Design and Research Institute Co., Ltd., Beijing 100084, China; songzhisheng@thad.com.cn
* Correspondence: dories_1012@tju.edu.cn; Tel.: +86-159-1065-0629

Abstract: The paper used technical parameters to investigate optimized solutions to protect the ecological environment and improve the microclimate adaptability among the traditional villages in Beijing. Shuiyu Village was used as a case study to analyze the coupling relationship between landscape patterns and the microclimate of traditional villages, with a focus on the ecological relationship between residents and the microclimate. This study also developed a climate index system, which includes computer numerical simulation and microclimate comprehensive analysis methods. The distinct types of landscape patterns were studied using the system. In addition, this paper studied the adaptive design mechanism in-depth, the form parameters of comfort evaluation controllability, and map expression technology of morphological parameters. The findings of this study include the optimized value of the environment based on landscape pattern and the map through the Rhino modeling platform. An interactive platform was developed, and a parametric-assisted optimization design process for traditional villages in the northern part of China was proposed. Moreover, this study concluded optimized strategies and technical guidelines for future planning of the rural areas in northern China with a goal to protect traditional villages and transform them into smart villages with microclimate adaptability.

Keywords: adaption; landscape pattern; local microclimate; comprehensive analysis; morphological parameters

Citation: Qi, L.; Liu, R.; Cui, Y.; Zhou, M.; Bonenberg, W.; Song, Z. Study of the Landscape Pattern of Shuiyu Village in Beijing, China: A Comprehensive Analysis of Adaptation to Local Microclimate. *Sustainability* **2022**, *14*, 375. https://doi.org/10.3390/su14010375

Academic Editor: Alejandro Rescia

Received: 31 October 2021
Accepted: 25 December 2021
Published: 30 December 2021

Publisher's Note: MDPI stays neutral with regard to jurisdictional claims in published maps and institutional affiliations.

1. Introduction

The environment of traditional Chinese villages has been severely impacted by globalization, urbanization, and new rural construction, causing its degradation and alienation. One focus of traditional village protection is to ensure the villages' landscape pattern of harmony between humans and nature. The frequently used strategies for quality control of the village environment include using site selection and traditional Fengshui models from ancient times to build landscape patterns. The ancient Chinese generated Fengshui models to make the best use of limited land and established optimal regions, cities and buildings without doing too much damage to the earth [1]. Therefore, their respect for the site, the comprehensive judgment based on "adapting to the situation", and the concept of harmony between human and nature should be critical in the field of sustainable climate adaptability design. Thus, to retain the sustainable development of cities in urban and rural areas, the protection and development of those traditional villages are necessary. It is critical to study the ecological system in environment construction.

The protection of traditional villages has been a concern in many countries. For example, various research methods and technologies have developed theoretical systems and studied traditional village renewal and spatial patterns in the United Kingdom, Germany, and Japan. Recently, the studies on the landscape pattern of rural settlements have been moved from physical properties to hidden factors in the settlement pattern. In addition, quantitative methods have been used to determine the degree of influence of hidden elements in the spatial distribution [2]. A lot of current research on the development and protection of traditional Chinese villages uses qualitative value research to develop an evaluation system and village spatial form and layout. Moreover, many studies use combined qualitative and quantitative research on building monomers [3,4], and most research studied climate adaptability, human comfort, urban planning and layout, air environment quality, and plant cultivation [3–28]. For example, Liu Binyi and his team used a combination of subjective and objective methods in a case study of the waterfront green space in Shanghai. In their research, green space was studied to improve the comfort feeling in summer. Urban streets were studied as the research object, and the research was conducted from three areas, including the influence of street space on microclimate elements, the evaluation of the comfort feeling, and the interaction between street space and comfort feeling. Comfort evaluation systems with professional characteristics of landscape architecture were established to integrate space elements, microclimate elements, and human feelings [10–13]. Fu Fan et al. undertook research on the indirect effect of urban green spaces on reducing the concentration of fine particulate matter in the air [14] and proposed an optimization plan to improve the thermal environment in Beijing's urban green space system. Dai Fei et al. also studied the impact of the internal spatial structure of green space on the thermal environment and its implementation in Wuhan by analyzing the morphological spatial patterns [15]. Moreover, Feng Xianhui discussed the correlation between plant communities, green space layouts, and microclimate effects in Guangzhou and used the correlation results to optimize site microclimate design strategies and plant community microclimate design [16,17]. Using the microclimate effect, Jin Hexian et al. applied their design strategies among Hangzhou streets and parks [18–20]. Other scholars have been researching the microclimate characteristics of scenic tourist areas, ancient towns, and traditional settlements [23,24]. However, most of the research has been conducted in urban areas [28,29], and limited research has been done on landscape patterns and microclimate adaptation design strategies.

A combination of qualitative and quantitative analysis methods hs been used, including numerical simulation, data comparison, and model construction at the micro-level. Moreover, quantitative, dynamic, and interdisciplinary research has been done in this field, and there are three main problems identified in current traditional village landscape pattern research. Firstly, although many studies are focused on historical and cultural value, settlement spatial distribution characteristics and structure, and architecture, little attention has been paid to the landscape pattern research. Therefore, research in this field is unbalanced, with much focus on macro-scales and micro-research on construction technology but little on the micro-scale research in the environment. Moreover, there is much research on the urban microclimate, but little in traditional villages.

Furthermore, there is limited research on the coupling control of traditional village landscape patterns and microclimate adaptation mechanism characteristics. Second, from the perspective of analysis methods, a lot of qualitative analyses and evaluations have been used, but not many quantitative analysis and design strategies have been used. Thus, there is a lack of comprehensive analysis methods used. Finally, from the perspective of model analysis, design mechanism, and application, most of the existing research conclusions are a single regularity or quantitative result, and the coupling relationship between the multi-factors of microclimate and the characteristics of landscape construction and predictive control methods have not been investigated further [30,31]. This causes a challenge to transform the evaluation results into practice efficiently and effectively.

To protect the ecological systems and improve the microclimate in the coordinated development of China's Beijing–Tianjin–Hebei coordinated development of urban and rural areas, this paper aimed to use various research methods to investigate internal and external factors in landscape construction and the microclimate of traditional villages that influence the overdevelopment of traditional villages and the destruction of the ideal landscape pattern caused by the misunderstanding of protection. In this study, a case study of Shuiyu Village was conducted in 2017.

2. Materials and Methods

2.1. Site Description

Shuiyu Village is located in Beigou, Nanjiao Township, Beijing, China (repainted in Figure 1a–c). The village is distributed along a northwest-southeast ditch rock formed in Shuiyugou. The terrain is high in the southwest and low in the northeast, surrounded by mountains. The village was built on a hillside. The landscaped environment of Shuiyu Village is aligned with Fengshui surroundings with mountains and rivers embracing Yin and Yang (Figure 2). The Zhongjiaoliang in the north is a natural barrier of the landscape. Shuiyu Village is high in the north and low in the south. The Nanpo Ridge is low and gentle, with Shamao Mountain in the east resembling Wu Shamao, and the mountains in the west resembling a throne. The tall mountains in the north block the northwest wind in winter, and with a low south slope, wind can be blown into the valley from the south. The overall ventilation environment is a good condition (Figure 3). The east part of Shuiyu Village preserves the pattern of villages during the Ming and Qing Dynasties, and the texture of streets and lanes is in the Yin-Yang-Bagua pattern. The Kun location is Changling Tuo (appreciating the moon); the trunk location is the big locust tree (Figure 4). the waterfront road is through the village.

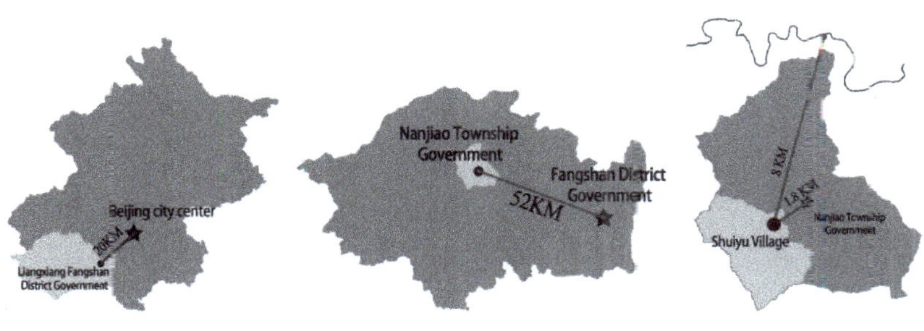

(a) (b) (c)

Figure 1. Shuiyu Village sitemap.

This paper considered the macroscopic combination space of all mountains and water in Shuiyu Village as the research object. It also quantified the coupling relationship between the "shape" of the landscape pattern and the "number" of the microclimate to construct a model [32] and a framework based on microclimate adaptability.

2.2. Data Resource

The data source came from three parts: (1) Field observation (measured data of the wind and thermal loop measured data locations of the typical landscape pattern of Shuiyu Village, and location photos with qualitative descriptions); (2) Simulation data (microclimate wind and thermal environment simulation mainly for the terrain of Shuiyu Village); and (3) Parametric terrain data (a digital model presented on the Rhino platform using programming software, such as Grasshopper). This model was developed based upon the CAD contours of the village and displayed the macroscopic landscape and spatial combination of Shuiyu Village.

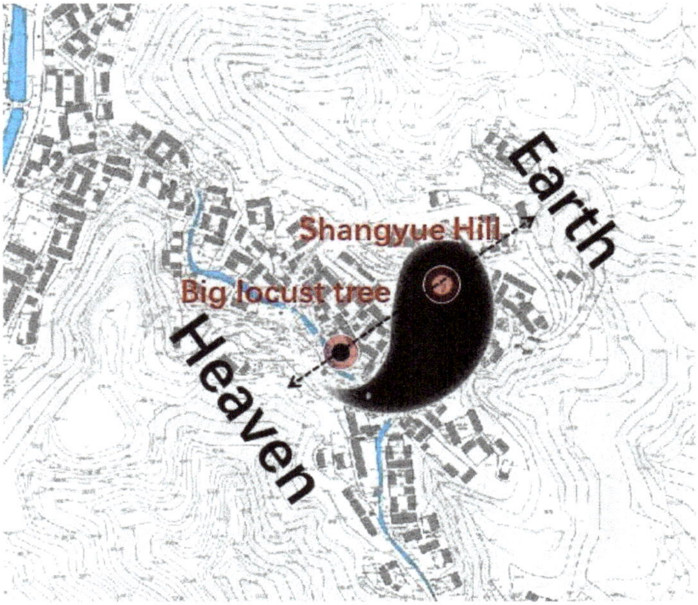

Figure 2. The trigram pattern of the east of Shuiyu Village.

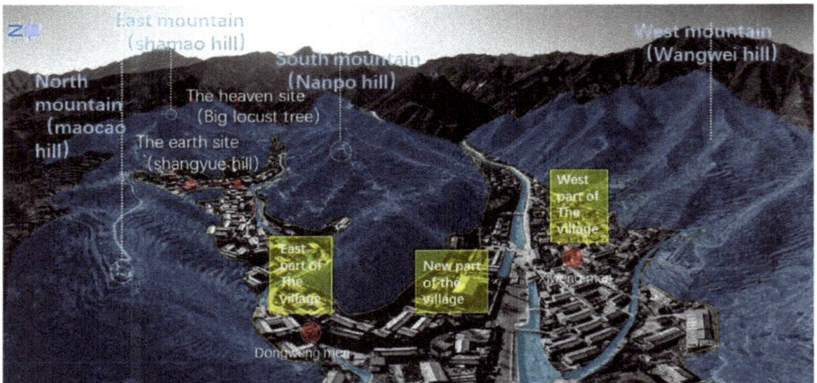

Figure 3. Schematic diagram of experimental observation locations.

2.3. *Methodology*

2.3.1. Field Observation

Field observation was conducted 1.5 m from the ground to observe the study object. Climate data, such as air temperature, wind speed and direction, relative humidity, and radiation temperature, at the experimental measurement locations, were analyzed.

(1) Research instruments: The equipment for the experiment included a mobile weather station (ZK-YD6A), hand-held heat-sensitive anemometer (TESTO405-V1, Germany), a temperature and humidity auto-logging instrument (Beijing Tianjianhua Instrument WSZY-1), and a black Bulb temperature recorder (Beijing Tianjianhua Instrument

HQZY-1). The mobile weather station collected data from the fixed space observation locations 24 h a day throughout the year. Due to the limited conditions of the field measurement, one measurement day in each of the 3 seasons and 10 space observation locations were used. The measurement dates were 11 March 2017, 5 July 2017, and 13 January 2018. All the collected data with reference to the data of China Meteorological Network among Beijing area and the data of small weather stations have been used as the input parameters of the developed model. In spring, 18 measuring locations were set up, with nine in each of the east and west villages. Observations were carried out in the mornings and afternoons, respectively. Optimized adjustments were made in summer and winter, and ten measuring locations were set up as well (Figure 4 and Table 1). The measuring locations were selected based on the experimental conditions, landscape environmental characteristics, height and slope orientation, and the principle of uniform distribution of locations. Additionally, the properties of different underlying surfaces, and the overall landscape pattern, buildings, vegetation, and other influencing factors had also been considered.

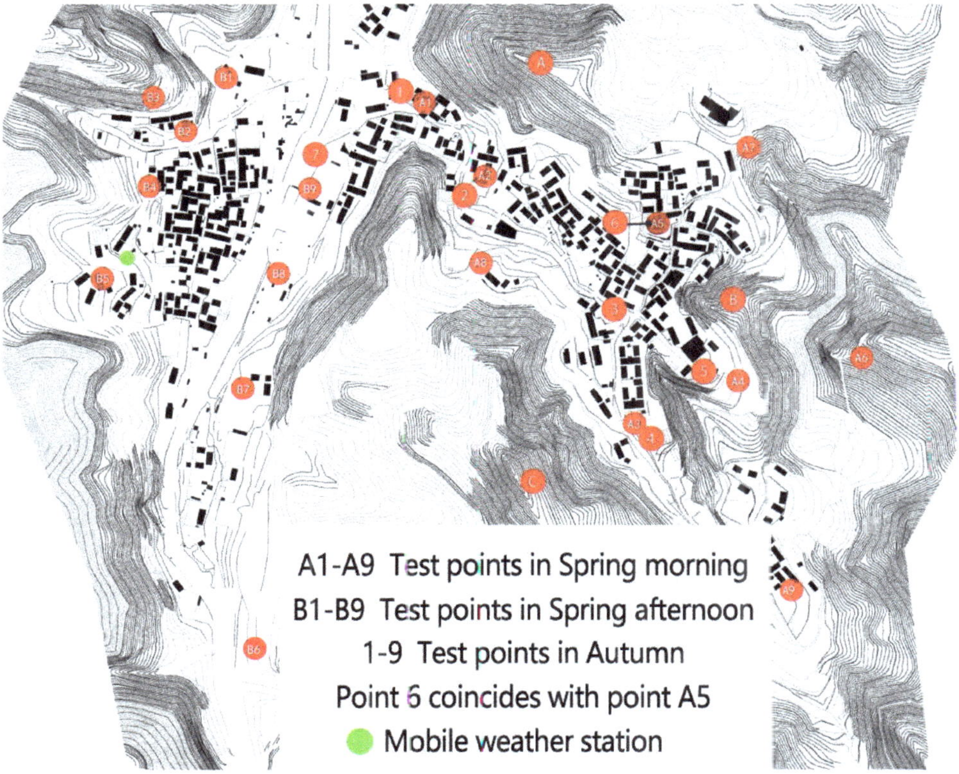

Figure 4. The landscape pattern of Shuiyu Village.

(2) Procedure: The research team completed the mobile weather station installation, conducted the 24-h observations at fixed locations, and monitored and collected data through the information platform. The recording frequency of the measured data was 10 min/time. The anemometer was set at the height of about 1.6 m from the ground, and the maximum wind speed was read, and the wind direction was recorded. The temperature and humidity probe was wrapped in tin foil to avoid direct sunlight and placed at a height of about 1.2 m above the ground.

Table 1. Experimental Observation Locations.

Test Locations	Spatial Structure	Environmental Characteristics	Real Scene of the Site Environment	Measuring Instrument	Floor Plan	Landscape Pattern Type
1	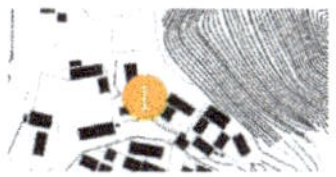	The stone road was paved with hills on both sides. The doorway and the valley are in the same direction.		Hand-held anemometer and temperature and humidity auto logging instrument	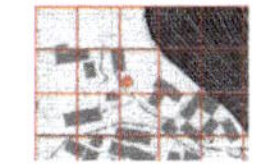	 Dependent Slope (east)
2		It was located on the hill into the village ramp. The terrain was relatively steep, with mountains and streams on both sides of the southwest and northeast and few dwellings and vegetation.		Hand-held anemometer and temperature and humidity auto-logging instrument	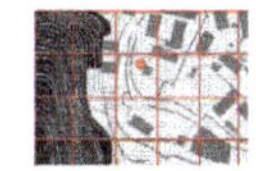	 Dependent Slope (West)
3		The river on the northwest side was frozen. Yang Family Courtyard was on the west side. The underlying surface was slate.		Hand-held anemometer and temperature and humidity auto-logging instrument	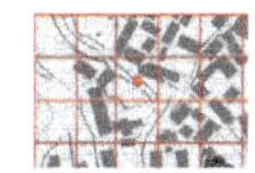	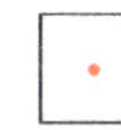 Open flat ground
4		Located at the bottom of the valley. The terrain was narrow, with mountains and buildings on both sides.		Hand-held anemometer and temperature and humidity auto-logging instrument		 Two sides off the valley (steep)
5	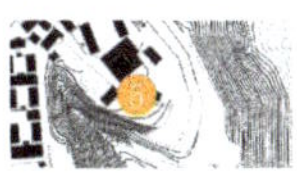	Deep in the valley, surrounded by high mountains. Narrow space.		Hand-held anemometer and temperature and humidity auto-logging instrument		 Three sides off the valley (steep)
6		Located on the Moon Viewing Hill, with open terrain, no mountain shelter, no water source, and no vegetation.		Hand-held anemometer and temperature and humidity auto-logging instrument		 Open flat ground

Table 1. *Cont.*

Test Locations	Spatial Structure	Environmental Characteristics	Real Scene of the Site Environment	Measuring Instrument	Floor Plan	Landscape Pattern Type
7		Flat square; adjacent to the main road; two houses with pavilions, surrounded by mountains, no water source, and limited vegetation.		Hand-held anemometer and temperature and humidity auto-logging instrument	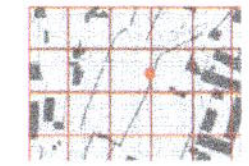	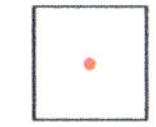 Open flat ground
A		Located on the top of the mountain to the northeast of the village. The terrain was high and sunny. Limited vegetation.		Hand-held anemometer and temperature and humidity auto-logging instrument		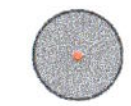 Open mountain top
B		The pavilion on the top of the mountain that overlooks the ancient village. The wind speed was relatively slow, and the wind direction was relatively stable.		Hand-held anemometer and temperature and humidity auto-logging instrument	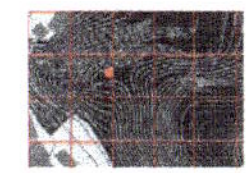	 Open mountain top
C		The pavilion to the south of Shuiyu Village was open without architectural vegetation.		Hand-held anemometer and temperature and humidity auto-logging instrument	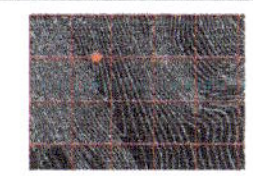	 Open mountain top

2.3.2. Numerical Simulation

The numerical simulation was undertaken through modeling, using the software of Ecotect and Phoenics. The simulation applied theoretical analysis to model calculations. Numerical simulation involves performing multiple simulation calculations on the same model, checking the measured data, calibrating the measured wind direction, and collecting parameter data, and providing data support for the subsequent parameterized model construction for planning and design.

2.3.3. Calculation Method of Microclimate Comfort Index

The layout of traditional village buildings and the landscape environment influences the surrounding microclimate environment and, consequently, the comfort level of the human living environment. The human comfort level in this paper was calculated by using the WBGT balance formula proposed by Dong Liang. The formula calculated the summer microclimate comfort value [33], TS-Givoni index [7], and THI index [34] to retrieve the microclimate comfort value during winter and in spring.

$$\mathrm{WBGT}_{autumn} = 0.8901t + 7.3771 \times 10^{-3}G + 13.8297a - 8.7284v^{-0.0551} \quad (1)$$

$$\mathrm{TS} - \mathrm{Givoni}_{\mathrm{winter}} = 1.7 + 0.1172t + 0.0019G - 0.322v - 0.0073a \quad (2)$$

$$\mathrm{THI}_{\mathrm{spring}} = t - (0.55 - 0.005a)(t - 14.5) \quad (3)$$

t: temperature in degrees Celsius, G: solar radiation, a: relative humidity of the air, v: wind speed.

2.3.4. Morphological Characterization Quantitative Method

The landscape pattern of traditional villages is affected by natural topographical conditions and human beings' interventional work. It has been continuously reshaped through interactions with the natural environment, forming an ideal landscape pattern. The microscopic morphological representation is reflected in slope, aspect, building orientation, elevation, vegetation coverage, and mountain occlusion. In this study, the morphological characterization of the landscape pattern was measured, and the main parameter factors were extracted by screening the parameterized factors of the geographic spatial model of the landscape pattern. They were presented in the southward space opening and closing degree X and the dominant wind direction opening and closing degree Y. The optimization factor extraction uses the sight analysis method to establish the parameterized logic and the space opening and closing degree visualization diagram (Figures 5 and 6). With reference to the related landscape design and parametric views [35,36], this study investigated the slope, aspect, and water body (inundation line) of the digital model of Shuiyu Village. This study also used the traditional village landscape pattern characterization influencing factors and the quantification of related factors to quantify the factors as the linear algebra relationship of the basic operation unit of parameterized programming.

2.3.5. Coupling Calculation Method of Landscape Pattern and Microclimate

In this paper, the multivariate linear panel data regression method was used to explore the coupling relationship between landscape pattern and microclimate. In multi-parameter analysis, human comfort had been used as a dependent variable, and landscape morphological characterization factors, such as the opening and closing degree of the southerly space and the opening and closing degree of the dominant wind direction, were taken as independent variables. The contribution of multiple variables to the microclimate was studied, and the regression equation is expressed as follows:

$$\mathrm{W}_{it} = \sum_{k=1}^{K} \alpha_{ki} x_{kit} + \sum_{k=1}^{K} \beta_{ki} y_{kit} + \mu_{it} \quad (4)$$

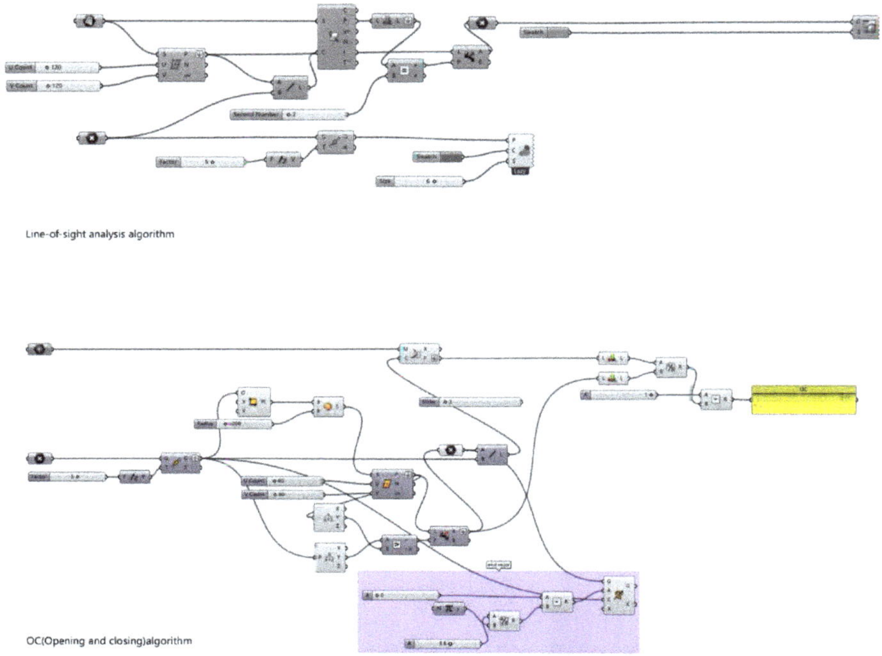

Figure 5. Landscape pattern factor algorithm.

Among them, i = 1, 2, 3, 4, ... , N, representing locations; t = 1, 2, 3, ... , T, representing the time location of the test. W_{it} is the explained variable (human comfort). When the observation value of location n is at t, x_{kit} and y_{kit} (south opening and closing degree, dominant wind direction opening and closing) are the k-th non-random explanatory variable; locations α_{ki} and β_{ki} are the parameters to be measured, and μ_{it} is a random error.

2.3.6. Comfort Evaluation and Visual Expression Method

The microclimate index was rated and evaluated (Figures 7 and 8) using the coupling verification and fitting equation of the landscape pattern factor and microclimate comfort in its microclimate environment and the evaluation standards of human comfort (WBGT index and TS-Givoni index). A visual map of Shuiyu Village's spring, summer, and winter comfort was generated based on climate adaptability.

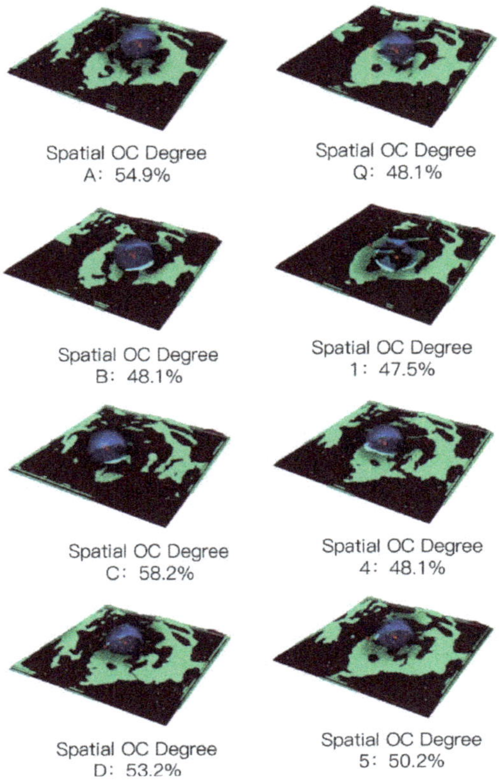

Figure 6. Data on the degree of space opening and closing and microclimate factors.

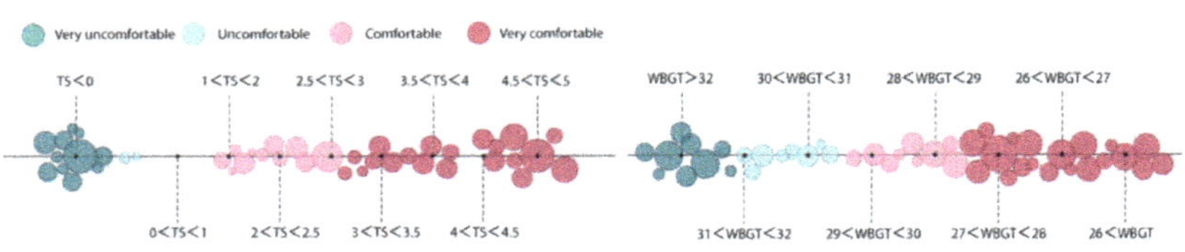

Figure 7. Comfort evaluation standard.

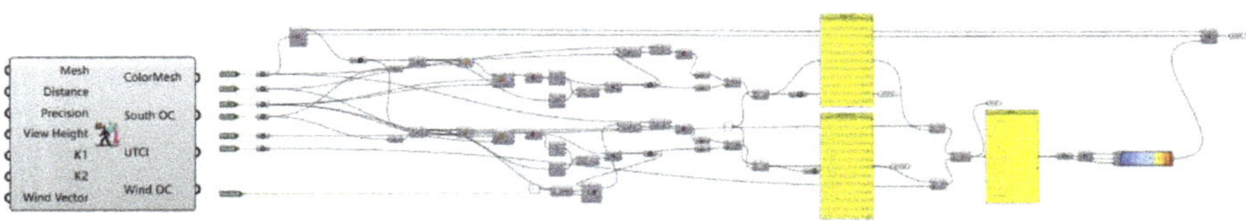

Figure 8. Comfort evaluation algorithm.

3. Results

3.1. Comparison of Microclimate Environment between Shuiyu Village and Beijing

This paper compared the measurement data of Shuiyu Village and the Beijing Meteorological Station. It was found that the average temperature of Shuiyu Village in all four seasons was slightly lower, its humidity was higher, and the wind speed of the village was lower than that of Beijing City (Table 2). It could be owing to the well-designed landscape and environment of Shuiyu Village.

Table 2. Comparison between Shuiyucun Village and Beijing City.

Time	Average Temperature (°C)		Average Humidity (RH %)		Average Wind Speed (m/s)	
	Weather Station	City	Weather Station	City	Weather Station	City
5.11–6.10	21.14	23.85	48.02	44.07	0.66	2.48
8.11–9.10	22.46	24.74	82.64	70.70	0.41	1.82
9.11–10.10	15.85	19.95	73.69	57.4	0.41	1.73
10.11–11.10	9.66	11.31	77.83	67.44	0.38	1.51

3.2. Microclimate Environment in Shuiyu Village

(1) Temperature: With no wind speed being considered, the higher the openness of the space is, the higher the temperature is. Using the four sets of data in three seasons, the temperature on the top of the mountain and the open area was found higher than the other areas. In spring and summer, the temperature of the location with greater openness raised quickly, and the average temperature was high. During the mornings (10:10 a.m.) in spring, the wind speeds at A6 and A1 were relatively high, causing the rise of temperature at these two locations to slow down. This phenomenon also occurred in summer, but the overall trend remains unchanged. However, in winter, excessive wind speed changed the temperature trend. Consequently, the temperature fluctuations at locations A, B, and C were higher than those of other locations, and Location 3 was located at the high location of the mountain system, and the wind speed was relatively high (Figure 9).

(2) Humidity: Open space affects the sunlight, ventilation, and humidity of the measuring locations and is negatively correlated with the humidity of the measuring locations. It was found that the influence of sunshine was more significant than the influence of wind speed when the spatial opening and closing degrees are similar (Figure 10). The lush degree of vegetation at the measuring location was negatively correlated with the fluctuation of humidity. In particular, the humidity at each measuring location in summer was affected by rainfall. Among the measuring locations, Location 3 was located in the core area of the village, with dense surrounding buildings and plants, which is more conducive to the absorption of rainwater. The humidity rose slowly accordingly.

(3) Wind speed: Wind direction changes when the wind passes through valleys or streets. The wind direction of the experimental locations in the village is parallel to its spatial direction. The wind direction at the top of the mountain in open space is more diverse than the other parts of the mountain. In terms of wind speed, taking Locations 2 and 3 as examples, the altitude and underlying surface of the two locations were similar. However, due to the fact that Location 3 was located at the intersection of north–south and east–west valleys and there was no high mountain around it, but Location 2 was on the south side and sheltered by high mountains, and the openness of the space was low, the average wind speed of Location 2 in the third season was much lower than Location 3. The prevailing wind direction on the measuring day of spring in Shuiyu Village was from the north, and the wind speeds were highest at Locations B1, B2, and B6 on the north windward side, where the building density was small.

The wind speeds at the B83 measuring location were relatively low. Taking the winter measurement results as an example, the wind direction was stable, and the terrain affected the wind speed. The wind speeds were high at the open locations in the north–south direction. For example, the wind speeds at six measuring locations, such as 1, 3, 7, A, B, and C, were relatively higher than the other locations. These locations had at least one opening in the north–south direction, while the wind speeds at other locations were low (Figure 11) The wind speed at the measuring location on the top of the mountain was higher than the wind speed at Location 6 inside the Village. In addition, Location B was the most elevated location in summer and winter, and the wind speed was higher than Locations A and C.

Figure 9. (**a**): spring morning temperature curve of Shuiyu village; (**b**): spring morning temperature analysis diagram of Shuiyu village; (**c**): spring afternoon temperature curve of Shuiyu village; (**d**): spring mafternoon temperature analysis diagram of Shuiyu village; (**e**): Autumn temperature curve of Shuiyu village; (**f**): Autumn temperature analysis diagram of Shuiyu village; (**g**): Winter temperature curve of Shuiyu village; (**h**): Winter temperature analysis diagram of Shuiyu village.

Figure 10. (**a**): spring morning humidity curve of Shuiyu village; (**b**): spring afternoon humidity curve of Shuiyu village; (**c**): spring mafternoon humidity analysis diagram of Shuiyu village; (**d**): spring afternoon humidity analysis diagram of Shuiyu village; (**e**): Autumn humidity curve of Shuiyu village; (**f**): Autumn humidity analysis diagram of Shuiyu village; (**g**): Winter humidity curve of Shuiyu village; (**h**): Winter humidity analysis diagram of Shuiyu village.

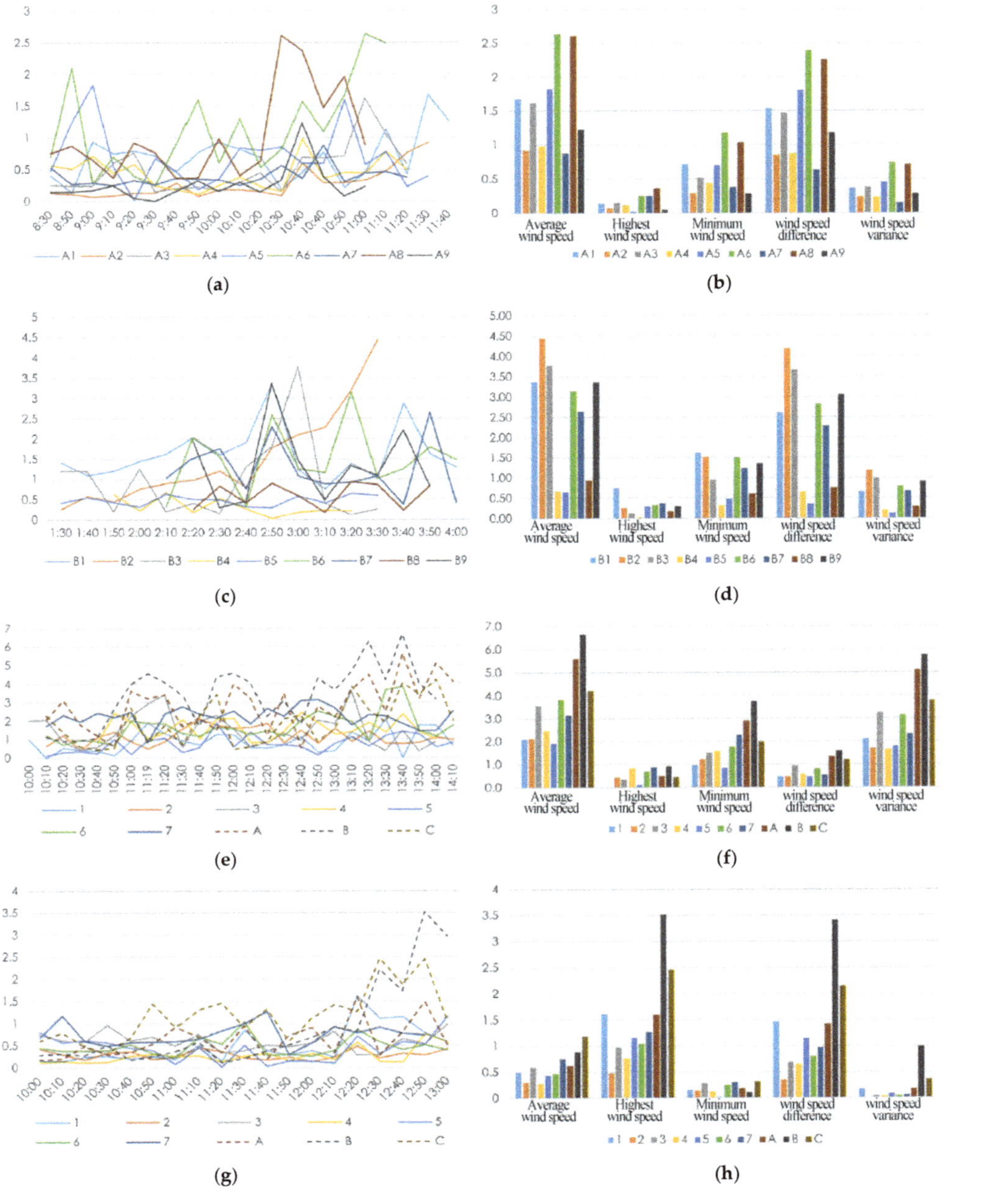

Figure 11. (**a**) spring morning wind speed curve of Shuiyu village; (**b**) spring morning wind speed analysis diagram of Shuiyu village; (**c**) spring afternoon wind speed curve of Shuiyu village; (**d**) spring mafternoon wind speed analysis diagram of Shuiyu village; (**e**) Autumn wind speed curve of Shuiyu village; (**f**) Autumn wind speed analysis diagram of Shuiyu village; (**g**) Winter wind speed curve of Shuiyu village; (**h**) Winter wind speed analysis diagram of Shuiyu village.

3.3. *Numerical Simulation Results*

3.3.1. Wind Environment Simulation

The overall wind environment of Shuiyu Village was characterized by high wind speed on the windward side and low wind speed on the leeward side. The East Village is located on the leeward side of the mountain, which is also in a small valley between the two mountains. The Nanshan and Beishan mountain ranges with large slopes pass through both sides of the village entrance. The peaks on the northwest side are barriers to the invasion of the northwest wind, and the south wind passes through the valley mouths from the east and west sides of the Nanshan Mountains. Xicun and Xincun are located in the north–south valley between the three mountains. The West Village was built on the mountains in the valley. The southeast opening of the valley is suitable for summer ventilation and is greatly affected by the valley wind. The mountains on the west side slow down the cold wind from the northwest. These regulate a microclimate environment in winter. However, because the valley is in the southeast direction, lower wind speeds appeared in some areas. Nevertheless, the wind speed at all locations in the east part of Shuiyu Village was balanced (between 2.2 and 2.6 m/s). According to the definition by the Beaufort index, it is within the range between a breeze (1.79 m/s) and a gentle breeze (3.58 m/s), the most suitable wind environment for human settlement. When the breeze entered the mountain valley from an open area, the cross-sectional area of the airflow decreased, and the airflow accelerated, thus forming a strong wind (Figures 11 and 12).

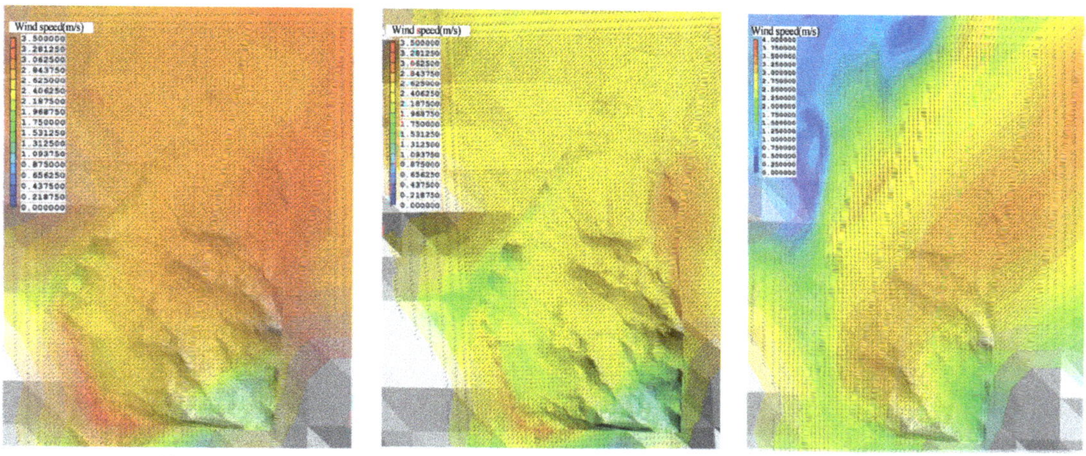

(a) (b) (c)

Figure 12. Simulation chart of the Spring (**a**), Summer (**b**), Winter (**c**) wind environment in Shuiyu Village.

3.3.2. Sunshine Environment Simulation

Locations A5 and A7 had the highest temperature because they were exposed to direct sunlight. In addition, the deeper the valley and the smaller the mouth are, the later the sunrise and the shorter the daily sunshine time are. For example, Location A4 was at the deepest part of the valley, the mountain was severely blocked, there was no sunshine before 9:30 am, and the temperature dropped significantly in the afternoon. The location of the mountain had a more significant impact on the sunshine of the area, while the sunlight had a critical influence on the temperature in the site (Figure 13).

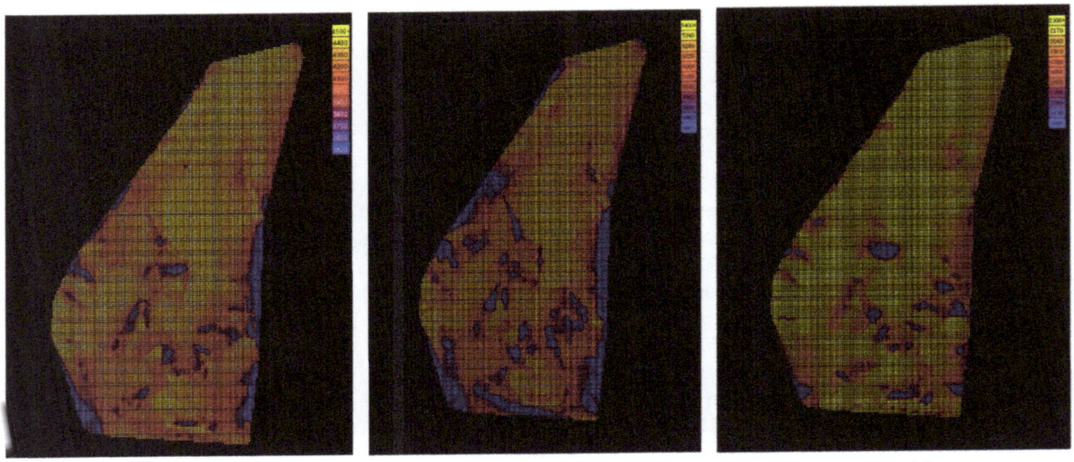

(a) (b) (c)

Figure 13. Thermal simulation chart of Spring (**a**), Summer (**b**), Winter (**c**) sunshine in Shuiyu Village.

The West Village and New Village are located north–south, and the mountains are on the east and west of the settlement, respectively. The closer they are to the bottom of the valley, the shorter the sunshine time they have. The mountains on the west side of the new village were relatively high, and it affected the sunshine environment of the west village in the afternoon.

3.4. Results on the Correlation between Landscape Pattern and Microclimate

Based on the spatial opening and closing degree and the influence degree of the mountain's south side, data from the field measurement location data were collected as the south opening degree and the dominant wind direction opening degree, using multiple linear regression of panel data. The *p*-value determines the significance of the model; that is, whether the landscape pattern has a significant impact on the microclimate environment, and the influence of each index was determined by the estimated parameters (Table 3).

The degree of space opening and closing had a greater impact on summer temperature, humidity, wind speed, and spring temperature. In particular, the microclimate factors in summer were significantly affected by the opening and closing of the space. Among them, the summer temperature was positively correlated with the influence degree of the south side of the mountains. The less sheltered by the mountains on the south side, the higher the temperature. It was negatively correlated with the opening and closing of the dominant wind direction. It was also found that the south side of the mountain had a strong influence on the wind speed in summer. The humidity in summer was positively correlated with the spatial opening and closing degree of the dominant wind direction and negatively correlated with the influence degree of influence of the south side of the mountains. The temperature in spring was not significantly affected by the southward mountain occlusion and was positively correlated with the spatial opening and closing of the dominant wind direction. It was hence affected by the dominant wind direction. Using data collation, the analysis of the landscape pattern was verified.

3.5. Expression of Comfort Degree in Shuiyu Village Based on Microclimate Adaptability

The modeling platform was developed by analyzing the relationship between the microclimate and the landscape pattern parameters. Moreover, the comfort grading standard was also included. The village landscape pattern information was used to collect from the locations in the village to develop the human body comfort map. The comfort degree of Shuiyu Village was found to be great in spring, summer, and winter. In the degree map (shown in Figure 14), values were used to distinguish between colors in spring. Moreover, the comfort evaluation standards were used for selection in winter and summer. The map provided guidance to the adaptive design for the microclimate of a local area.

Table 3. Regression models between microclimate and the comfort index.

Dependent Variable		Independent Variable	Regression Equation	R^2	*p*-Value
the comfort index (W)	WBGT (summer) TS-Givoni (winter) THI (spring)	Southward openness = *x* dominant wind direction openness = *y* Southward openness = *x* dominant wind direction openness = *y* Southward openness = *x* dominant wind direction openness = *y*	W = 7.1516X + 33.4031 W = −1.0072Y + 36.2758 W = −6.7148X + 4.6134 W = 0.5019Y + 2.1881 W = −0.3009X + 16.1620 W = 2.5765Y + 15.1103	0.18 0.01 0.15 0.12 0.01 0.08	<0.001 0.718 0.142 0.825 0.706 0.017 <0.1
Temperature (t)	summer winters pring	Southward openness = *x* dominant wind direction openness = *y*	t = 15.5769X − 9.3404Y + 28.2636 t = −88.2864X + 27.1348Y + 24.86 t = −0.8629X + 4.5002Y + 15.8701	0.25 0.13 0.08	<0.05 0.1834 <0.1
Humidity (a)	summer winter spring	Southward openness = *x* dominant wind direction openness = *y*	a = −46.5144X + 22.0111Y + 73.1636 a = 54.0313X + 12.8910Y + 5.9255 a = 3.1929X + 0.0702Y + 22.1475	0.27 0.12 0.15	<0.05 0.4820 0.3583
Wind speed (v)	summer winter spring	Southward openness = *x* dominant wind direction openness = *y*	v = 6.6988X + 0.0753Y − 0.8955 v = −0.7197X − 0.3989Y + 1.1853 v = 0.4590X + 2.5161Y−0.3912	0.37 0.08 0.05	<0.05 0.6910 0.49
Combined		WBGT (summer) TS-Givoni (winter) THI (spring)	W = 9.5550X − 4.0073Y + 33.9075 W = −10.0626X − 3.0953Y + 4.7461 W = −0.4472X + 2.6499Y + 15.2445	0.22 0.13 0.09	<0.001 0.1998 0.0547 <0.1

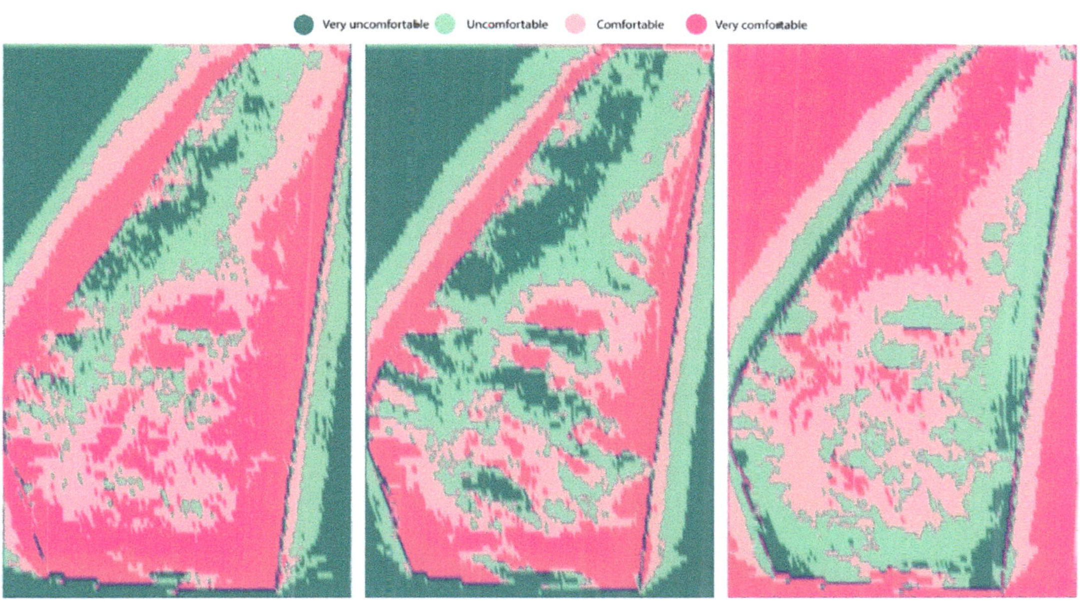

(a) (b) (c)

Figure 14. Shuiyu Village comfort map in Autumn (**a**), Winter (**b**), Spring (**c**).

4. Discussion

4.1. Analysis and Discussion

Four conclusions were drawn by using the actual observation measurements, numerical simulation, and discussion of the correlation between microclimate and landscape environment.

(1) It was found that the landscape pattern had a strong relationship with the wind environment simulation results at various locations. The wind was easy to form in the narrow valley area. The wind speed was slow on the leeward of the mountain, but vortex wind might form (e.g., at Location B7).

(2) By simulating and analyzing the heat gain from radiation, it was found that the orientation of the slope of the mountain has a great influence on the heat radiation of the site.

(3) This study found that the environment of Shuiyu Village was suitable for living in spring, summer, and winter, using the data of the sunshine and numerical wind simulation. The simulation indicators of the east part of the village (the ancient village area) show more comfort than the west part. The wind speed in the east was relatively low in spring and winter but high in summer. The intensity of sunshine in the east part of the village was found average in spring, while in summer it was found that more areas were shaded.

(4) The comprehensive analysis of the results of multiple linear regression shows that summer and winter landscape patterns had similar effects on the microclimate comfort when the mountain shelter was considered. Moreover, the southward opening and closing degree and the dominant wind direction opening and closing degree were found correlated to each other. With human comfort in mind, when the summer comfort classification standard was taken into consideration, the greater the opening and closing degree of the dominant wind direction was, the greater the required south opening and closing degree were. The results indicated that Shuiyu Village is a valley type in landscape patterns. When this landscape pattern is open to the south, the dominant wind direction is low (three-sided valley type). Only when the dominant wind direction was high and southward was low (two-sided valleys or four-sided valleys) was it not suitable as a residential location, and the human body comfort was low. When the opening and closing degree of the southward mountain was large, the opening and closing degree of the dominant wind direction would, consequently, be large (such as flat open type or slope-dependent landscape pattern). Then the degree of the southward opening and closing and the magnitude of the dominant wind direction wind speed needed to be considered. When the large mountain environment meets the human comfort index, a reasonable layout of vegetation and buildings can be appropriately used to adjust the human comfort and form a pleasant small site. It was also found that most of the types of landscape patterns met the standards of human comfort. This finding is aligned with the spring environment in North China, when the temperature is suitable for human comfort.

In summary, the microclimate characteristics of different landscape pattern areas in traditional villages were initially affected by the climate and environmental characteristics of the large area; that is, the spatial correlation of the climate environment. North China has a typical temperate monsoon climate with hot and dry summers. In the cold winter, when the environmental characteristics of large areas were similar, the influence of the traditional village landscape pattern was focused. When the general landscape pattern and site selection met the comfort index, the influence of the layout and suitable constructions of buildings were paid attention to, which created the residential environment of the village. Although this model was developed based on the field measuring data of Shuiyu Village and collected the quantitative indicators of the landscape pattern, it provided a certain level of guidance in selecting suitable settlements and the construction and layout of human settlements in the village. Its specific coefficient indicators showed it was not applicable to other villages in non-Beijing–Tianjin–Hebei regions. However, through the verification process of the validity of the model results, the approximate parameterization relationship in the method can be generalized to a certain extent.

Through the algorithmic coupling of "number" (comprehensive microclimate index) and "shape" (optimization of landscape patterns, such as spatial opening and closing), this study found the landscape pattern characteristics of traditional villages (such as the shape of the terrain, the slope of the mountain, the shelter of plants, and the structure of the building) were closely related to local microclimate related factors (such as temperature, humidity, and solar radiation).

This study has some limitations. This study has measured only a number of field locations. This has a limited application to a broader representation of the village landscape patterns and their characterized microclimates for village planning. Moreover, it was found that the impact of some village layouts on microclimate environmental indicators was complex and non-linear. The model was a simplification of the real natural system under certain experimental conditions. In future research, the data richness of the measured database will be increased, with various types of typical landscape pattern locations and different graphic databases of traditional village landscape patterns in different macroclimate environments. Further, different types of microclimate indicator calculation equations will be used to improve the model and collect the parametric coupling state of landscape pattern and microclimate indicators under ideal conditions.

4.2. Preliminary Study on Parametric Aided Design Process Based on Climate Adaptability

The study of the parametric relationship between the landscape pattern of traditional villages and microclimate indicators was conducive to digitizing the experience of the original planners in practice. The wisdom of the ancient ancestors and the experience of the planners, such as "Bearing the Yin and Embracing the Yang" and "backing the mountains and facing the water", has been transformed into scientific guidance. The combination of the obtained model and the modeling platform has a broader application prospect, enabling planners to conduct comprehensive research from two dimensions to three dimensions. This paper investigated the applicability and practicability of the model and discussed the application scenarios of the parametric model.

This paper adopted the comfort map expression algorithm combined with the digital model of the site. It was conducted by using the map expression method, guiding the adaptive design of local microclimates, and combining relevant microclimate design experience to formulate optimization strategies and save planning efficiency. This model can be extended in the actual planning of the village. A three-dimensional site model of the village has been used to obtain data from the human body comfort, temperature, wind speed, and slope of the village site. The purpose was to achieve clear, descriptive, and testable primary conditions for the site. It was also aligned with design requirements. Appropriate parameter conditions were set, using the Rhino modeling platform and Grasshopper to iteratively calculate and establish a combination map of morphological relationships to meet the design requirements. A parametric simulation-aided design site selection process was used (Figure 15). The specific process is as follows:

(1) The construction site is investigated to select representative check locations. Then, the microclimate index information is used to generate a digital model of the landscape pattern.

(2) The factors and generated site analytical models are used to analyze the landscape pattern characteristics of the site from the perspectives of elevation, slope, aspect, water environment, and microclimate environment.

(3) According to the construction needs and related conditions (e.g., slope $\leq 10°$, the southeast is relatively open, the wind environment is between the breeze and gentle, and winter and summer conditions are comfortable), the parameters influencing conditions are adjusted to comprehensively analyze factors in the system performance sites. Multiple cycles of adjustment of the parameters are needed to generate and calculate the recommended site selection. It provides an optimized auxiliary design strategy for future detailed traditional village protection and transformation design.

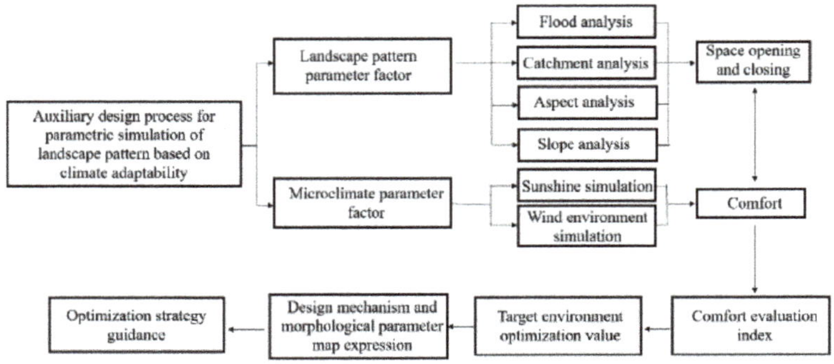

Figure 15. Parameterized model construction of traditional village landscape pattern design framework.

5. Conclusions

To sum up, this paper developed a parametric model of landscape pattern and discussed the feasibility from the two dimensions of model results and research methods. This model was established based upon the previous case studies of traditional villages and observation measurement and simulated data analysis in Shuiyu Village. Moreover, the algorithmic coupling has been conducted with the parameterized landscape pattern factor morphology generation and the climate adaptability mechanism.

This paper has also made the following contributions. First, it combines the general human comfort evaluation standards and uses Grasshopper to map hierarchical expression. Secondly, it compiles the logic construction process of village comfort evaluation through Grasshopper and R language programming design software. By merging the algorithms, it also tried to build an interactive platform interface schematic and summarize the parametric auxiliary design process in the village planning and site reconstruction. Additionally, it demonstrates the parametric simulation design process from setting parameter conditions to align with the design requirements to evaluate the parametric sites, and finally, generate site recommendations.

The research method used belongs to the "black box" theoretical research; that is, to trace the source from the results. It is a verification study under certain experimental conditions on the ideal landscape pattern of traditional villages. It simulates and quantifies perceptual cognition and inherits traditional villages through quantitative methods. The ecological system provides a scientific underpinning for the improvement of its human settlement environment based upon inheriting and carrying forward traditional culture and greatly improves the efficiency of planning to meet a number of rural planning needs. It also provides a scientific basis for the development of a rural revitalization strategy. The model's applicability has certain limitations due to the limited number of measurement locations. However, the model can be used and applied to any regional characteristics and landscape patterns similar to those of Shuiyu Village.

Although the parametric design strategy was extended based on this model quantifies the "hidden" relationship between the microclimate environment and the landscape pattern through parametric assisted design, it is only limited to the single-line logic of "parameter A→evaluation→screening" [37]. It can be used by designers as an auxiliary technical reference to improve design efficiency. The planning and design of landscape architecture under the human settlement environment science is complicated but systematic. The key is to coordinate and work with the relationship between human beings and nature. Our values determine the position and design process, which consequently determines our design attitude and influences our design method. This ultimately affects the design form and content, designer's values, and ability to make judgments. Our respect for the site and comprehensive judgment based on "adapting to the situation" and "adapting measures to local conditions" should be critical to our design. Thus, the interactions among the various parameter factors will be investigated in further research.

Author Contributions: Conceptualization, L.Q.; Data curation, L.Q., R.L. and Y.C.; Formal analysis, L.Q., R.L. and Y.C.; Funding acquisition, L.Q., M.Z., W.B. and Z.S.; Investigation, L.Q., R.L., Y.C., M.Z. and Z.S.; Methodology, L.Q. and R.L.; Project administration, L.Q., M.Z., W.B. and Z.S.; Resources, L.Q., R.L. and Y.C.; Software, R.L. and Y.C.; Supervision, L.Q., M.Z. and W.B.; Validation, R.L. and Y.C.; Visualization, Y.C.; Writing—original draft, L.Q., R.L.; Writing—review & editing, L.Q. and R.L. All authors have read and agreed to the published version of the manuscript.

Funding: This research was sponsored by the International Research Cooperation Talent Introduction and Cultivation Project of Beijing University of Technology(No. 2021C10), the Youth Program of National Natural Science Foundation of China on the project "Based on microclimatological adaptive design of the Beijing-Tianjin-Hebei traditional village landscape pattern research" (No. 51608012), Beijing Municipal Social Science Foundation project (key project) "Building information model technology in the Beijing core area of public facilities design collaborative optimization research" (No. 19YTB050), the National Natural Science Foundation of China on the project "Urban site not permeable form of hydro-ecological effects and its updated design research" (No. 52078006).

Institutional Review Board Statement: Not applicable.

Informed Consent Statement: Not applicable.

Data Availability Statement: Data is contained within the article.

Conflicts of Interest: The authors declare no conflict of interest.

References

1. Skinner, S. *The Living Earth Manual of Feng-Shui: Chinese Geomancy*; Graham Brash Party, Limited (SI): Singapore, 1983.
2. Liu, H.; Liao, B. Summary of researches on the landscape pattern of rural settlements at home and abroad. *Mod. Urban Res.* **2014**, *11*, 30–35+74. (In Chinese)
3. Yi, C.Y.; Peng, C. Microclimate change outdoor and indoor coupled simulation for passive building adaptation design. *Procedia Comput. Sci.* **2014**, *32*, 691–698. [CrossRef]
4. Zhang, D. Research in Space Analysis and Practice of Beijing Ancient Villages. Ph.D. Thesis, Tianjin University, Tianjin, China, 2014.
5. Brown, R.D.; Gillespie, T.J. *Microclimatic Landscape Design: Creating Thermal Comfort and Energy Efficiency*; Wiley: Hoboken, NJ, USA, 1995.
6. Noguchi, M.; Givoni, B. Outdoor comfort as a factor in sustainable towns. In Proceedings of the Second International Conference for Teachers in Architecture, Florence, Italy, 16–18 October 1997.
7. Givoni, B.; Noguchi, M.; Saaroni, H.; Pochter, O.; Yaacov, Y.; Feller, N.; Becker, S. Outdoor comfort research issues. *Energy Build.* **2003**, *35*, 77–86. [CrossRef]
8. Roth, M. Review of urban climate research in (sub) tropical regions. *Int. J. Climatol.* **2007**, *27*, 1859–1873. [CrossRef]
9. Stewart, I.D. A systematic review and scientific critique of methodology in modern urban heat island literature. *Int. J. Climatol.* **2011**, *31*, 200–217. [CrossRef]
10. Xue, S.; Liu, B. Analyses on microclimatic factors and human thermal comfort of different types of greenbelt and non-greenbelt in riparian zone of Suzhou River in Shanghai City in summer. *J. Plant Resour. Environ.* **2018**, *27*, 108–116.
11. Liu, B.; Peng, X. The process and Enlightenment of Research on Microclimate Comfort in Urban Streets. *Chin. Landsc. Archit.* **2019**, *35*, 57–62.
12. Wei, D.; Liu, B. The Analysis and evaluation of Thermal comfort at Shanghai Knowledge & Innovation Community Square. *Chin. Landsc. Archit.* **2018**, *34*, 5–12.
13. Klemm, W.; Lenzholzer, S.; van den Brink, A. Developing green infrastructure design guidelines for urban climate adaptation. *J. Landsc. Archit.* **2017**, *12*, 60–71. [CrossRef]
14. Fu, F.; Zhao, C.; Sun, Y.; Liu, D.; Li, Y. Study on the Relationship between the Temperature, Humidity and the Concentration of Fine Particulate Matter (PM2.5) in Green spaces in Autumn: A Case Study of Beijing Yuyuantan Park. *Chin. Landsc. Archit.* **2018**, *34*, 36–40.
15. Dai, F.; Chen, M.; Wang, M.; Zhu, S.; Fu, F. Effect of Urban Block Form on Reducing Particulate Matter: A case study of Wuhan. *Chin. Landsc. Archit.* **2020**, *36*, 109–114.
16. Feng, X.; Chu, Y. Research on Local Microclimate Effect of Urban Green Space Based on Aerodynamic Simulation. *Chin. Landsc. Archit.* **2017**, *33*, 29–34.
17. Lai, H.; Feng, X. Correlation Study of Flora and Microclimate effects based on canopy shade and canopy Enclosure: The Case of field observation of Guangzhou Academic of Forestry. *Urban. Archit.* **2018**, *33*, 98–102.

18. Rusinga, O.; Chapungu, L.; Moyo, P.; Stigter, K. Perceptions of climate change and adaptation to microclimate change and variability among smallholder farmers in Mhakwe communal area, Manicaland province, Zimbabwe. *Ethiop. J. Environ. Stud. Manag.* **2014**, *7*, 310–318. [CrossRef]
19. Yang, X.; Jin, H.; Peng, H.; Chen, C. Research on Adaptive Design Strategies of Hangzhou Streets Based on Summer Microclimate Effects. *Landsc. Archit.* **2019**, *26*, 100–104.
20. Zhuang, X.; Duan, Y.; Jin, H. Research Review on urban landscape Micro-climate. *Chin. Landsc. Archit.* **2017**, *33*, 23–28.
21. Vecellio, D.J.; Bardenhagen, E.K.; Lerman, B.; Brown, R.D. The role of outdoor microclimatic features at long-term care facilities in advancing the health of its residents: An integrative review and future strategies. *Environ. Res.* **2021**, *201*, 111583. [CrossRef]
22. Blank, D.A. Using microclimate of arid landscape as a resource in goitered gazelle comfort behavior. *J. Arid Environ.* **2020**, *180*, 104201. [CrossRef]
23. Cheng, H.; You, D.; Liu, Q.; Li, X.; Lan, S. A study on the Relationship between Microclimate and spatialselection of Tour Path in the Ancient Town of Songkou. *Chin. Landsc. Archit.* **2020**, *36*, 83–88.
24. Du, C.; Lin, L. Analysis of Microclimatic Environment Characteristics of Yi traditional Settlements in Yunnan Province. *Chin. Landsc. Archit.* **2020**, *36*, 43–48.
25. El-Bardisy, W.M.; Fahmy, M.; El-Gohary, G.F. Climatic Sensitive Landscape Design: Towards a Better Microclimate through Plantation in Public Schools, Cairo, Egypt. *Procedia Soc. Behav. Sci.* **2016**, *216*, 206–216. [CrossRef]
26. Lin, J.; Brown, R.D. Integrating Microclimate into Landscape Architecture for Outdoor Thermal Comfort: A Systematic Review. *Land* **2021**, *10*, 196. [CrossRef]
27. Chen, J.; Saunders, S.C.; Crow, T.R.; Naiman, R.J.; Brosofske, K.D.; Mroz, G.D.; Brookshire, B.L.; Franklin, J.F. Microclimate in Forest Ecosystem and Landscape Ecology: Variations in local climate can be used to monitor and compare the effects of different management regimes. *BioScience* **1999**, *49*, 288–297. [CrossRef]
28. Schettini, E.; Blanco, I.; Campiotti, C.A.; Bibbiani, C.; Fantozzi, F.; Vox, G. Green Control of Microclimate in Buildings. *Agric. Agric. Sci. Procedia* **2016**, *8*, 576–582. [CrossRef]
29. Ragheb, A.A.; El-Darwish, I.I.; Ahmed, S. Microclimate and human comfort considerations in planning a historic urban quarter. *Int. J. Sustain. Built Environ.* **2016**, *5*, 156–167. [CrossRef]
30. Qi, L.; Ma, Z.; Guo, Y.; Liu, J.; Song, Z. Analysis of landscape pattern of Xijingyu Village, Jizhou District, Tianjin City based on microclimate adaptability design. *Chin. Gard.* **2018**, *266*, 34–41.
31. Qi, L.; Ma, Z.; Zhang, Y.; Liu, J. Study on the landscape pattern of Shuiyu Village, Jiao Township, Southwest Beijing based on microclimate adaptability design. *Landsc. Archit.* **2018**, *159*, 38–44.
32. Yuan, Y.; Cheng, Y. Process, logic and Model: Research on Parametric Landscape Architecture planning and design. *Chin. Landsc. Archit.* **2018**, *34*, 77–82.
33. Chen, R.; Dong, L. Brief introduction and enlightenment of foreign research on microclimate comfort. *Chin. Gard.* **2009**, *11*, 81–83. (In Chinese)
34. Manos, N.E. Discomfort Index. *Science* **1959**, *130*, 1624. [CrossRef]
35. Cai, J. The Concept Hierarchy and Procedure Diagram of Digital Planning and Design for Landscape Architecture. *Landsc. Archit.* **2013**, *20*, 48–57.
36. Cai, J. Discussing on the Digital Strategies of Landscape Architecture Planning and Design. *Chin. Landsc. Archit.* **2012**, *28*, 14–19.
37. Kuang, W. Overview and Reflection on the Development of Parametric Landscape Planning and Design. *Landsc. Archit.* **2013**, *2*, 58–64.

sustainability

MDPI

Article

Developing the Urban Blue-Green Infrastructure as a Tool for Urban Air Quality Management

Joanna Badach [1,*], Jakub Szczepański [2], Wojciech Bonenberg [3], Jacek Gębicki [4] and Lucyna Nyka [1]

1 Department of Urban Architecture and Waterscapes, Faculty of Architecture, Gdańsk University of Technology, 11/12 Narutowicza Street, 80-233 Gdańsk, Poland
2 Department of History, Theory of Architecture and Conservation of Monuments, Faculty of Architecture, Gdańsk University of Technology, 11/12 Narutowicza Street, 80-233 Gdańsk, Poland
3 Institute of Architecture and Spatial Planning, Faculty of Architecture, Poznań University of Technology, 2 Jacka Rychlewskiego Street, 61-131 Poznań, Poland
4 Department of Process Engineering and Chemical Technology, Faculty of Chemistry, Gdańsk University of Technology, 11/12 Narutowicza Street, 80-233 Gdańsk, Poland
* Correspondence: joanna.badach@pg.edu.pl

Abstract: Urban structure is an important factor that shapes the process of urban ventilation and pollution dispersion. With proper planning of the urban spatial layout, city breathability can be effectively regulated, contributing to urban air quality improvement. This paper investigates the development and current management of urban systems of green and open spaces in four Polish cities: Gdańsk, Warsaw, Poznań and Wrocław, with a particular focus on the planning aspects of urban ventilation and air quality management. The initial GIS-based comparison of historical plans and the current spatial layouts of the cities show that these systems, consciously shaped at the beginning of the twentieth century, remain clearly identifiable. However, in some locations, the continuance of these systems was interrupted by later investments. The next step was to develop GIS procedures to effectively map the spatial distribution of selected urban form indicators that are related to urban ventilation, especially the frontal area index. The results made it possible to determine the main features of the current ventilation systems and to identify some of the local problem areas. The last phase of the study was to conduct a local-scale analysis of these problem areas. With this study, the applicability of various analysis and simulation tools for the purpose of improving city breathability by appropriate integrated planning and design decisions was demonstrated. The presented approach, taking into account the city- and micro-scale interactions, should be used in current planning practice to preserve the historically developed ventilation systems.

Keywords: city breathability; urban ventilation; urban air quality management; blue-green infrastructure; integrated urban planning; sustainable development

Citation: Badach, J.; Szczepański, J.; Bonenberg, W.; Gębicki, J.; Nyka, L. Developing the Urban Blue-Green Infrastructure as a Tool for Urban Air Quality Management. *Sustainability* **2022**, *14*, 9688. https://doi.org/10.3390/su14159688

Academic Editor: Elena Cristina Rada

Received: 30 June 2022
Accepted: 31 July 2022
Published: 6 August 2022

Publisher's Note: MDPI stays neutral with regard to jurisdictional claims in published maps and institutional affiliations.

1. Introduction

It is estimated that over four million people worldwide die prematurely every year as a result of exposure to outdoor air pollution. Almost all the global population breathes air that does not meet the World Health Organization (WHO) standards [1]. Being listed among the 10 major threats to global health, air pollution and the strategies for its mitigation are high on the worldwide agenda [1,2]. Although in Europe, air pollution emissions have been reduced in the past several years due to the implementation of many policies, such as the development of sustainable transport systems, a significant proportion of European residents still live in places where the pollutant levels are above the set standards [3]. In Poland, environmental and air quality management is particularly ineffective in comparison to many European Union (EU) countries [3,4].

The process of urban ventilation affects air pollution dispersion and, thus, contributes to air quality improvement. Urban ventilation efficiency, in turn, is significantly affected by urban structure [5–7]. This aspect should be comprehensively integrated into urban planning [8–11]. The consideration of urban airflow in relation to the spatial characteristics of urban form should follow the integration of different urban scales—see Figure 1. At On the scale of the entire city, or even on a regional scale, urban ventilation corridors should be designated and preserved to improve air exchange between the polluted city centre and the surrounding areas [6,12–16]. On the neighbourhood scale, the appropriate layout of the urban blocks can ensure the development of secondary ventilation pathways, ideally being connected with the main ventilation corridors [9,10,17]. Moreover, not only the 2D spatial configuration but also the three-dimensional structure should be considered in relation to effective airflow within the urban built environment [18–23]. Finally, on a local scale, the orientation of buildings and open spaces, their configurations and the typology of building complexes in relation to the local wind environment can be also analysed [24–29]. Such a comprehensive approach toward urban air quality and ventilation management is consistent with the recommended practice for scale-sensitive integration in urban planning [30,31] and environmental management [32,33].

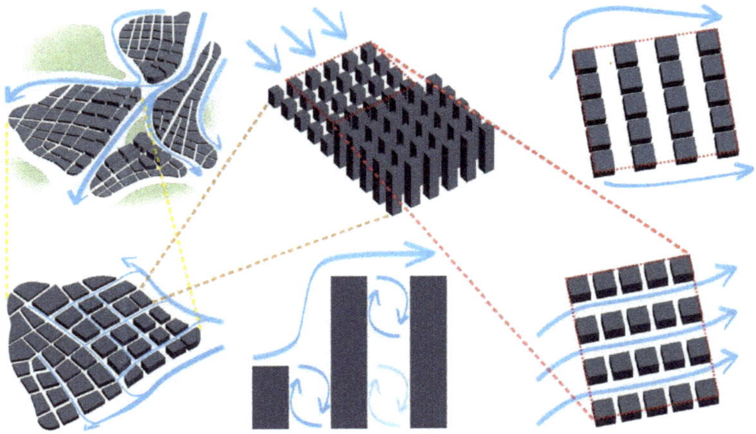

Figure 1. Impact of urban form on urban ventilation at various levels of urban planning and design (authors' own elaboration, based on the referenced literature).

Improving urban ventilation efficiency contributes not only to air quality improvement but also mitigates the urban heat island (UHI) effect. The UHI is the effect of an increase in ground-level atmospheric temperature within a city, relative to the air temperature outside the city. It was first identified at the beginning of the nineteenth century in London and was subsequently reported in other industrial cities [34]. Depending on the structure of the city, the heat island takes different spatial forms. The general pattern is an increase in temperature from suburban areas to dense urban centres, where temperatures reach a maximum. The UHI is the result of changes in the land use of urbanised areas. Such changes distort the natural state of urban climates, making them much warmer than rural climates. As studies show, the UHI effect intensifies under conditions of low wind speed [35]. This is confirmed by meteorological observations that indicate that urban climates are warming due to disturbed airflow mechanisms, as a result of tall and densely situated buildings [36]. These observations reveal that above certain wind speeds, the UHI is significantly reduced. This situation occurs in well-ventilated areas with low building heights [37].

Inquiries into how urban planning may contribute to healthier cities and improve urban air quality began to appear in the nineteenth century. The idea was that certain designated areas should remain natural and allow unobstructed airflow—providing clean air and blowing pollutants away. Such concepts as John Claudius Loudon's plans for green belts for London, the idea of the Garden City developed by Ebenezer Howard, waterside parks by Frederick Law Olmsted and many other schemes were developed in response to the poor quality of life in industrial cities. Today, the pursuit of making cities more sustainable, healthier, more closely related to nature and resilient to climate change has triggered diversified research studies on the policies and instruments that are necessary to control urban transformations [38–41]. The availability of advanced digital tools and complex databases opens new paths toward developing innovative methods of investigation into the effectiveness of urban ventilation.

The planning aspects of the development and maintenance of urban ventilation systems in four Polish cities (Gdańsk, Poznań, Warsaw and Wrocław) were addressed in this research. The aim was to trace how the current systems of green and open spaces in these cities were shaped by their historical spatial development. Geographic information system (GIS) tools were then used to assess the degree of preservation and to identify the current features of the ventilation systems in these cities. Mapped urban form indicators were also used to identify local problem areas, in which alternative design solutions should be applied to avoid inhibiting the ventilation process. Finally, it was demonstrated that the sensitivity of GIS-based analysis on the city scale is not sufficient to identify all the spatial issues related to reduced ventilation efficiency. On the contrary, after a preliminary GIS analysis is completed, further investigation of the local urban layout is required. Moreover, computational fluid dynamics (CFD) simulations should be used on the local scale to fully understand how urban planning and design decisions can impact the efficiency of the ventilation process at the local level.

2. Materials and Methods

2.1. Study Areas

Four municipalities were selected for the study: Gdańsk, Warsaw, Poznań and Wrocław. Gdańsk (54.35° N, 18.65° E), a city on the Baltic coast and the largest in the Pomeranian Voivodeship, incorporates the largest Polish seaport. Poznań (52.41° N, 16.92° E), a city on the Warta River, in west-central Poland, is the fifth-largest city in Poland, in terms of population. Warsaw (52.23° N, 21.01° E), located on the Vistula River, is the capital city and the largest city in Poland. Wrocław (51.11° N, 17.04° E), a city in southwestern Poland on the Odra River, is the fourth most populated city in Poland.

The cities are largely of comparable size: Gdańsk has an area of 262 square kilometres and a population of 470,621; Poznań has an area of 262 square kilometres and a population of 529,410; Warsaw has an area of 517 square kilometres and a population of 1,795,569; Wrocław has an area of 293 square kilometres and a population of 642,687 [42,43]. However, they exhibit heterogenic parameters in terms of topography, urban morphology, the distribution of built-up structures, land use, and vegetation systems. In particular, these cities have distinctive ventilation systems. These systems began to be consciously shaped at the beginning of the twentiethcentury.

In Gdańsk, the first modern urban plan covering the entire city was not developed until 1929–1930. It was the Development Plan of Greater Gdańsk (German: Bebauungsplan Groß-Danzig)—see Figure 2. Its authors were Hugo Althoff, Karl Fehlhaber, and Karl Hell [44]. Part of this plan was the green system plan (Grünflächenplan). The essence of the solution was the creation of two green belts—forests on hills, with post-fortification areas and seaside parks. These belts were to be connected by wedges of greenery stretching along the streams. In the plans developed in the 1930s, the green system project proposed by Althoff was partially continued; however, some original assumptions were abandoned [45].

In 1916, a group of architects, led by Tadeusz Tołwiński, created the "Sketch of the preliminary regulatory plan of Warsaw" [46]. Green wedges appeared in this project, which were to perform hygienic functions and at the same time constitute the framework for the future development of the city as a radial structure [47]. In 1931, the "General Plan of Greater Warsaw for 3 million inhabitants" made by Stanisław Różański's group was finally approved. Work on this document took several years, but the basic assumptions were published as early as 1928 [48]. In the "General Plan", the concept of green wedges was developed. These consist of areas covering four categories: the Vistula valley, parks and sports grounds, cemeteries and "reserves", i.e., garden and agricultural areas where building is prohibited [49]—see Figure 3. Green wedges connect to the green insulation strip surrounding the entire city. The authors of the plan emphasised the particular importance of the river valley for the ventilation of the city. They also pointed out the treatment of cemeteries as an essential element of the ecosystem, which is consistent with contemporary research [50].

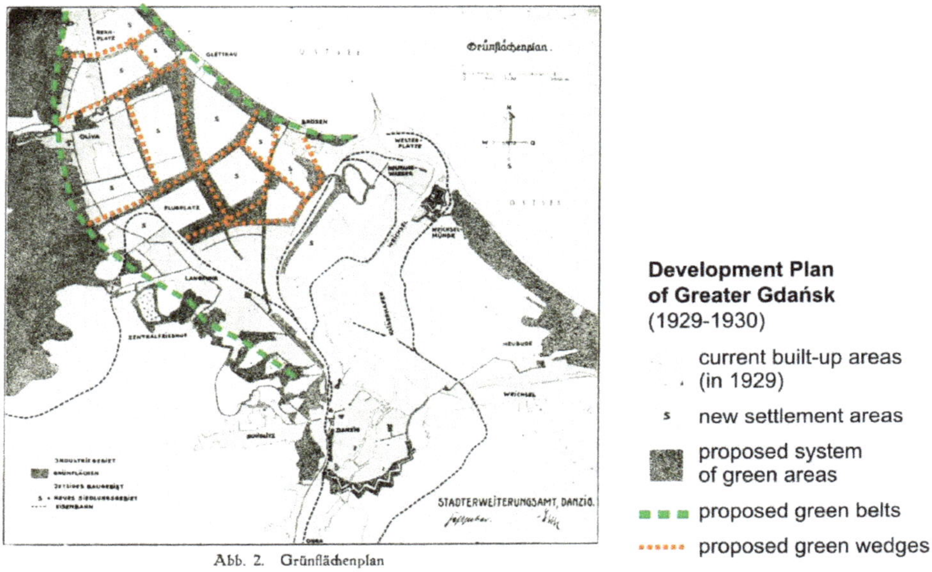

Figure 2. The system of green areas proposed in the Development Plan of Greater Gdańsk [44].

In 1903, Josef Stübben prepared the first of a series of detailed development and zoning plans for Poznań. These plans were modified and supplemented over the next 10 years. In Stübben's plans, the green system was formed by three half-rings surrounding the city centre , supplemented by radial bands along the river valleys. Stübben's plans in the 1930s were further developed by Władysław Czarnecki and his Town Planning Studio of the City of Poznań [51]. A consistent, coherent radial-radial green system was created, and the overall green area in Poznań increased significantly during the 20th century [52]. Initially, 10 wedges were proposed by Czarnecki in his plan from 1932—as depicted in Figure 4—but their number was reduced to 4 main wedges in subsequent municipal plans [53].

From 1921 to 1922, an urban competition was held in Wrocław, the aim of which was to re-organise the existing city and create a spatial framework for future development. The winner of the competition was the architect and town planner, Ernst May, acting in cooperation with Herbert Boehm [54]. May believed that the previously used concentric city systems and the more modern radial city system with green wedges were not appropriate solutions to the problem of providing access to greenery. The architect proposed a project called Trabanten ("satellites" in German)—see Figure 5. The city was to be decentralised, meaning that the urbanised central city zone would be surrounded by self-sufficient satellite housing estates. The boundaries between the estates would be fixed using urban greenery [55,56].

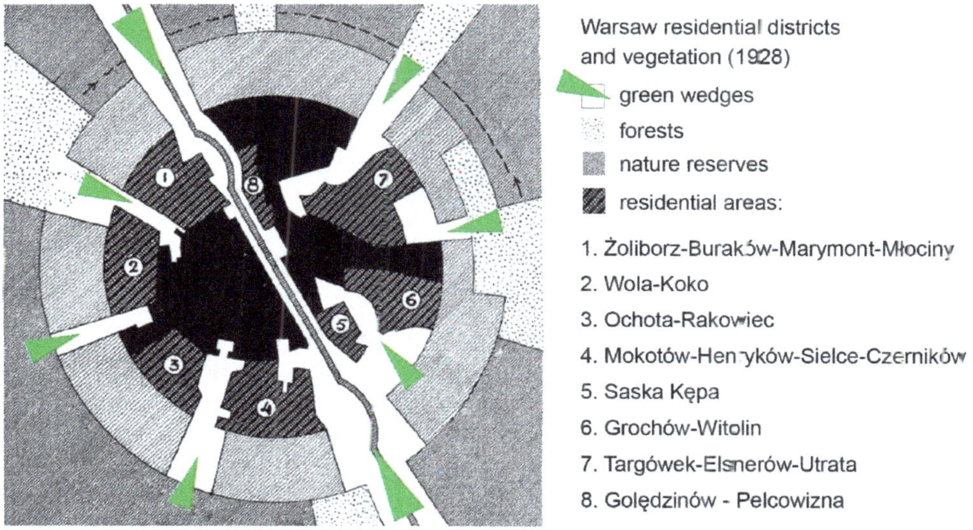

Figure 3. Scheme of the green wedges in relation to the residential districts in the city of Warsaw, as presented by Maria Buckiewiczówna (1928) [49].

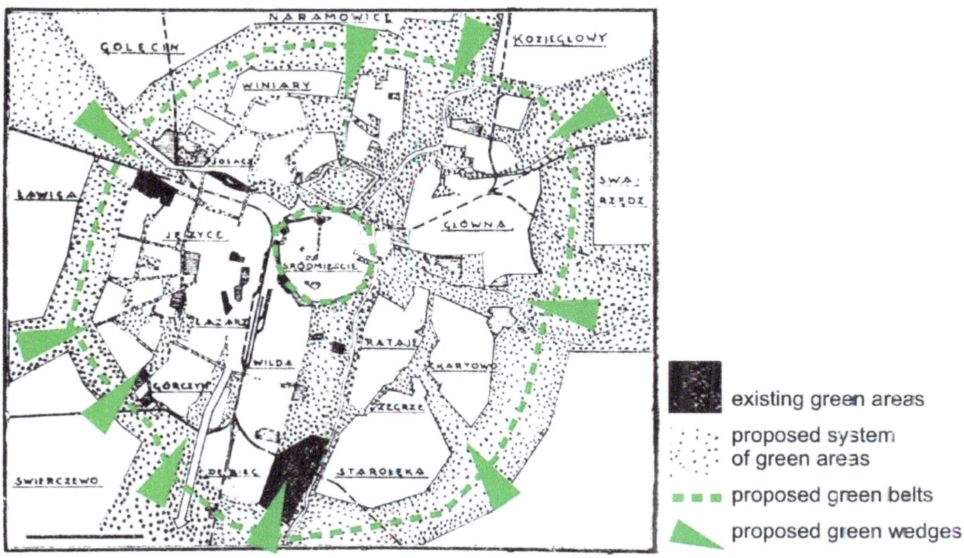

Figure 4. Scheme of green rings and wedges in Poznań by Władysław Czarnecki—concept from 1932, published in his book from the 1960s [57].

The political changes that took place after the Second World War broke the continuity of urban planning in two of the analysed cities. Wrocław and Gdańsk ceased to belong to Germany; there was an almost complete replacement of the population in these cities. The conflict of national identities meant that the architectural past of these cities was negated [58]. Urban concepts that were developed before 1945 had no continuators. The situation was different in Warsaw and Poznań, where the Polish designers of urban plans continued their work and where the implementation of green and ventilation systems that were designed in the 1930s continued. However, nowadays, insufficient planning provisions to protect these systems and pressure from developers have led to their degradation [59,60]. The current configuration of these systems and the initial planning concepts that shaped them are shown in Figure 6.

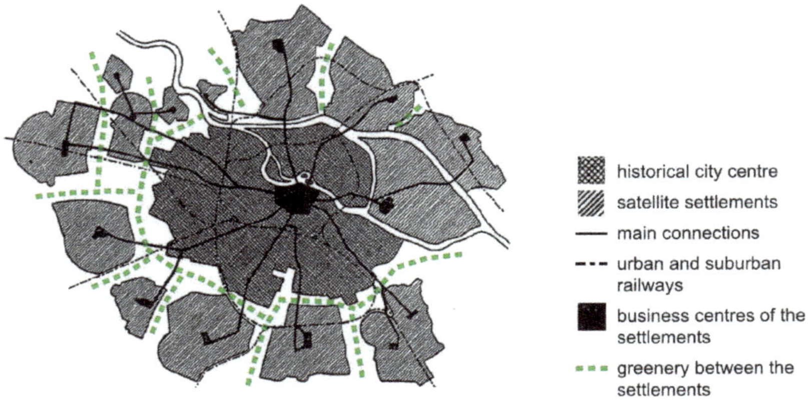

Figure 5. Concept of satellite settlements with open spaces and connections between them for the city of Wrocław, developed in the 1920s by Ernst May [54].

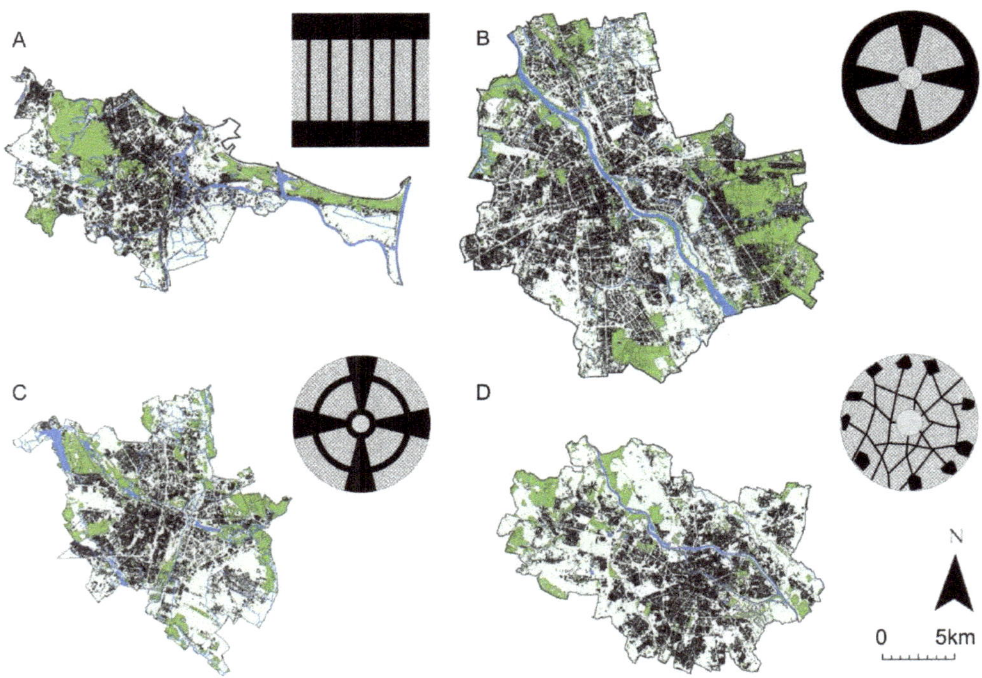

Figure 6. System of green and open spaces in (**A**) Gdańsk, (**B**) Warsaw, (**C**) Poznań and (**D**) Wrocław (maps—authors' own elaboration, based on data from the Head Office of Geodesy and Cartography [61], schemes—adapted based on the work in [62]).

Air quality is monitored within the four cities on a daily basis: in Gdańsk by the ARMAAG Foundation [63], and in Warsaw, Poznań and Wrocław by the WIOŚ environmental authority (Voivodeship Inspectorate of Environmental Protection) [64–66]. Air quality is also modelled and estimated in various studies on the municipal scale with the use of dispersion models (see, e.g., [67–69]). However, so far, the research into the existing urban structure and planning policies with respect to urban ventilation and air quality monitoring is still insufficient.

2.2. Analysis of the Historical Context

In the first stage of the research, all available source materials were identified and juxtaposed with the current spatial layouts of the cities. The workflow for the analysis and classification of the source materials is presented in Figure 7. After graphical amendments, the historical plans were georeferenced according to the current spatial datasets. It is important to note that due to the low degree of accuracy of the available historical plans, some state-of-the-art methods [70–72] were not available and the historical plans were adjusted manually, using the characteristic control points of the water system. However, when dealing with historical geographical materials, high accuracy is not always achievable [73].

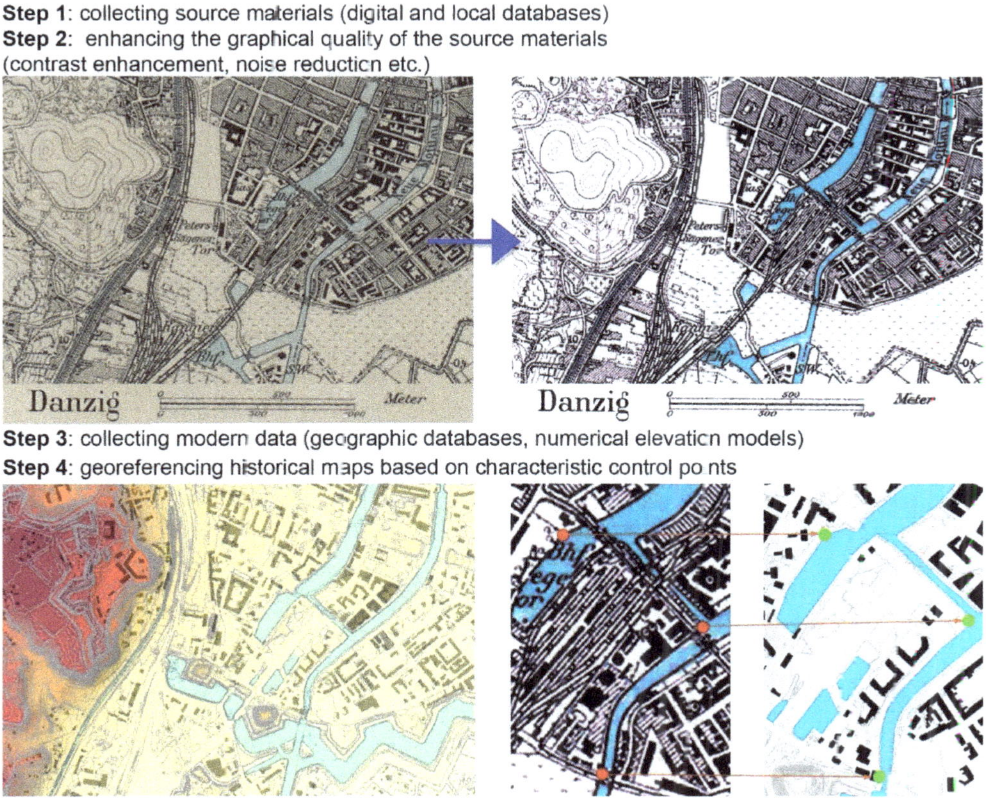

Figure 7. Workflow for the analysis of the source materials and for monitoring the degree of preservation of the urban systems of green and open spaces.

2.3. GIS Analysis—Identification of the Current Elements of the Ventilation Systems

In the second phase of the research, the urban forms of the four cities were quantified using GIS-based tools to identify the main elements of the current ventilation systems. Although the relationship between the parameters of urban form and urban ventilation or air quality is extremely complex and it is not always possible to definitively determine it [74], some urban form indicators, particularly those related to the effectiveness of the ventilation process, have been identified in previous research [8,9]. The following urban form indicators were selected, based on the previous research on the impact of urban structure on air quality and ventilation conditions—see Table 1.

Table 1. Urban form indicators used in the GIS study to identify the key elements of the ventilation system.

Category	Indicator	Definition	Ref.
Land use	**Dominant land use (l_U) (-)**	- the dominant land use in the grid cell (that use occupying the largest area) from the land use categories defined in the municipal database: **LU1**—built-up land, **LU2**—transport infrastructure (combined road network areas, urban squares, and railway network and airport areas feature classes), **LU3**—woodland and land covered with high greenery, **LU4**—land covered with shrubs, **LU5**—grassland and agricultural land, **LU6**—surface water, **LU7**—municipal landfill, **LU8**—unutilised land, **LU9**—undeveloped land, **LU10**—excavation and heap land	[75,76]
Vegetation system	**Biologically active cover area density (λ_{BA}) (%)**	$\lambda_{BA} = A_{BA}/A_c$ where A_{BA} is the total footprint area of all biologically active areas (woodland and land covered with high greenery, land covered with shrubs, grassland and agricultural land) in each grid cell, and where A_c is the area of the grid cell	[8,77,78]
Built-up intensity	**Plan area density (λ_P) (%)**	$\lambda_P = A_p/A_c$ where A_p is the total footprint area of all buildings in each grid cell, and where A_c is the area of the grid cell	[8,19,24,79]
	Gross floor area ratio (λ_{GFA}) (-)	$\lambda_{GFA} = A_{GFA}/A_c$ where A_{GFA} is the total building gross floor area in each grid cell, and where A_c is the area of the grid cell	[8,80,81]
Height structure	**Height variability (σ_H) (m)**	- calculated as the standard deviation of the buildings' height in the grid cell	[8,19,79]
	Average height (μ_H) (m)	- calculated as the average buildings' height in the grid cell (area weighted)	
Urban structure porosity	**Frontal area index (λ_{FAI}) (-)**	$\lambda_{FAI} = A_{FA}/A_c$ where A_{FA} is the area of the buildings' façades facing the given wind direction in each grid cell, and where A_c is the area of the grid cell—the λ_{FAI} was calculated for the cardinal wind directions (North–South and East–West) and averaged for each grid cell	[15,79,82,83]

All indicators were calculated with the ArcGIS Pro 2.9 software (Esri Inc., Redlands, CA, USA). For calculating each indicator, geoprocessing models were designed and automated using the ArcGIS 2.9 Model Builder©. The data was drawn from a database of topographical objects corresponding to a 1:10,000 topographic map, provided by the Head Office of Geodesy and Cartography [61]. It is a vector database containing the spatial location of the topographical objects, together with their characteristics, developed in the plane coordinates of the PL-1992 system (EPSG code: 2180). The data were retrieved in May 2022. A grid approach was adopted for the GIS calculations—see Figure 8. They were based on a 100 × 100-metre rectangular grid, which is often used in studies on the city scale regarding urban ventilation conditions, e.g., [12,15].

The following indicators—plan area density (λ_P), gross floor area ratio (λ_{GFA}), height variability (σ_H) and average height (μ_H)—were calculated using the methods described by Badach et al. (2020) [8]. For the frontal area index (λ_{FAI}), which is the indicator describing the relationship between urban structure and wind conditions, a set of methods was developed based on tools used in previous studies, adapting them to the available data describing the built-up structures. The concept of the frontal area index was developed by Grimmond and Oke (1999) [84] and Burian et al. (2002) [85] for the purposes of local climate zone and urban ventilation studies; it is defined as the sum of the frontal façades

(façades facing the selected wind direction), divided by the site area. This definition was re-defined and modified for various studies, depending on the urban context and the focus of the analysis.

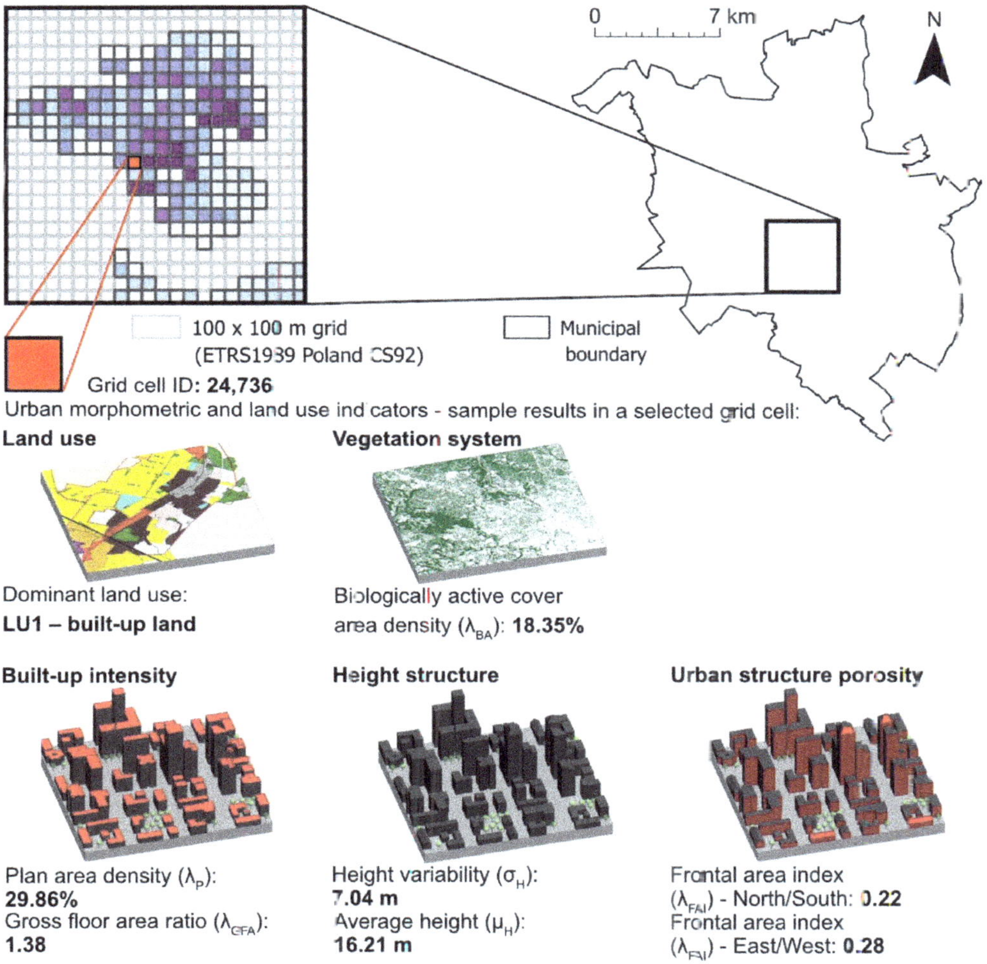

Figure 8. Grid approach used in the GIS study.

When calculating the frontal area index, two approaches are most commonly employed. In the first one, described, e.g., in [15,86], a set of lines, aligned with the selected wind direction and with a predefined spacing, is generated. Then the set of lines is intersected with the building layer. The lines intersecting with the buildings are counted and are then multiplied by the area resulting from the predefined line spacing and the building heights. If the effect of the imposing façades is to be eliminated, the given line is taken into account only once; it is counted only when it comes into contact with a building for the first time for a given site or grid cell—see Figure 9A. In another approach, described, e.g., in [60], the lines of the building's external walls are created by projecting the width of the façade facing the selected wind direction to the grid cell boundary or plane, perpendicular to the wind direction. Each building height is assigned to the line representing its projected width. Then, the line widths are multiplied by the height parameter and summed up—see

Figure 9B. By using this approach, the effect of imposing façades can be eliminated as well. Moreover, the frontal area index indicator can be calculated using three-dimensional GIS building data, e.g., using a dedicated application based on the ArcGIS Pro software called the Urban Morphology Extractor (UME) [82,87].

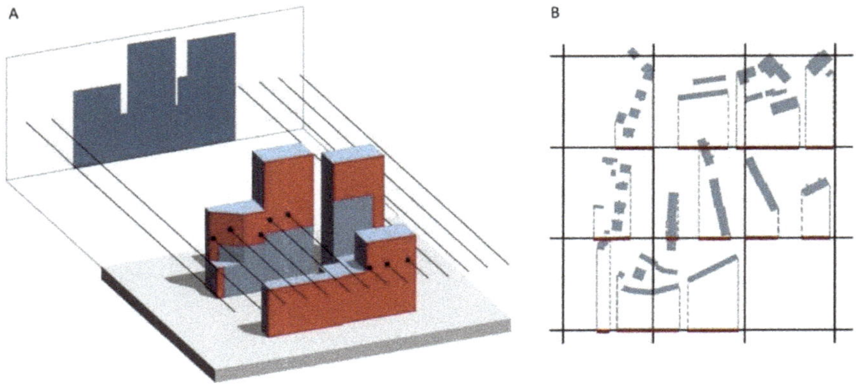

Figure 9. (**A**,**B**) two approaches to calculating the frontal area index. (Authors' own elaboration, based on: [15,60,86]).

An approach based on the one proposed by Burghardt (2014) [88] for the ArcMap software (Esri Inc., Redlands, CA, USA) was adapted for the ArcGIS Pro software. Instead of creating Thiessen polygons for site partition, a grid approach was used, similarly to the approach of Wong et al. (2010) [15]. The same 100 × 100 metre grid was used, as with the previous indicators. In the first step, layers with lines representing the cardinal and inter-cardinal wind directions were generated, with 2 metres spacing. Then, buildings with the same heights and with a common edge were merged. Next, the wind lines were intersected with the buildings, resulting in information about the buildings' heights being assigned to the wind lines. For each line intersecting the buildings, a new attribute was created, which was calculated by multiplying the building height by the line spacing (2 metres). Finally, this attribute was summed up for each grid cell, based on the unique cell ID, and divided by the grid cell area. For some high-density urban areas, a correction is often required to account for the effect of imposing buildings (see, e.g., [15,89]). However, given the relatively low built-up density of the selected cities, the lack of such a correction did not lead to the overestimation of this indicator, so it was not taken into account.

2.4. Local Case Studies—Design Recommendation

Analysis of the λ_{FAI} spatial distribution and its values in particular areas and even grid cells was used to designate some local problem areas—examples of urban development scenarios that interrupted the continuity of the ventilation systems. Subsequently, CFD simulations were used to illustrate the finding that city-scale GIS-based analysis is not sufficiently sensitive to identify improper design solutions that inhibit the city's ventilation efficiency due to micro-scale wind environment effects. A CFD case study of a particular problem area was performed to demonstrate that the integration of various analytical and simulation tools is required for effective urban ventilation management.

The ventilation simulations were conducted using the Autodesk CFD 2021 software (Autodesk, San Rafael, CA, USA). An area with a radius of 400 m around a particular point of interest was modelled in CAD/CAM software Autodesk AutoCAD 2021 (Autodesk, San Rafael, CA, USA), based on data from the geographic database, and was then imported into the CFD software. The simulation parameters and domain were set according to the Architectural Institute of Japan's guidelines for CFD predictions of the urban wind environment [90]. As discussed extensively in many previous studies, the CFD technique is currently the most accurate tool for such an analysis on a local scale [91]. However, high simulation accuracy was not pursued in this case. The simulations were performed to illustrate how this technique, already well-established or even required in the planning and design process in various cities (see, e.g., [92,93]), could be a basis for design solutions that are more beneficial for the local wind environment.

3. Results

An introductory evaluation of the degree of preservation of the ventilation corridors was performed by superimposing the georeferenced historical maps and contemporary urban maps. Since maps contain only two-dimensionalinformation, the analysis of such a superimposition only makes it possible to gain an approximate insight into the researched problem. However, it indicates that the ventilation corridors are often blocked by later built-up structures. Figure 10 presents fragments of Gdańsk's large-scale Zaspa and Przymorze housing developments, built in the 1970s, clearly showing that the originally designed ventilation system was, generally, not taken into consideration in urban development plans of that time. A striking example of such negligence are the elongated and wavy, 11-story buildings that were placed across the ventilation corridor. Similarly, the hexagonal forms of blocks in the Zaspa estate were built at the intersection of the two main ventilation corridors, heavily impairing the airflow. In other cases, the buildings are located along the direction of ventilation, causing less damage to the functioning of the whole system.

Similarly, in Warsaw, several examples of improper design solutions can be found, many of which were erected in recent years when an awareness of the issues of city breathability and air quality management was more widespread. The massive bulk of the recently built Galeria Bemowo shopping centre was constructed across one of the ventilation wedges—see Figure 11. In this case, a more fragmented building complex, oriented along the direction of ventilation, would have been more beneficial.

In Poznań, in recent years, a new multi-storey residential and commercial development was permitted in the area of the Droga Dębińska, with the Bielniki and Piastowska streets within the southern wedge—see Figure 12.

Finally, in Wrocław, the spatial layout of the complex of the Military Clinic Hospital, with its elongated façade facing the direction of the ventilation corridor, was not properly shaped—see Figure 13.

Contemporary GIS tools provide more in-depth insights into the continuity of ventilation systems by enabling the integration of two-dimensional data and land-use characteristics. Figures 14–17 show the distribution of the urban form indicators related to the process of urban ventilation. They also reveal the differences in the urban morphology of the investigated cities. The average values of specific indicators (λ_P, c_H, λ_{FAI}) for each city, provided in the legends for Figures 14–17, confirm their spatial heterogeneity. In particular, the spatial distribution of λ_{FAI} was considered in relation to the urban ventilation systems. By mapping the λ_{FAI} values below the average value in each city (the areas marked with blue dots in map 4, in Figures 14–17) and λ_{BA} (green color), the major ventilation pathways and some secondary paths were identified.

A close-up analysis of the λ_{FAI} values shows how city-scale mapping can rapidly identify local problem areas—see Figures 18 and 19. The first case shows the previously discussed residential and commercial development in Poznań. In the second case, the λ_{FAI} mapping pointed to one of the Gdańsk University of Technology campus buildings, which was constructed post-WWII and was built perpendicularly to the direction of a local ventilation pathway, alongside one of the waterways (currently underground).

However, in some cases, this data needs to be analysed in relation to the existing urban layout, which can be illustrated using the example of the Manhattan shopping centre in Gdańsk (indicated in red in Figure 20). It can be noticed that even though the highest λ_{FAI} values indicate other problem areas, the more serious problem related to the inhibited ventilation efficiency is the large bulk of this building. This is because it is blocking a secondary ventilation pathway along Partyzantów street, which is connected to the main ventilation corridors in this area.

Although using GIS-based tools makes it possible to identify the existing elements of ventilation systems and ventilation pathways, the results of CFD modelling more accurately recognises the characteristics of air movement disturbances, especially in the urban micro-scale. The CFD simulations conducted for the Manhattan shopping centre case study—see Figure 21—confirm that the GIS-based analysis was not sufficient to identify this issue. Analysis of the results of the CFD modelling provides more in-depth insights into how particular building arrangements constructed in the paths of ventilation corridors negatively affect air movement and ventilation efficiency. The continuity of the ventilation pathway in this area could be maintained, e.g., by dividing the complex into two parts, with a shopping street between them, as an extension of Partyzantów street—see Figure 21.

Figure 10. Juxtaposition of the historical plan [44] and the current spatial distribution of built-up and open areas in Gdańsk.

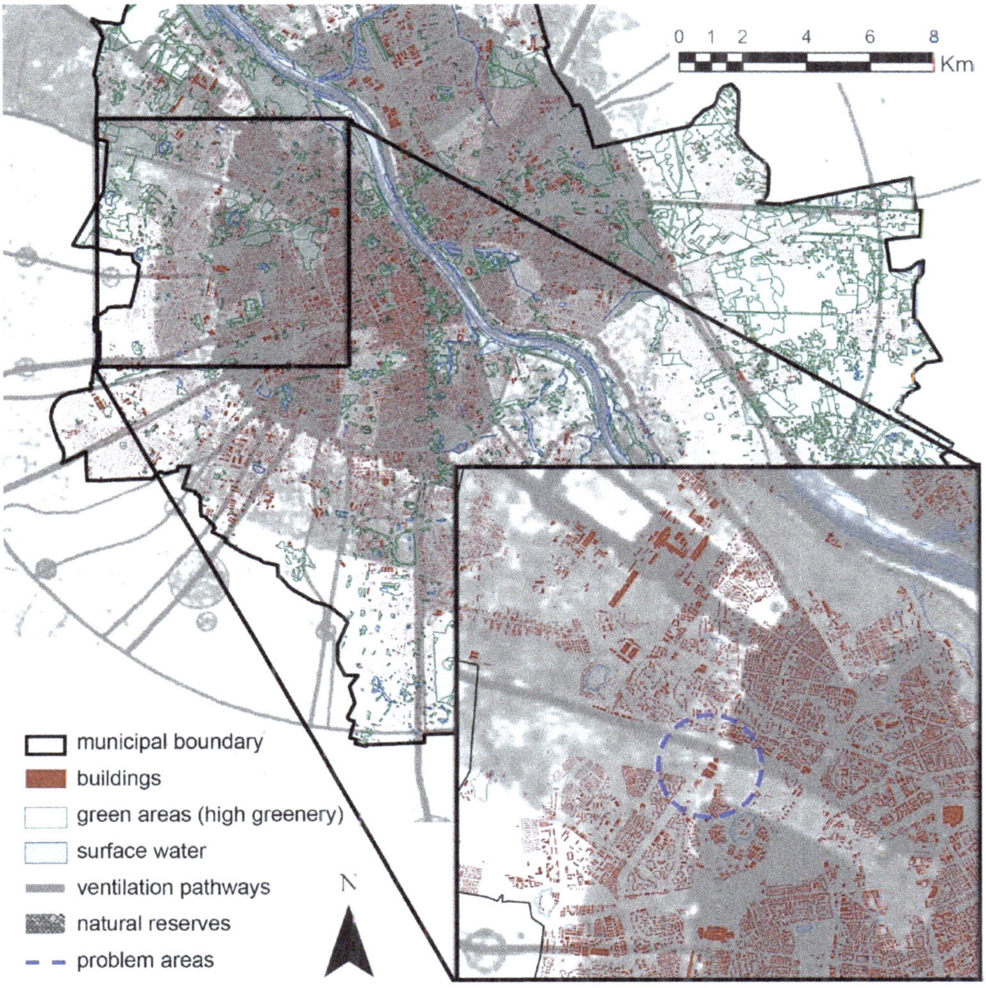

Figure 11. Juxtaposition of the historical plan [49] and the current spatial distribution of built-up and open areas in Warsaw.

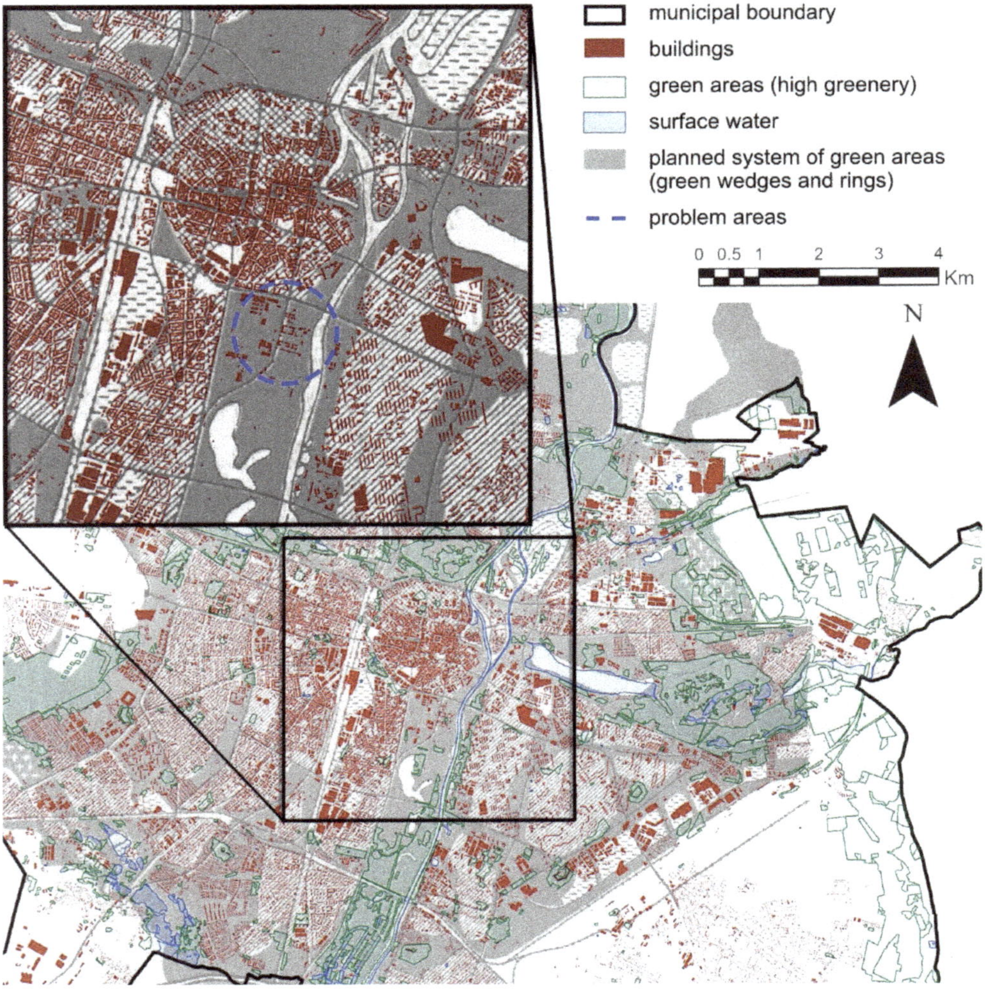

Figure 12. Juxtaposition of the historical plan (as presented in [62]) and the current spatial distribution of built-up and open areas in Poznań.

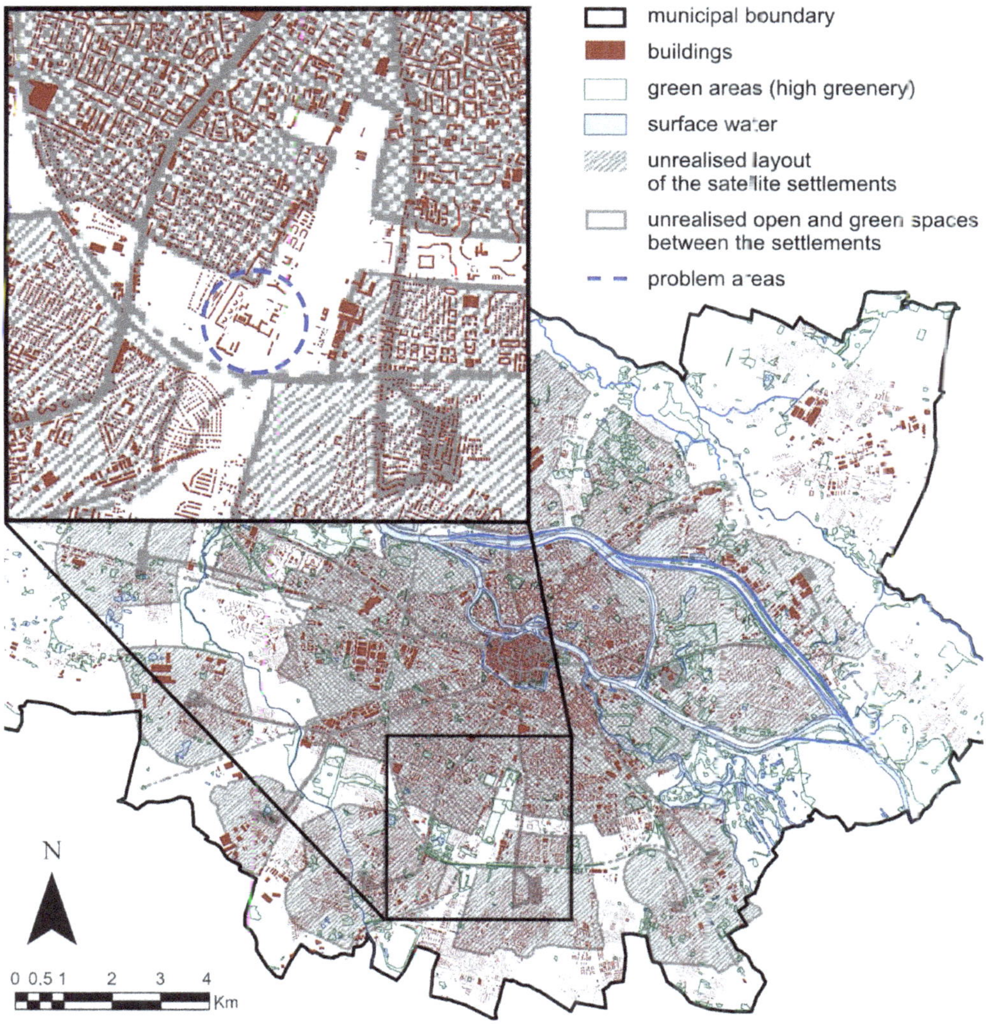

Figure 13. Juxtaposition of the historical plan [54] and the current spatial distribution of built-up and open areas in Wrocław.

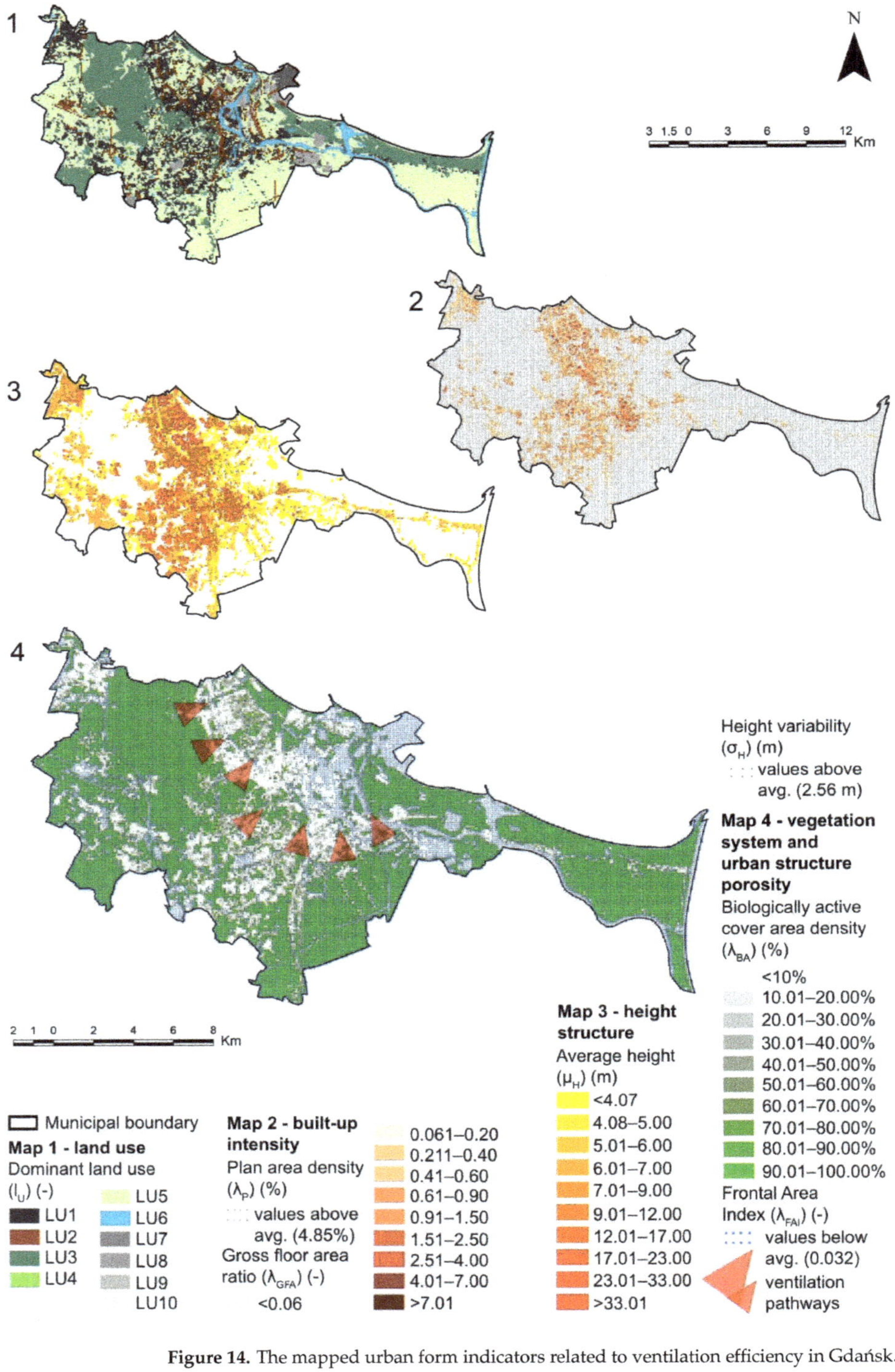

Figure 14. The mapped urban form indicators related to ventilation efficiency in Gdańsk.

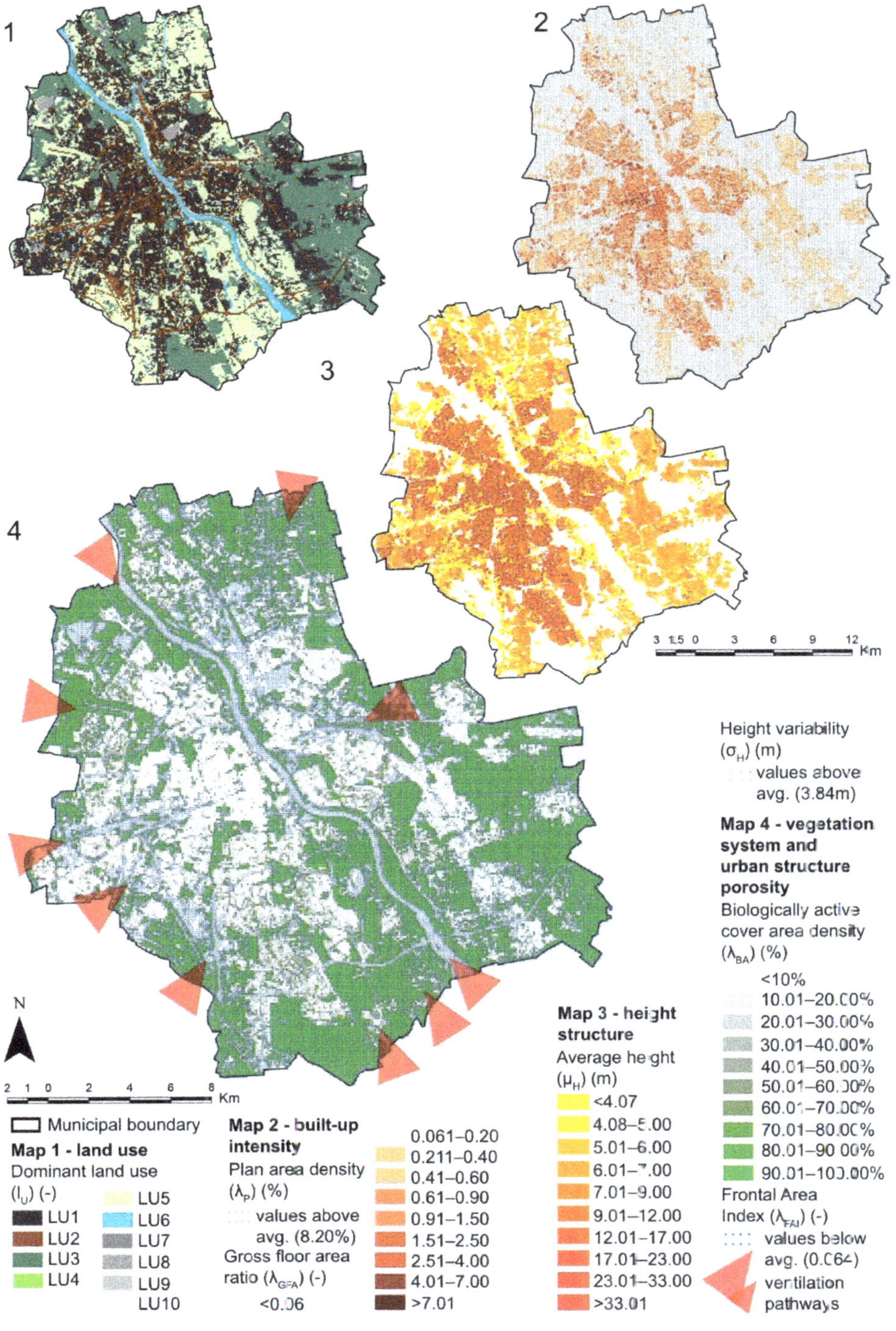

Figure 15. The mapped urban form indicators related to ventilation efficiency in Warsaw.

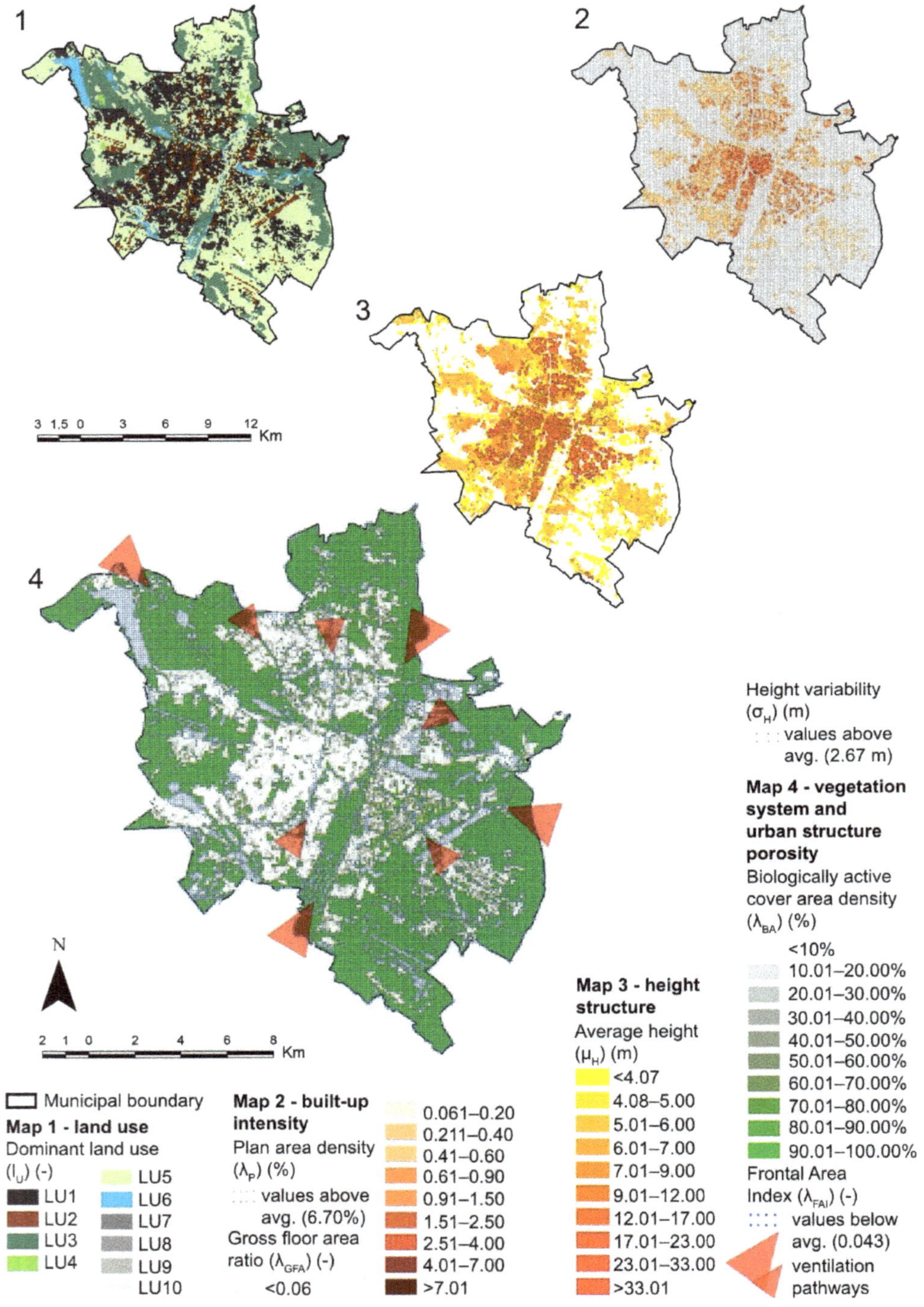

Figure 16. The mapped urban form indicators related to ventilation efficiency in Poznań.

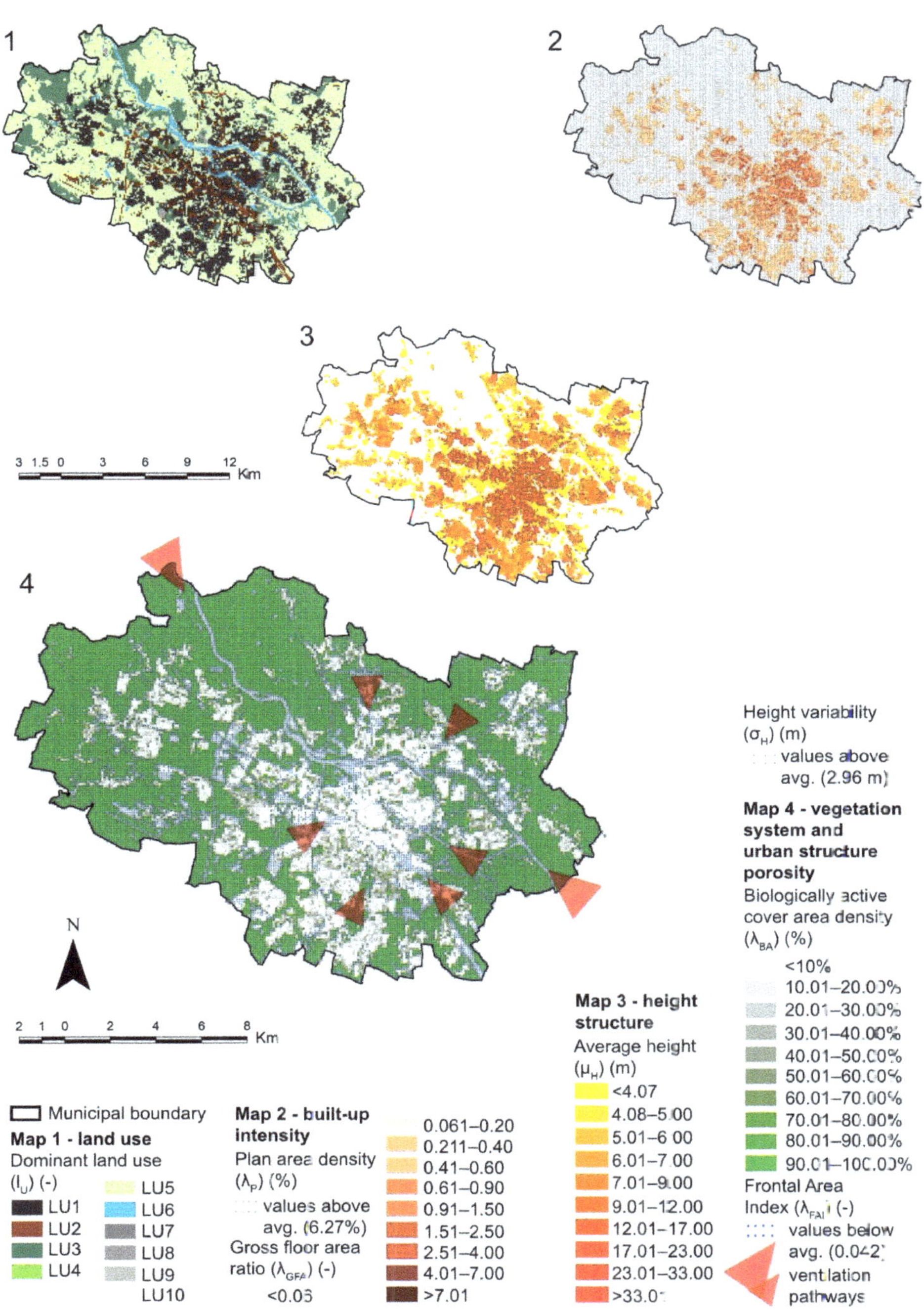

Figure 17. The mapped urban form indicators related to ventilation efficiency in Wrocław.

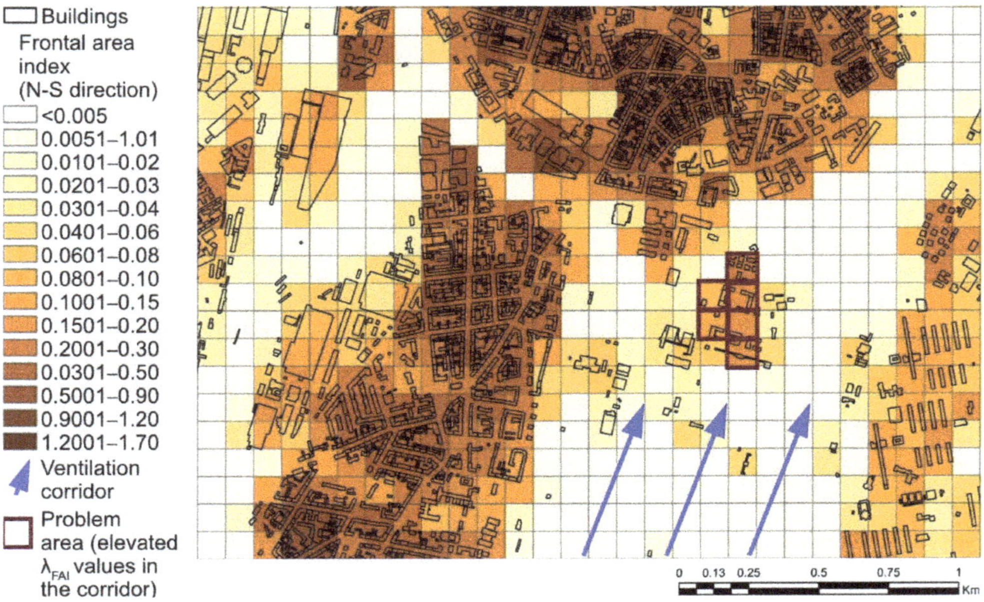

Figure 18. Local case study of the ventilation corridor (wedge) in Poznań.

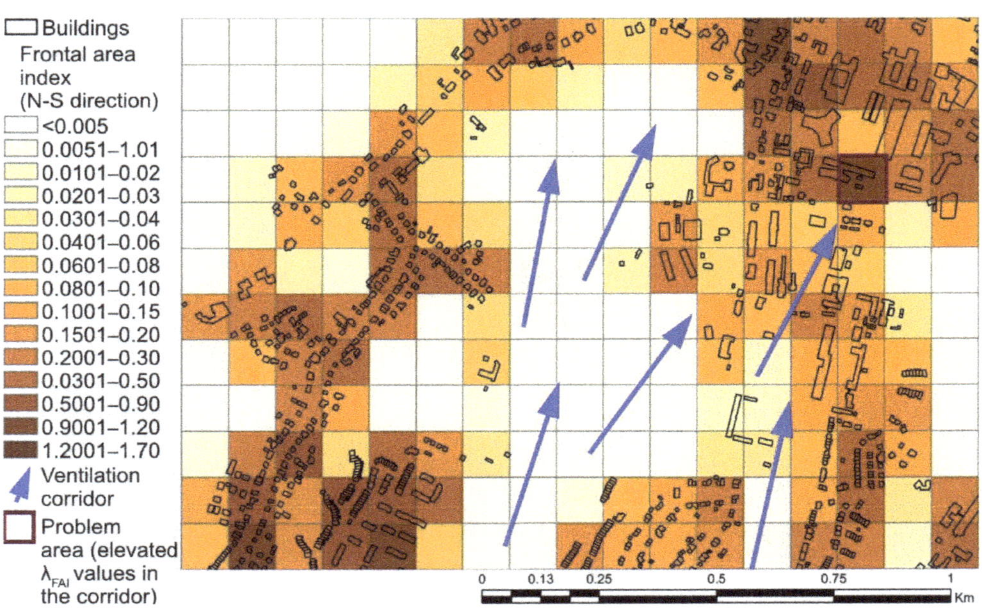

Figure 19. Local case study of the ventilation pathway alongside one of the waterways in Gdańsk.

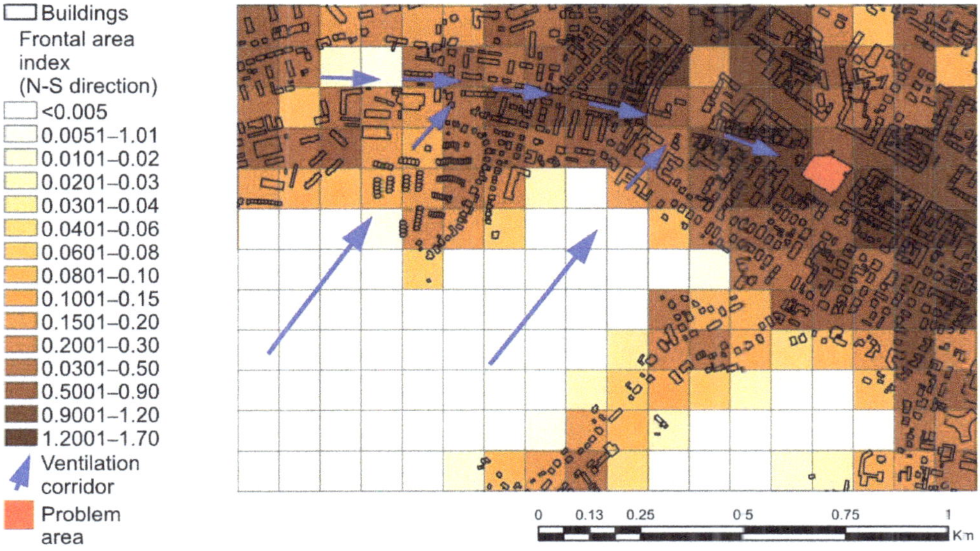

Figure 20. Local case study of the ventilation pathway alongside Partyzantów street in Gdańsk

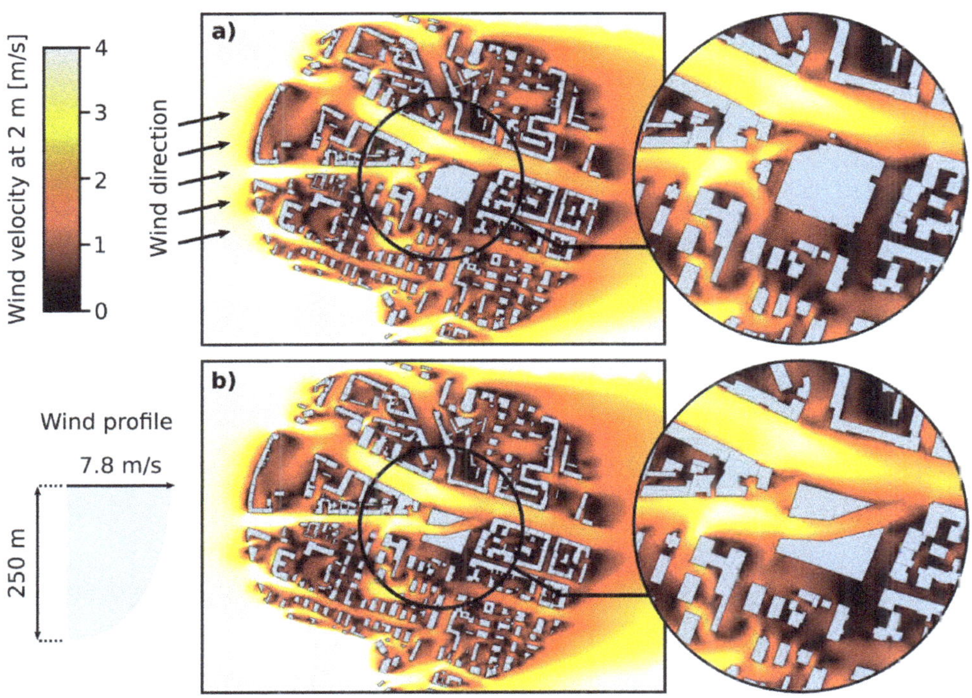

Figure 21. CFD study of the ventilation pathway alongside Partyzantów street in Gdańsk: (**a**) existing scenario, (**b**) alternative design scenario for ventilation efficiency improvement.

4. Discussion and Recommendations

A comparative analysis of urban ventilation conditions in four Polish cities demonstrates, in general, the significant potential of early twentieth-century planning concepts. They were designed without the use of advanced analytical tools, but they took into account local knowledge of the main wind directions, local topography, the location of watercourses, the natural disposition of green areas, and the characteristics of the urban structure. The main assumptions of these early urban ventilation systems are still visible in all the analysed cities. However, GIS-based analysis made it possible to correct some compromised airflow directions to the prevailing ones through an examination of the current city structure and the land-use characteristics. Additionally, the presented method led to the identification of problem areas with major disruptions to airflow. While analysis using GIS-based tools confirms the disturbances on the scale of the whole city, CFD modelling offers detailed insights into the process causing impaired airflow due to the improper location of specific buildings.

With an increasing awareness of the value of the quality of life in healthy cities and the ongoing pursuit toward sustainable development goals, after decades of negligence, the value of effective urban ventilation is re-appearing. This is, however, happening at a different pace in each of the four analysed cities. In the case of the city of Gdańsk, the issue of urban ventilation is not even mentioned in strategic planning documents. This extreme situation could be due to the massive exchange of population after 1945, which brought about a radical break in the continuity of urban design concepts. In the strategic general planning documents for Gdańsk, there are no provisions protecting the ventilation corridors. Even so, the issue of air quality and urban ventilation corridors is slowly beginning to appear in the individual local development plans. In 2021, the Gdańsk Development Office commenced a study regarding the definition of the Gdańsk Green Policy [94], where city ventilation corridors will be designated and then gradually introduced to local planning acts. To protect the local ventilation pathways, regulations concerning building height, density and the percentage of biologically active areas will be appropriately adjusted in the local spatial development plans. As this study indicates, such provisions are still incomplete, and other urban form indicators should be also taken into consideration. For example, the results of the CFD study prove that low buildings do not guarantee undisturbed airflow—their layout and fragmentation in relation to the local wind conditions are also crucial.

After decades of inattention, the wedged-shaped system of ventilation corridors in Warsaw that complement the main ventilation corridor of the Vistula River started to be discussed again and was recommended for protection in 1992, when the issue of air exchange and regeneration appeared in the "General Spatial Development Plan for the Capital City of Warsaw" [95]. This document named nine ventilation corridors running radially from the city outskirts toward its centre. These are consistent with the findings of this study, as well as those of previous studies [96,97]. This identification, however, did not halt the process of intensive housing development and construction that largely reduced the functionality of important ventilation corridors, the large-scale development in Bemowo being a striking example. In 2006, the document "Eco-physiographic Study" was issued [98], including ventilation recommendations to inform planning decisions and guide the directions of the spatial development of Warsaw. Despite these recommendations, in numerous cases, urban pressure and the gaps in the local planning regulations have led to further disturbances in the functionality of the ventilation system, as revealed by this study.

In the city of Poznań, the latest planning documents, which are very clearly defined and act as a continuation of the early modern concepts, define the strategy for better ventilation of the city. The protection of the system of green rings and wedges is included in the Study on Conditions and Spatial Development Directions for Poznań [99]. The provisions of the new study not only protect the green wedges as ventilation corridors but also call for new green belts, especially along the communication routes. The strategy regarding blue-green networks explicitly refers to the Stübben Ring and the belt along the outer ring of fortifications. However, the integrity of this system is still not properly protected in the planning and development practice of the area [100].

In the Study of Conditions and Directions of Spatial Development of Wrocław [101], the question of urban ventilation is also discussed; however, there is no reference to the earlier concepts—most probably, due to the very vague politics of urban ventilation and the reduced degree of its implementation in the first decades of the twentieth century. To protect the natural environment against air, water and soil pollution, the study defines sectoral policies, including air quality policy, among other factors. According to the study, to reduce air pollution, the green wedges that cut into the city structure should be re-established and protected. This is also consistent with the findings of our study, in which we identified a centric layout of the ventilation system. However, in other studies, different concepts were also considered (see, e.g., [102]). Additionally, the connections between existing green areas will be shaped to transform the scattered green spaces into networks.

All these provisions in the recent planning documents confirm that the issue of urban ventilation has become an essential element of urban policy. Advanced tools and methods are needed to support the intentions expressed in these general planning acts and facilitate their transfer into operational decisions, along with their integration into more detailed local spatial development plans. For example, as research has shown, there are large differences in λ_{FAI} depending on the orientation of the buildings. This means that in most airflow-sensitive urban areas, restrictions should not only concern building heights but also their orientations and the distances between them, which may vary depending on the prevailing wind conditions. Similarly, it may happen that the building heights and intensities are low but they still block the airflow, which confirms the importance of CFD research on the urban micro-scale.

For the purpose of sustaining or improving the ventilation efficiency in cities, the designation of new networks of open public spaces [103], blue-green connections [104,105] and the introduction of high-density urban structures [28,92] should be subjected to careful consideration of the research results and recommendations. Some design aspects are particularly relevant in this process—for example, the re-development of waterfront areas. This is because ventilation corridors along the riverbeds are very often a core part of urban ventilation systems, as it is in the case of the studied cities. Therefore, the waterfront areas, in particular, should be considered in future studies of urban ventilation systems [92,106].

The design and maintenance of the urban green infrastructure represent another crucial aspect that should be integrated into urban air quality and ventilation management. Future research and planning practice based on the proposed workflow can be broadened by using more indicators to comprehensively describe the parameters of the vegetation systems. In this study, the vegetation systems in the four cities were studied in relation to their historical development; then, λ_{BA} was calculated to account for the current spatial distribution of the green areas. The structure of urban vegetation, both at the city- and micro-scale, is relevant to urban air quality and ventilation management [8,107], but its proper design to improve ventilation conditions is not straightforward [108]. Therefore, more scale-sensitive parameters describing the vegetation parameters should be included in further research, e.g., crown and plant height or vegetation density and porosity [109], while accounting for the variations in the vegetation forms [110].

Finally, it is of particular importance that the scientific knowledge regarding effective urban ventilation becomes widespread among practitioners, policymakers and authorities. However, as previously investigated by Badach et al. [11], this is not always the case in the Polish context. Therefore, the awareness of the local stakeholders should be subjected to further investigation. More attention should be given to the dissemination of the results of current research results on improving city breathability using planning and design solutions.

5. Conclusions

It is becoming evident that effective urban ventilation studies are of key importance in the drive to improve urban air quality. At the same time, in many Polish cities, including the four that have been analysed in this study, intensive urban transformation processes have been taking place—in the majority of cases, without any ventilation monitoring tools. The high pace of urban spatial development results in changes in the land use characteristics of suburban areas that may affect the efficiency of ventilation corridors. The rise in building heights and their frontal index in city centres, when uncontrolled, may lead to serious disturbances in airflow.

The methods proposed in this paper based on GIS tools, including the analysis of historical source materials, supported by CFD modelling make it possible to monitor the changes in urban structure and to verify the ventilation paths. The system enables the monitoring and shaping of airflow systems and evaluation of urban ventilation efficiency. Moreover, it is worth pointing out that the set of urban form indicators proposed in this study, including the indicators describing the buildings' density and height structure, are commonly used in urban planning, for example, to determine the parameters for new developments in local spatial plans. We demonstrated that the same parameters are useful for urban air quality management, which further facilitates the integration of urban ventilation control within the process of urban planning.

However, other parameters used in this study, for example, λ_{FAI}, are rarely incorporated into local and city-scale planning guidelines and directives in Poland. It is our recommendation that this parameter should be more commonly used as a tool to control the parameters of new developments, especially in the more densely built-up urban areas and in those areas that are part of ventilation systems. The proposed GIS-based workflow may be used as a supporting planning and design tool, especially on the city scale. The urban form indicators can be calculated and updated on an ongoing basis, to examine whether the designed layout impedes effective ventilation. As such, this method also allows for updating the current ventilation systems to account for the ongoing spatial development of the city and designing new ventilation corridors. Based on the analysis of these indicators, it is possible to identify problem areas and study them in detail.

Moreover, we wanted to show that the use of GIS tools can account for the geometrical features of the urban form but it does not account for local phenomena related to the ventilation process. Calculating the urban form indicators and, in particular, λ_{FAI}, is useful for the effective evaluation of the ventilation systems at the city scale, but more suitable numerical simulations, supported by an analysis of the local conditions, should be used to support urban design on a local scale (the scale of building complexes or even single buildings). It would be beneficial if the use of this tool was required by local planning regulations, at least in the most problematic areas. As discussed above, there are already some examples of such requirements in other countries.

It is not possible to define one overall strategy for better air quality in all of the four cities, due to their different geographical and urban conditions, as well as the introduction of different urban ventilation concepts in the first decades of the twentieth century. The tools presented in this paper, based on the use of GIS-based analysis and CFD modelling, make it possible to identify the existing elements of urban ventilation systems, designate problem areas, and simulate changes in ventilation efficiency and scenarios as an effect of new urban transformations. To formulate detailed recommendations for any specific city, more detailed on-site local studies are needed, supported by the tools and procedures proposed in this paper.

Author Contributions: Conceptualization, J.B., J.S., W.B., J.G. and L.N.; methodology, J.B.; investigation, J.B.; writing—original draft preparation, J.B., J.S., W.B., J.G. and L.N.; writing—review and editing, J.B., J.S., W.B., J.G. and L.N.; visualization, J.B. All authors have read and agreed to the published version of the manuscript.

Funding: This research was financially supported by the National Science Centre in Poland (Grant no. 2019/33/N/HS4/00978).

Institutional Review Board Statement: Not applicable.

Informed Consent Statement: Not applicable.

Data Availability Statement: The relevant data underlying this study are fully available under the details provided in the references. For further questions, please contact the corresponding author

Acknowledgments: The authors would like to thank Wojciech Wojnowski for his help with the CFD simulations.

Conflicts of Interest: The authors declare no conflict of interest.

References

1. World Health Organization. Air Pollution. Available online: https://www.who.int/health-topics/air-pollution#tab=tab_1 (accessed on 30 June 2022).
2. Tan, X.; Han, L.; Zhang, X.; Zhou, W.; Li, W.; Qian, Y. A review of current air quality indexes and improvements under the multi-contaminant air pollution exposure. *J. Environ. Manag.* **2021**, *279*, 111681. [CrossRef]
3. European Environment Agency. Exceedance of Air Quality Standards in Europe. Available online: https://www.eea.europa.eu/ims/exceedance-of-air-quality-standards (accessed on 20 November 2021).
4. Court of Justice of the European Union. *Judgement in Case C-336/16*; Court of Justice of the European Union: Luxembourg, 2018.
5. Mirzaei, P.A.; Haghighat, F. A procedure to quantify the impact of mitigation techniques on the urban ventilation. *Build. Environ.* **2012**, *47*, 410–420. [CrossRef]
6. Ren, C.; Yang, R.; Cheng, C.; Xing, P.; Fang, X.; Zhang, S.; Wang, H.; Shi, Y.; Zhang, X.; Kwok, Y.T.; et al. Creating breathing cities by adopting urban ventilation assessment and wind corridor plan—The implementation in Chinese cities. *J. Wind Eng. Ind. Aerodyn.* **2018**, *182*, 170–188. [CrossRef]
7. Yang, J.; Shi, B.; Shi, Y.; Marvin, S.; Zheng, Y.; Xia, G. Air pollution dispersal in high density urban areas: Research on the triadic relation of wind, air pollution, and urban form. *Sustain. Cities Soc.* **2020**, *54*, 101941. [CrossRef]
8. Badach, J.; Voordeckers, D.; Nyka, L.; Van Acker, M. A framework for Air Quality Management Zones-useful GIS-based tool for urban planning: Case studies in Antwerp and Gdańsk. *Build. Environ.* **2020**, *174*, 106743. [CrossRef]
9. He, B.-J.; Ding, L.; Prasad, D. Enhancing urban ventilation performance through the development of precinct ventilation zones: A case study based on the Greater Sydney, Australia. *Sustain. Cities Soc.* **2019**, *47*, 101472. [CrossRef]
10. Kurppa, M.; Hellsten, A.; Auvinen, M.; Raasch, S.; Vesala, T.; Järvi, L. Ventilation and Air Quality in City Blocks Using Large-Eddy Simulation—Urban Planning Perspective. *Atmosphere* **2018**, *9*, 65. [CrossRef]
11. Badach, J.; Dymnicka, M.; Załęcki, J.; Brosz, M.; Voordeckers, D.; Van Acker, M. Exploring the Institutional and Bottom-Up Actions for Urban Air Quality Improvement: Case Studies in Antwerp and Gdańsk. *Sustainability* **2021**, *13*, 11790. [CrossRef]
12. Guo, F.; Zhang, H.; Fan, Y.; Zhu, P.; Wang, S.; Lu, X.; Jin, Y. Detection and evaluation of a ventilation path in a mountainous city for a sea breeze: The case of Dalian. *Build. Environ.* **2018**, *145*, 177–195. [CrossRef]
13. Lou, B.; Barbieri, D.M.; Passavanti, M.; Hui, C.; Gupta, A.; Hoff, I.; Lessa, D.A.; Sikka, G.; Chang, K.; Fang, K.; et al. Air pollution perception in ten countries during the COVID-19 pandemic. *Ambio* **2021**, *51*, 531–545. [CrossRef]
14. Wang, W.; Yang, T.; Li, Y.; Xu, Y.; Chang, M.; Wang, X. Identification of pedestrian-level ventilation corridors in downtown Beijing using large-eddy simulations. *Build. Environ.* **2020**, *182*, 107169. [CrossRef]
15. Wong, M.S.; Nichol, J.E.; To, P.H.; Wang, J. A simple method for designation of urban ventilation corridors and its application to urban heat island analysis. *Build. Environ.* **2010**, *45*, 1880–1889. [CrossRef]
16. Zheng, Z.; Ren, G.; Gao, H.; Yang, Y. Urban ventilation planning and its associated benefits based on numerical experiments: A case study in beijing, China. *Build. Environ.* **2022**, *222*, 109383. [CrossRef]
17. Azizi, M.M.; Javanmardi, K. The Effects of Urban Block Forms on the Patterns of Wind and Natural Ventilation. *Procedia Eng.* **2017**, *180*, 541–549. [CrossRef]
18. Zhang, Y.; Gu, Z. Air quality by urban design. *Nat. Geosci.* **2013**, *6*, 506. [CrossRef]
19. Chen, L.; Hang, J.; Sandberg, M.; Claesson, L.; Di Sabatino, S.; Wigo, H. The impacts of building height variations and building packing densities on flow adjustment and city breathability in idealized urban models. *Build. Environ.* **2017**, *118*, 344–361. [CrossRef]
20. Hang, J.; Li, Y.; Sandberg, M.; Buccolieri, R.; Di Sabatino, S. The influence of building height variability on pollutant dispersion and pedestrian ventilation in idealized high-rise urban areas. *Build. Environ.* **2012**, *56*, 346–360. [CrossRef]
21. Lee, R.X.; Jusuf, S.K.; Wong, N.H. The study of height variation on outdoor ventilation for Singapore's high-rise residential housing estates. *Int. J. Low-Carbon Technol.* **2015**, *10*, 15–33. [CrossRef]
22. Liao, W.; Hong, T.; Heo, Y. The effect of spatial heterogeneity in urban morphology on surface urban heat islands. *Energy Build.* **2021**, *244*, 111027. [CrossRef]

23. Maing, M. Superblock transformation in Seoul Megacity: Effects of block densification on urban ventilation patterns. *Landsc. Urban Plan.* **2022**, *222*, 104401. [CrossRef]
24. Mei, S.J.; Hu, J.T.; Liu, D.; Zhao, F.Y.; Li, Y.; Wang, Y.; Wang, H.Q. Wind driven natural ventilation in the idealized building block arrays with multiple urban morphologies and unique package building density. *Energy Build.* **2017**, *155*, 324–338. [CrossRef]
25. Ramponi, R.; Blocken, B.; de Coo, L.B.; Janssen, W.D. CFD simulation of outdoor ventilation of generic urban configurations with different urban densities and equal and unequal street widths. *Build. Environ.* **2015**, *92*, 152–166. [CrossRef]
26. Trindade da Silva, F.; Reis, N.C.; Santos, J.M.; Goulart, E.V.; Engel de Alvarez, C. The impact of urban block typology on pollutant dispersion. *J. Wind Eng. Ind. Aerodyn.* **2021**, *210*, 104524. [CrossRef]
27. Wen, H.; Malki-Epshtein, L. A parametric study of the effect of roof height and morphology on air pollution dispersion in street canyons. *J. Wind Eng. Ind. Aerodyn.* **2018**, *175*, 328–341. [CrossRef]
28. Ng, E. (Ed.) *Designing High-Density Cities For Social and Environmental Sustainability*, 2nd ed.; Routledge: Hong Kong, China, 2015.
29. Xu, F.; Gao, Z.; Zhang, J.; Hu, Y.; Ding, W. Influence of typical street-side public building morphologies on the ventilation performance of streets and squares. *Build. Environ.* **2022**, *221*, 109331. [CrossRef]
30. Faehnle, M.; Söderman, T.; Schulman, H.; Lehvävirta, S. Scale-sensitive integration of ecosystem services in urban planning. *GeoJournal* **2015**, *80*, 411–425. [CrossRef]
31. Li, F.; Hu, D.; Liu, X.; Wang, R.; Yang, W.; Paulussen, J. Comprehensive urban planning and management at multiple scales based on ecological principles: A case study in Beijing, China. *Int. J. Sustain. Dev. World Ecol.* **2008**, *15*, 524–533. [CrossRef]
32. Lim, T.K.; Ignatius, M.; Miguel, M.; Wong, N.H.; Juang, H.-M.H. Multi-scale urban system modeling for sustainable planning and design. *Energy Build.* **2017**, *157*, 78–91. [CrossRef]
33. Song, Y.; Song, X.; Shao, G. Effects of Green Space Patterns on Urban Thermal Environment at Multiple Spatial–Temporal Scales. *Sustainability* **2020**, *12*, 6850. [CrossRef]
34. AlKhaled, S.; Coseo, P.; Brazel, A.; Cheng, C.; Sailor, D. Between aspiration and actuality: A systematic review of morphological heat mitigation strategies in hot urban deserts. *Urban Clim.* **2020**, *31*, 100570. [CrossRef]
35. Ngarambe, J.; Nganyiyimana, J.; Kim, I.; Santamouris, M.; Yun, G.Y. Synergies between urban heat island and heat waves in Seoul: The role of wind speed and land use characteristics. *PLoS ONE* **2020**, *15*, e0243571. [CrossRef] [PubMed]
36. Gaffin, S.R.; Rosenzweig, C.; Khanbilvardi, R.; Parshall, L.; Mahani, S.; Glickman, H.; Goldberg, R.; Blake, R.; Slosberg, R.B.; Hillel, D. Variations in New York city's urban heat island strength over time and space. *Theor. Appl. Climatol.* **2008**, *94*, 1–11. [CrossRef]
37. He, B.-J.; Ding, L.; Prasad, D. Urban ventilation and its potential for local warming mitigation: A field experiment in an open low-rise gridiron precinct. *Sustain. Cities Soc.* **2020**, *55*, 102028. [CrossRef]
38. Ptak-Wojciechowska, A.; Januchta-Szostak, A.; Gawlak, A.; Matuszewska, M. The Importance of Water and Climate-Related Aspects in the Quality of Urban Life Assessment. *Sustainability* **2021**, *13*, 6573. [CrossRef]
39. Rynska, E.D.; Solarek, K. Adaptive Urban Transformation: Cities in Changing Health and Wellbeing Conditions. In *Sustainable Development and Planning X*; Passerini, G., Marchettini, N., Eds.; WIT Transactions on Ecology and the Environment; WIT Press: Rome, Italy, 2018; pp. 247–256.
40. Nyka, L. From Structures to Landscapes–towards Re-Conceptualization of The Urban Condition. In *Architectural Research Addressing Societal Challenges*; CRC Press: Boca Raton, FL, USA, 2017; pp. 509–515.
41. Christiaanse, K. Urban development as landscape. In *Changing Places, Contemporary German Landscape Architecture*; BDLA, Ed.; Birkhauser: Basel, Switzerland; Berlin, Germany; Boston, MA, USA, 2005; pp. 51–58.
42. Statistics Poland. Category K1 Territorrial Division. Group G441 Geodetic Area (Data of the Head Office Of Geodesy and Cartography). Subgroup P1410 Area. Available online: https://bdl.stat.gov.pl/bdl/metadane/cechy/1410 (accessed on 14 June 2022).
43. Statistics Poland. Category K3 Population. Group G7 Population. Subgroup P2462. Available online: https://bdl.stat.gov.pl/bdl/metadane/cechy/2462 (accessed on 14 June 2022).
44. Althoff, H. Siedlungsarbeit in der Freien Stadt Danzig 1920–1930, H. 2. 1930. Available online: https://pbc.gda.pl/dlibra/publication/2817/edition/3460?language=pl (accessed on 14 June 2022).
45. Lorens, P. Rozwój Urbanistyczny Gdańska w Latach 1918–1945 [Urban Development of Gdańsk in the Years 1918–1945]. In *100 Lat Nowoczesnej Urbanistyki w Gdańsku [100 Years of Modern Urban Planning in Gdańsk]*; Postawka, M., Lorens, P., Eds.; Politechnika Gdańska: Gdańsk, Poland, 2009; pp. 72–104.
46. Pawłowski, K.K. Początki polskiej nowoczesnej myśli urbanistycznej [The beginnings of Polish modern urban thought]. In *Sztuka około 1900 [Art around 1900]*; Białostocki, J., Ed.; Państwowe Wydawnictwo Naukowe: Warsaw, Poland, 1969; pp. 13–25.
47. Rudnicki, C. Regulacja Warszawy w okresie wieku XIX i początku XX [Regulation of Warsaw in the nineteenth and early twentieth centuries]. *Archit. I Bud.* **1928**, *11*, 404–409.
48. Różański, S.; Filipkowski, S.; Buckiewiczówna, M. Plan ogólny Wielkiej Warszawy [The general plan of the Great Warsaw]. *Archit. I Bud.* **1928**, *11–12*, 410–438.
49. Buckiewiczówna, M. Higjena urbanistyczna [Urban hygiene]. *Archit. I Bud.* **1928**, *11*, 433–438.
50. Myślińska, A.; Szczepański, J.; Dłubakowski, W. The Impact of Decommissioning Cemeteries on the Urban Ecosystem. *Sustainability* **2021**, *13*, 9303. [CrossRef]
51. Kodym-Kozaczko, G. *Urbanistyka Poznania w XX wieku [Town Planning of Poznań in the 20th Century]*; Wydział Architektury Politechniki Poznańskiej: Poznań, Poland, 2017.

52. Macias, A.; Dryjer, M. Forest cover dynamics in the city of Poznań from 1830 to 2004. *Quaest. Geogr.* **2010**, *29*, 47–57. [CrossRef]
53. Urbański, P.; Szpakowska, B.; Raszeja, E. Walory rekreacyjne zieleni Poznania [Recreational value of the city of Poznan]. *Nauk. Przyr. Technol.* **2008**, *2*, 27.
54. Fuchs, M.; Behrendt, F. *Die Stadt Breslau und die Eingemeindung ihres Erweiterungs Gebietes*; Friedrichdruck Grass, Barth & Comp.: Breslau, Poland, 1925.
55. Kononowicz, W. Accomplishments of urbanist and architect Ernst May in Wrocław in the years 1919–1925—A stage in the process towards functional Frankfurt. *Kwart. Archit. I Urban.* **2010**, *55*, 3–38.
56. Masztalski, R.; Kryczka, P. The Wrocław urban planning from general plans to studies on land use planning. *Teka Kom. Archit. Urban. I Stud. Kraj.* **1970**, *14*, 119–136. [CrossRef]
57. Czarnecki, W. *Planowanie Miast i Osiedli [Planning Cities and Urban Settlements]*; Państwowe Wydawnictwo Naukowe: Warszawa/Poznań, Poland, 1961.
58. Dymnicka, M.; Szczepański, J. Dilemmas of Identity in Contemporary Cities. The City of Gdansk as an Example. *Procedia Eng.* **2016**, *161*, 1225–1229. [CrossRef]
59. Osińska-Skotak, K.; Zawalich, J. Analysis of land use changes of urban ventilation corridors in warsaw in 1992–2015. *Geogr. Pol.* **2016**, *89*, 345–358. [CrossRef]
60. Bogucka, E.P. The use of GIS tools in the study of ventilation corridors in Warsaw—The example of frontal area index method. The case of Mokotowski Ventilation Corridor. In Proceedings of the EARSeL 34th Symposium, Warsaw, Poland, 16–20 June 2014.
61. Główny Urząd Geodezji i Kartografii [Head Office of Geodesy and Cartography] Geoportal. Available online: https://geoportal.gov.pl/ (accessed on 1 May 2022).
62. Ptaszycka, A. *Przestrzenie Zielone w Miastach [Green Spaces in Cities]*; Ludowa Spółdzielnia Wydawnicza: Warsaw, Poland, 1950.
63. ARMAAG Foundation. Available online: http://armaag.gda.pl/en/index.htm (accessed on 1 May 2022).
64. WIOŚ Poznań Air Quality Monitoring System. Available online: https://powietrze.poznan.wios.gov.pl/ (accessed on 1 May 2022).
65. WIOŚ Wrocław Air Quality Monitoring System. Available online: https://www.wroclaw.pios.gov.pl/ (accessed on 1 May 2022).
66. WIOŚ Warszawa Air Quality Monitoring System. Available online: http://wios.warszawa.pl/ (accessed on 1 May 2022).
67. Czernecki, B.; Półrolniczak, M.; Kolendowicz, L.; Marosz, M.; Kendzierski, S.; Pilguj, N. Influence of the atmospheric conditions on PM10 concentrations in Poznań, Poland. *J. Atmos. Chem.* **2017**, *74*, 115–139. [CrossRef]
68. Holnicki, P.; Nahorski, Z. Air quality modeling in Warsaw Metropolitan Area. *J. Theor. Appl. Comput. Sci.* **2013**, *7*, 56–69.
69. Majewski, J.G.; Przewoźniczuk, W. Study of particulate matter pollution in Warsaw area. *Pol. J. Environ. Stud.* **2009**, *18*, 293–300.
70. Yilmaz, I.; Gullu, M. Georeferencing of Historical Maps Using Back Propagation Artificial Neural Network. *Exp. Tech.* **2012**, *36*, 15–19. [CrossRef]
71. Baiocchi, V.; Lelo, K.; Milone, M.V.; Mormile, M. Accuracy of different georeferencing strategies on historical maps of Rome. *Geogr. Tech.* **2013**, *1*, 10–16.
72. Brigante, R.; Radicioni, F. Georeferencing of historical maps: Gis technology for urban analysis. *Geogr. Tech.* **2014**, *9*, 10–19.
73. Rumsey, D.; Williams, M. Historical maps in GIS. In *Past Time, Past Place: GIS for History*; ESRI Press: Redlands, CA, USA, 2002; pp. 1–18.
74. Antoniou, N.; Montazeri, H.; Wigo, H.; Neophytou, M.K.A.K.-A.; Blocken, B.; Sandberg, M. CFD and wind-tunnel analysis of outdoor ventilation in a real compact heterogeneous urban area: Evaluation using "air delay". *Build. Environ.* **2017**, *126*, 355–372. [CrossRef]
75. Janssen, W.D.; Blocken, B.; van Hooff, T. Pedestrian wind comfort around buildings: Comparison of wind comfort criteria based on whole-flow field data for a complex case study. *Build. Environ.* **2013**, *59*, 547–562. [CrossRef]
76. Righini, G.; Cappelletti, A.; Ciucci, A.; Cremona, G.; Piersanti, A.; Vitali, L.; Ciancarella, L. GIS based assessment of the spatial representativeness of air quality monitoring stations using pollutant emissions data. *Atmos. Environ.* **2014**, *97*, 121–129. [CrossRef]
77. Nowak, D.J.; Greenfield, E.J. Tree and impervious cover change in U.S. cities. *Urban For. Urban Green.* **2012**, *11*, 21–30. [CrossRef]
78. Tallis, M.; Taylor, G.; Sinnett, D.; Freer-Smith, P. Estimating the removal of atmospheric particulate pollution by the urban tree canopy of London, under current and future environments. *Landsc. Urban Plan.* **2011**, *103*, 129–138. [CrossRef]
79. Shi, Y.; Xie, X.; Fung, J.C.H.; Ng, E. Identifying critical building morphological design factors of street-level air pollution dispersion in high-density built environment using mobile monitoring. *Build. Environ.* **2018**, *128*, 248–259. [CrossRef]
80. Kubota, T.; Miura, M.; Tominaga, Y.; Mochida, A. Wind tunnel tests on the relationship between building density and pedestrian-level wind velocity: Development of guidelines for realizing acceptable wind environment in residential neighborhoods. *Build. Environ.* **2008**, *43*, 1699–1708. [CrossRef]
81. Peng, Y.; Gao, Z.; Ding, W. An Approach on the Correlation between Urban Morphological Parameters and Ventilation Performance. *Energy Procedia* **2017**, *142*, 2884–2891. [CrossRef]
82. Jhaldiyal, A. Automatic estimation of Urban Roughness Parameters for Microclimatic Analysis. Master's Thesis, Indian Institute of Remote Sensing, ISRO, Dehradun, Uttarakhand, India, 2015.
83. Wong, M.S.; Nichol, J.E.; Ng, E.Y.Y.; Guilbert, E.; Hei, K.; Kwokl, P.H.T.; Wang, J.Z.; Kwok, K.H.; To, P.H.; Wang, J.Z. GIS techniques for mapping urban ventilation, using frontal area index and least cost path analysis. *Int. Arch. Photogramm. Remote Sens. Spat. Inf. Sci.—ISPRS Arch.* **2002**, *38*, 586–591.

84. Grimmond, C.S.B.; Oke, T.R. Aerodynamic Properties of Urban Areas Derived from Analysis of Surface Form. *J. Appl. Meteorol.* **1999**, *38*, 1262–1292. [CrossRef]
85. Burian, S.J.; Brown, M.J.; Linger, S.P. *Morphological Analyses Using 3D Building Databases: Los Angeles, California. LA-UR-02-0781*; Los Alamos National Laboratory: Los Alamos, NM, USA, 2002.
86. Xie, P.; Liu, D.; Liu, Y.Y.; Liu, Y.Y. A Least Cumulative Ventilation Cost Method for Urban Ventilation Environment Analysis. *Complexity* **2020**, *2020*, 9015923. [CrossRef]
87. Jhaldiyal, A.; Gupta, K.; Gupta, P.K.; Thakur, P.; Kumar, P. Urban Morphology Extractor: A spatial tool for characterizing urban morphology. *Urban Clim.* **2018**, *24*, 237–246. [CrossRef]
88. Burghardt, R. *Development of an ArcGIS Extension to Model Urban Climate Factors. A Method of Automatic and Interactive Analysis to Capture the Influencing Factors on Urban Climate*; University of Kassel: Kassel, Germany, 2014.
89. Wong, M.S.; Nichol, J.E. Spatial variability of frontal area index and its relationship with urban heat island intensity. *Int. J. Remote Sens.* **2013**, *34*, 885–896. [CrossRef]
90. Architectural Institute of Japan (Ed.) *AIJ Benchmarks for Validation of CFD Simulations Applied to Pedestrian Wind Environment around Buildings*; Architectural Institute of Japan: Tokyo, Japan, 2016; ISBN 978-4-8189-5001-6.
91. Blocken, B.; Tominaga, Y.; Stathopoulos, T. CFD simulation of micro-scale pollutant dispersion in the built environment. *Build. Environ.* **2013**, *64*, 225–230. [CrossRef]
92. Ng, E. Policies and technical guidelines for urban planning of high-density cities-air ventilation assessment (AVA) of Hong Kong. *Build. Environ.* **2009**, *44*, 1478–1488. [CrossRef] [PubMed]
93. City of London Wind Microclimate Guidelines for Developments in the City of London. Available online: https://www.cityoflondon.gov.uk/services/planning/microclimate-guidelines (accessed on 14 June 2022).
94. Referat Prasowy Urzędu Miejskiego w Gdańsku [Press Office of the Gdańsk Municipal Office] Zielona Polityka Gdańska [Gdańsk Green Policy]. Available online: https://media.gdansk.pl/komunikaty/689663/zielona-polityka-gdanska (accessed on 20 May 2022).
95. Biuro Architektury i Planowania Przestrzennego m.st. Warszawy [Architecture and Spatial Planning Office of the Capital City of Warsaw]. Miejscowy Plan Ogólny Zagospodarowania Przestrzennego m.st. Warszawy [General Spatial Development Plan for the Capital City of Warsaw]. Available online: https://architektura.um.warszawa.pl/-/plan1992 (accessed on 20 May 2022).
96. Wicht, M.; Wicht, A.; Osińska-Skotak, K. Detection of ventilation corridors using a spatio-temporal approach aided by remote sensing data. *Eur. J. Remote Sens.* **2017**, *50*, 254–267. [CrossRef]
97. Wicht, M.; Wicht, A. LiDAR-Based Approach for Urban Ventilation Corridors Mapping. *IEEE J. Sel. Top. Appl. Earth Obs. Remote Sens.* **2018**, *11*, 2742–2751. [CrossRef]
98. Urząd Miasta Stołecznego Warszawy [Office of the Capital City of Warsaw]; Biuro Naczelnego Architekta Miasta [Office of the Chief Architect of Warsaw]; Miejska Pracowania Planowania Przestrzennego i Strategii Rozwoju [Municipal Department of Spatial Plann Opracowanie]. Ekofizjograficzne do Studium Uwarunkowań i Kierunków Zagospodarowania Przestrzennego m.st. Warszawy [Eco-Physiographic Study for the Study on Conditions and Spatial Development Directions for Warsaw]. Available online: https://architektura.um.warszawa.pl/-/opracowanie-ekofizjograficzne-do-studium-2006- (accessed on 20 May 2022).
99. Miejska Pracownia Urbanistyczna [Poznań Municipal Urban Planning Office]. Studium Uwarunkowań i Kierunków Zagospodarowania Przestrzennego Miasta Poznania [The Study on Conditions and Spatial Development Directions for Poznań]. Available online: http://sip.geopoz.pl/sip/studium/pliki/id_rap/39 (accessed on 20 May 2022).
100. Dymek, D.; Wilkaniec, A.; Bednorz, L.; Szczepańska, M. Significance of Allotment Gardens in Urban Green Space Systems and Their Classification for Spatial Planning Purposes: A Case Study of Poznań, Poland. *Sustainability* **2021**, *13*, 11044. [CrossRef]
101. Biuro Rozwoju Wrocławia [Wrocław City Development Office]. Studium Uwarunkowań i Kierunków Zagospodarowania Przestrzennego Wrocławia [The Study of the Conditions and Directions of the Spatial Development of the City of Wrocław]. Available online: http://gis.um.wroc.pl/www/pliki/studium-2017/studium_wroclawia (accessed on 20 May 2022).
102. Suder, A.; Szymanowski, M. Determination of Ventilation Channels In Urban Area: A Case Study of Wrocław (Poland). *Pure Appl. Geophys.* **2014**, *171*, 965–975. [CrossRef]
103. Burda, I.M.; Nyka, L. Providing Public Space Continuities in Post-Industrial Areas through Remodelling Land/Water Connections. *IOP Conf. Ser. Mater. Sci. Eng.* **2017**, *245*, 082037. [CrossRef]
104. Alexandra, J. The city as nature and the nature of the city-climate adaptation using living infrastructure: Governance and integration challenges. *Australas. J. Water Resour.* **2017**, *21*, 63–76. [CrossRef]
105. Burda, I.; Nyka, L. Re-Shaping the Land and Water Connections and its Role in Achieving Landscape and Ecological Systems' Continuity on The Post-Industrial Territories. In Proceedings of the International Multidisciplinary Scientific Conferences on Social Sciences and Arts, Albena, Bulgaria, 24–30 August 2016; pp. 533–537.
106. Lan, H.; Lau, K.K.-L.; Shi, Y.; Ren, C. Improved urban heat island mitigation using bioclimatic redevelopment along an urban waterfront at Victoria Dockside, Hong Kong. *Sustain. Cities Soc.* **2021**, *74*, 103172. [CrossRef]
107. Badach, J.; Dymnicka, M.; Baranowski, A. Urban vegetation in air quality management: A review and policy framework. *Sustainability* **2020**, *12*, 1258. [CrossRef]
108. Vos, P.E.J.; Maiheu, B.; Vankerkom, J.; Janssen, S. Improving local air quality in cities: To tree or not to tree? *Environ. Pollut.* **2013**, *183*, 113–122. [CrossRef]

109. Tomson, M.; Kumar, P.; Barwise, Y.; Perez, P.; Forehead, H.; French, K.; Morawska, L.; Watts, J.F. Green infrastructure for air quality improvement in street canyons. *Environ. Int.* **2021**, *146*, 106288. [CrossRef]
110. Chen, H.-S.; Lin, Y.-C.; Chiueh, P.-T. High-resolution spatial analysis for the air quality regulation service from urban vegetation: A case study of Taipei City. *Sustain. Cities Soc.* **2022**, *83*, 103976. [CrossRef]

sustainability

MDPI

Article

Prefabrication Implementation Potential Evaluation in Rural Housing Based on Entropy Weighted TOPSIS Model: A Case Study of Counties in Chongqing, China

Jingyuan Shi * and Jiaqing Sun *

School of Architecture and Urban Planning, Chongqing Jiaotong University, Chongqing 400074, China
* Correspondence: 990201900031@cqjtu.edu.cn (J.S.); jiaqing.sun@foxmail.com (J.S.)

Abstract: Prefabrication as a sustainable construction method has become a trend for use in house construction. However, the construction of rural houses in China still mainly adopts on-site construction, which also raises wasteful resources and environmental problems. Previous studies lack an evaluation system for the implementation potential of prefabricated rural housing in counties, and thus cannot provide references for the government to formulate implementation strategies. This study uses PEST analysis to establish an evaluation index system for the implementation potential of prefabricated rural housing and then evaluates 32 counties in Chongqing with urbanization rates below 90% based on the entropy weighted TOPSIS model. The results show that the weight values of the four evaluation subsystems of political, economic, social, and technological are 0.4516, 0.3152, 0.0684, and 0.1648, respectively; the nearness degrees of Dianjiang, Yubei, Jiangjin, and Rongchang are 0.5475, 0.4439, 0.4312, and 0.4103, respectively, ranking in the top four in Chongqing. The results indicate that the potential of implementing prefabricated rural housing in Chongqing is closely related to policy orientation and construction industrialization; Dianjiang, Yubei, Jiangjin, and Rongchang have the relative advantage of implementing prefabricated rural housing. Finally, this paper proposes political, economic, social, and technological suggestions for the implementation of prefabricated rural housing in Chongqing.

Keywords: prefabricated rural housing; PEST; entropy weighted TOPSIS; implementation potential; Chongqing

Citation: Shi, J.; Sun, J. Prefabrication Implementation Potential Evaluation in Rural Housing Based on Entropy Weighted TOPSIS Model: A Case Study of Counties in Chongqing, China. *Sustainability* **2023**, *15*, 4906. https://doi.org/10.3390/su15064906

Academic Editors: Oleg Kapliński, Wojciech Bonenberg, Agata Bonenberg and Lili Dong

Received: 29 January 2023
Revised: 3 March 2023
Accepted: 5 March 2023
Published: 9 March 2023

1. Introduction

1.1. Research Background

Sustainability is a core issue currently in focus in the construction industry. The 2022 Global Status Report for Buildings and Construction (Buildings-GSR) shows that in 2021, the building and construction sector accounted for around 37% of energy- and process-related CO_2 emissions and over 34% of energy demand globally; the building sector's operational energy-related CO_2 emissions reached an all-time high of around 10 $GtCO_2$ [1]. This is because on-site construction has been a common construction method in the building industry for the past decades [2,3]. In addition to serious environmental damage, traditional construction methods can lead to economic and social problems, such as long construction cycles, low labor productivity, and frequent safety accidents [4]. Traditional on-site construction lacks sustainability [5]. To solve the above problems, prefabricated construction (PC) has been introduced into the construction industry [6]. Prefabricated buildings are those based on industrial production methods, where all or some parts of the building structure and the building interior are built in an integrated manner using assembly [7]. Compared with traditional on-site construction, prefabricated technology can reduce 50% of construction waste [8], save 35.82% of resources, reduce 6.61% of health damage, and reduce 3.47% of ecosystem damage [9]. Prefabricated buildings can achieve 15.6%

actual carbon reduction and 3.2% operational carbon reduction [10], effectively reducing the carbon emissions and environmental impact of the construction industry [11,12].

Against emission peak and carbon neutrality, low-carbon construction to achieve sustainability has become the focus of most countries throughout the world. In 2020, Chinese President Xi Jinping announced at the 75th UN General Assembly that China aims to peak CO_2 emissions by 2030 and work toward achieving its carbon neutrality goal by 2060 [13]. However, relevant studies show that China's construction sector is likely to reach peak carbon by 2035, five years later than the national plan [10]. Therefore, China's construction industry urgently needs to choose green and low-carbon construction methods to achieve sustainable development of the construction industry as well as contribute to slowing down climate deterioration and saving natural resources [14]. Currently, the promotion of PC has become the focus of China's construction industry [15]. However, in China, it is still concentrated in urban areas, with rural areas are obviously lagging behind [16]. Under the national strategy of China's new rural construction and rural revitalization, the material living standard of China's countryside is increasing, and a large number of rural houses are being built in a short period of time [17,18]. However, over 70% of China's rural buildings are traditional brick-and-mortar and brick-and-timber structures. Meanwhile, the remaining 30% are mainly cast-in-place reinforced concrete structures [19], which means the current building quality and environmental performance of China's countryside is not meeting the requirements of contemporary sustainable development [20]. In addition, over 90% of new rural houses in China are built by villagers on their own initiative [21], with village builders invited to build them, lacking unified planning and professional technological guidance [22], and suffering from unregulated construction [23]. The construction mode of villagers' self-built houses leads to environmental pollution [24], poor structural safety [20], poor insulation, and high energy consumption [25], which restricts the sustainable development of China's countryside [26].

The Chinese government is also aware of the urgency of solving rural construction problems, and building green and livable rural housing has become an important task [27]. Prefabrication does not only realize the unified planning of rural residential design and address the need for safety and comfort in rural housing but also realizes energy savings and emission reduction throughout the life cycle of the housing. In May 2022, the Chinese government introduced the Action Plan on Rural Construction [28], noting the implementation of the project is to improve the quality and safety of rural housing. This plan explicitly requires promoting prefabricated steel, wood, and bamboo structures. However, in rural areas, due to the high cost of prefabricated buildings and the low awareness of farmers, prefabrication implementation still needs to be improved by using government projects. China's 14th Five-Year Plan is based on the goal of carbon peaking and carbon neutrality. Rural construction is a significant area of carbon emissions in the countryside, so it has become consensus to vigorously promote the construction of prefabrication housing there. As a result, there is an urgent need to establish a scientific and practical regional evaluation system to provide a reference for implementing strategies to promote prefabricated rural housing in the region.

1.2. Methodology and Purpose

The regional evaluation system of the construction industry has been studied by scholars. Liu et al. [29] established the evaluation system of regional prefabricated development level from five dimensions–technology, economy, sustainability, enterprise development, and development environment–and then used the AHP method to determine the weights of each indicator and evaluated Jiangsu province as an example; Dou et al. [30] explored the data collection method and quantification of evaluation indicators by using the advantage of new media data collection. Wang et al. [31] extracted 33 different indicators affecting industrialized buildings from existing literature and evaluated the development level of regional industrialized buildings based on the cloud model, taking Guangzhou as an example; Jin et al. [32] determined indicators from four levels–economic, social, technological innovation, and environmental resources–and used the AHP method to determine the weight values of each indicator establishing a gray comprehensive evaluation model to assess the sustainable development level of construction industrialization in the Beijing-Tianjin-Hebei region. From this, it can be seen that the existing relevant studies on the establishment of construction industrialization index systems and evaluation methods have been more mature, and multi-criteria decision-making methods (MCDMs) are more often used for evaluation.

Multi-criteria decision-making methods (MCDM) are an important part of modern decision science and are designed to support decision-makers who are faced with multiple decision criteria and multiple decision options [33]. Currently, MCDM methods have been applied by many decision-makers and researchers to solve complex problems, for example, AHP, ELECTRE, PROMETHEE, VIKOR, TOPSIS, etc. have been proposed and extended successively [34]. One of the most widely used multi-criteria decision-making methods is AHP, but the method requires an accurate distinction between the values of the decision problem, the included factors, and their intrinsic relationships [35]. The ELECTRE method eliminates inferior solutions by constructing a series of weakly dominant relationships, which sequentially reduce the number of alternatives without affecting the results by considering fewer data [36]. However, this method lacks objective data to help further understand the differences between alternatives and cannot fully utilize the information in the decision problem [37]. In both design and implementation, PROMETHEE is relatively simple [38] and this method has gradually evolved from a single method to include I-VI, PROMETHEE GDSS, PROMETHEE TRI, and other method families. However, it has shortcomings in problem design and weight determination [39] Hwang and Yoon Introduced TOPSIS to specify the most suitable solution based on the nearness degree to the ideal solution [40]. The VIKOR method was developed by Opricovic to solve MCDM problems containing different units and conflicting criteria so as to determine compromise solutions [41]. The last two methods are widely used in MCDM problems; TOPSIS uses vector normalization while VIKOR uses linear normalization, thus the former has higher accuracy [42]. Various studies in the context of MCDM emphasize the use of simple and understandable techniques to deal with MCDM problems and the computations should be simple and easy to perform. From the above, it can be seen that the TOPSIS method is practical in dealing with the MCDM problem compared to the other methods proposed above.

In the evaluation of MCDM, due to the diversity of raw data, the assignment of weight value can be divided into subjective and objective assignment methods [34]. The first is the method in which the decision maker independently assigns values to indicators based on their importance and usually relies on the subjective experience or judgment of people, representing only the decision maker's judgment of the importance of the indicator, which is more subjective and arbitrary. The objective assignment method is that which assigns weights to indicators through scientific calculation algorithms, and is not influenced by the subjective judgment of decision-makers, with the original data being derived from the attributes of the decision scheme [43].

Entropy in information theory is used to quantify the information content of a certain message [44]. The decision matrix for a set of alternatives contains a certain amount of information, so entropy can be used as a tool in weight value evaluation [45]. The entropy and TOPSIS methods are combined into a completely objective decision-making method, where the determination of weights and decision outcomes does not involve any subjective preferences but relies entirely on objective data of alternatives. Previous studies have shown the importance of the entropy weighted TOPSIS method for the study of problems related to development strategies. Huang et al. [46] evaluated the operational performance of urban rail transit systems by the entropy weighted TOPSIS method based on 34 months of initial data from the Chengdu subway. Yu et al. [47] used the entropy weighted TOPSIS method to evaluate industrial wastewater treatment projects. Bhowmik et al. [48] used the entropy weighted TOPSIS method to select the best green energy from multiple alternatives for sustainable planning. Kaynak et al. [49] used the entropy weighted TOPSIS method to evaluate the innovation performance of four EU candidate countries. These show that the entropy TOPSIS method is widely used in various fields.

Currently, China has recognized the need to implement prefabrication in rural areas, but no research provides quantitative models to assist in the development of prefabrication implementation strategies. Governments at all levels need to formulate rural prefabrication implementation policies based on regional development potential, but there is a lack of research aimed at evaluating regional rural prefabrication implementation potential. To fill this gap, this study proposes a method for evaluating regional rural prefabricated implementation potential, aiming to identify political, economic, social, and technological potential problems to better promote regional prefabricated rural housing implementation. The research purpose is to:

1. Determine the evaluation index system of county rural prefabrication implementation potential by PEST analysis and literature analysis;
2. To propose an entropy weighted TOPSIS evaluation method of rural prefabrication implementation potential, by improving the evaluation object and the formula of taking positive and negative ideal solutions, to further match the evaluation results with the real situation.
3. Chongqing Municipality was selected for empirical analysis to analyze the advantages and disadvantages of implementing rural prefabrication in its subordinate counties, which could provide a reference for other regions.

2. Materials and Methods

2.1. Selection of Evaluation Indicators

The selection of evaluation indicators is based on the principles of measurability and easy access to data. Based on a PEST analysis and a literature analysis, the elements affecting the implementation of rural prefabrication are extracted. The evaluation indicator system is constructed in conjunction with China's prefabrication industry. The PEST analysis effectively analyzes macro-environmental factors, where P, E, S, and T represent political, economic, social, and technological factors, respectively [50]. Currently, it is difficult to obtain endogenous factors in China's rural prefabrication industry, so this method understands the macro environment of rural prefabrication implementation by studying external environmental factors. The process used the Google Scholar search engine, with the keywords "prefab", "precast", and "off-site", and filtered the literature for research relevance [16]. By using the principle of measurable indicators and reading the literature 16 indicators affecting the implementation of prefabricated rural housing were extracted from four dimensions, as shown in Table 1, so that the evaluation results genuinely reflect the actual local situation.

The political layer reflects the policy conditions for the implementation of rural prefabrication. As the implementation of rural prefabrication in China is currently government-led, policy information is essential. P1 represents the strength and importance that governments at all levels attach to implementing rural prefabrication. P2 represents the mandatory share of new buildings applying prefabricated technology in government industrial planning documents by county. P3 represents the current implementation of rural prefabrication and P4 represents the practical support for implementing the prefabrication industry in each county.

The economic layer reflects the material basis for developing rural prefabrication in each county. E1 represents the level of economic development of each county, while E2–E4 reflect the capacity of each county to produce prefabricated components. E5 represents the road transport conditions of each county; if the road network is denser, then the rural access rate is higher and the transport cost lower.

Table 1. Indicator system for evaluating the implementation potential of prefabricated rural housing.

Criterion Layer	Indicator Layer	Code	Unit	Property	References
Political	Number of policies to incentivize the construction of prefabricated rural housing	P1	/	–	[51–56]
	Policy targets for the proportion of prefabricated buildings	P2	%	+	[52,54–57]
	Area of the prefabricated rural housing demonstration project	P3	m^2	+	[52,54–56,58]
	Target output value of the prefabricated component	P4	100 million yuan	+	[52,54,56,58–60]
Economic	GDP per capita	E1	Yuan	+	[51,52,59,61]
	Production capacity of prefabricated concrete components	E2	10,000 m^3	+	[51–56,58,62,63]
	Production capacity of prefabricated wall panels	E3	10,000 m^3	+	[51,56,58,62,63]
	Production capacity of prefabricated steel components	E4	10,000 tons	+	[51,56,58,62,63]
	Road network density	E5	/	+	[52,53,55,59,62–64]
Social	Year-end residential completions	S1	10,000 m^2	+	[7,52,55,65]
	Number of rural population	S2	10,000 people	+	[7,52,53,55]
	Disposable income per resident in rural areas	S3	Yuan	+	[7,51–53,55]
Technological	Number of prefabrication industrial bases	T1	/	+	[54,56,58,60,63,64,66,67]
	Number of people working in the construction industry	T2	10,000 people	+	[53,55,59,62,66,67]
	Number of construction general contract enterprises	T3	/	+	[53,56,58,64,65,68]
	Number of construction enterprises	T4	/	+	[53,56,58–60,65,66]

The social layer reflects the current state of the housing market in each county, which is the direct driver of rural prefabrication. S1 represents the housing demand in each county, S2 represents the number of potential consumers of prefabricated rural housing, and S3 represents the level of purchasing power of potential rural consumers.

The technological layer reflects the technological support for the implementation of rural prefabrication. T1 represents the technological conditions of rural prefabrication in each county. T2 represents the number of specialized construction workers in each county. T3 represents the number of leading enterprises in each county in terms of construction technology and scale and T4 represents the degree of perfection of the construction industry chain in each county.

2.2. Entropy Weighted TOPSIS Model

The entropy weighted TOPSIS model is an improvement on the traditional TOPSIS model, where the weights of evaluation indicators are determined by the entropy weighted method, then the ranking of evaluation objects is determined by the TOPSIS model using the method of approximating the ideal solution. The entropy weight method is based on information provided by each evaluation indicator to objectively determine its weight, which not only objectively reflects the importance of a specific indicator in the indicator system at the time of decision making, but also prominently reflects the change in the weight of the indicator over time, and is, therefore, suitable for regional implementation potential evaluation research. The core idea of the TOPSIS method is to define the distance between the optimal and inferior solutions of a decision problem, calculate the relative nearness degree of each evaluation object to the ideal solution, and rank the solutions' superiority. Determining the weights is an essential aspect of the TOPSIS method, and using the information entropy method can effectively eliminate the influence of subjective factors [69]. The main calculation steps of the entropy weighted TOPSIS method follow.

Step 1: Standardize the indicators. The indicators for evaluating the implementation potential of prefabricated rural housing are all positive, with larger values of positive indicators indicating higher potentials of implementation, and smaller values of negative indicators higher implementation potentials. In the event that both positive and negative indicators are present they need to be standardized in a dimensionless way, with the values of the indicators being in the range (0, 1). The standardization formulas are

$$\text{Positive indicators}: \ x_{ij}' = (x_{ij} - \min x_{ij}) / (\max x_{ij} - \min x_{ij}) \tag{1}$$

$$\text{Negative indicators}: \ x_{ij}' = (\max x_{ij} - x_{ij}) / (\max x_{ij} - \min x_{ij}) \tag{2}$$

where x_{ij} indicates the original value of the evaluation indicator, x_{ij}' is the standard value, and. $\max x_{ij}$ and $\min x_{ij}$ indicate the maximum and minimum values of the jth indicator for ith regions, respectively.

Step 2: H_j is determined as the information entropy value. $(1 - H_j)$ The greater the information utility value of an indicator, the greater the weight of that indicator in the evaluation and the more critical it is. Where, p_{ij} is the weight of the jth indicator in the ith region.

$$H_j = -\frac{1}{\ln m}\sum_{i=1}^{m} p_{ij} \ln p_{ij} \tag{3}$$

Step 3: Determine the weight of the jth indicator W_j

$$W_j = (1 - H_j) / \sum_{j=1}^{n} (1 - H_j) \tag{4}$$

Step 4: Construct a weighted decision matrix V

$$V = w_i * x_{ij}' \tag{5}$$

Step 5: Determine the positive and negative ideal solutions for the indicator. Let V^+ denote the best of all solutions, called the positive ideal solution, and V^- denote the least desirable solution, called the negative ideal solution.

$$V^+ = \{\max v_{ij} | i = 1, 2, \cdots, m\} \tag{6}$$

$$V^- = \{\min v_{ij} | i = 1, 2, \cdots, m\} \tag{7}$$

Step 6: The Euclidean distance is calculated. Let the distances of each evaluation object vector to the positive and negative ideal solutions be D^+ and D^-, respectively, then

$$D^+ = \sqrt{\sum_{j=1}^{m} \left(V_{ij} - V_j^+\right)^2} (i = 1, 2, \cdots, n) \tag{8}$$

$$D^- = \sqrt{\sum_{j=1}^{m} \left(V_{ij} - V_j^-\right)^2} (i = 1, 2, \cdots, n) \tag{9}$$

Step 7: Calculate the nearness degree C_j as

$$C_j = \frac{D^-}{D^+ + D^-} \tag{10}$$

The nearness degree indicates how close the rating object is to the positive ideal solution, i.e., the optimal solution, and is expressed by C_j. Obviously, for $C_j \in (0,1)$, the closer C_j is to 1, the closer the implementation potential value of prefabricated rural housing in the region is to the optimal level, and the promotion of prefabricated rural housing can be prioritized; conversely, the closer C_j is to 0, the further the potential implementation value of prefabricated rural housing in the region is to the optimal level, and the promotion of prefabricated rural housing can be carried out after the relevant industrial base is perfected.

2.3. Study Region

Chongqing is located in southwestern China, with 38 counties under its jurisdiction, as shown in Figure 1, a total area of 82,402 square kilometers, and a resident population of 32 million, of which 9.79 million people (30.54%) live in the countryside. As the only municipality directly under the central government in western China, Chongqing has many counties under its jurisdiction, a large rural population, and a typical dual economic structure. Moreover, there are significant differences among counties in terms of economy, geography, and culture. For example, counties in western Chongqing have convenient transportation and good economic development under the radius of Chongqing's urban area; southeastern Chongqing and northeastern Chongqing are relatively backward in economy and industry due to the obstruction of mountains and rivers and poor transportation; southeastern Chongqing is also a region inhabited by China's ethnic minorities.

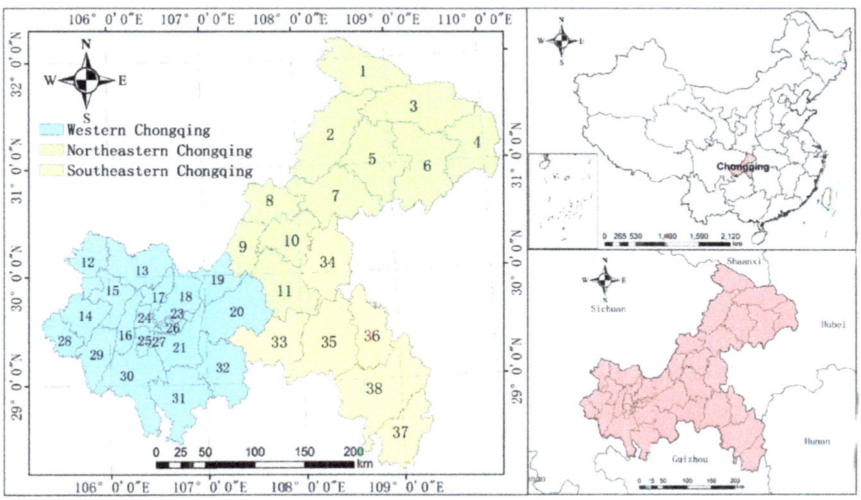

Figure 1. Chongqing Location Map. 1. Chengkou; 2. Kaizhou; 3. Wuxi; 4. Wushan; 5. Yunyang; 6. Fengjie; 7. Wanzhou; 8. Liangping; 9. Dianjiang; 10. Zhongxian; 11. Fengdu; 12. Tongnan; 13. Hechuan; 14. Dazu; 15. Tongliang; 16. Bishan; 17. Beibei; 18. Yubei; 19. Changshou; 20. Fuling; 21. Banan; 22. Yuzhong; 23. Jiangbei; 24. Shapingba; 25. Jiulongpo; 26. Qijiang; 37. Xiushan; 38. Youyang.

Therefore, the villages in Chongqing are complex and can better reflect the characteristics of Chinese villages. Simultaneously, Chongqing has also begun to pay attention to the implementation of prefabricated rural housing, which is the premise for this study being conducted. Therefore, the empirical evidence in this paper takes Chongqing as an example to measure the implementation potential of prefabricated rural housing in 32 of Chongqing's counties, excluding its jurisdictions with an urbanization rate of more than 90%, to provide a reference for the government to develop rural prefabrication implementation strategies.

2.4. Data Sources

The main sources of research data are the Chongqing Statistical Yearbook 2022, Chongqing Government Documents, the "14th Five-Year Plan" for the development of the modern construction industry in Chongqing, the vector data of Chongqing's road network, Chongqing's prefabricated components production enterprise list, and government websites. Except for the statistical yearbook, all data are as of January 2023.

3. Results

3.1. Evaluation Process

The original evaluation index matrix composed of raw data was standardized by Formulas (1) to (2), and then the weights of each evaluation index of the prefabricated implementation potential of Chongqing counties were calculated according to Formulas (3) to (5), and the results are shown in Table 2. Next, the positive and negative ideal solutions were determined according to Formulas (6) to (7). Then the weighted normalized evaluation matrix V is substituted into the Formulas (8) to (9) to derive the distance between the rural prefabricated implementation potential and the positive and negative ideal solutions of 32 counties in Chongqing, the results of which are shown in Table 3. Based on the distance between the prefabricated rural housing implementation potential and the positive and negative ideal solutions of each county in Table 3, the nearness degree of the rural prefabricated implementation potential of 32 counties in Chongqing can be obtained by Formula (10). Meanwhile, the evaluation results of the four evaluation subsystems of political, economic, social, and technological are calculated by the same method, and the final results are shown in Table 4.

The results of the evaluation of the implementation potential in 32 counties are divided into five levels based on the nearness degree: Level I (0.0, 0.2), Level II (0.2, 0.4), Level III (0.4, 0.6), Level IV (0.6, 0.8), and Level V (0.8, 1), as shown in Table 5 and Figure 2. When the nearness degree is closer to 1, the greater the potential for implementing prefabricated rural housing in the county. Based on the values of the overall nearness degree, it can be seen that 28 counties in Chongqing have an overall nearness degree concentrated between 0.0 and 0.4. There are significant difficulties in promoting prefabricated rural housing in Chongqing.

Table 2. The weighting of indicators.

Criterion Layer	Weight	Code	Indicator Layer	Weight
Political	0.4516	P1	Number of policies to incentivize the construction of prefabricated rural housing	0.1246
		P2	Policy targets for the proportion of prefabricated buildings	0.0547
		P3	Area of the prefabricated rural housing demonstration project	0.1817
		P4	The target output value of the prefabricated component	0.0907
Economic	0.3152	E1	GDP per capita	0.0158
		E2	Production capacity of prefabricated concrete components	0.0765
		E3	Production capacity of prefabricated wall panels	0.0858
		E4	Production capacity of prefabricated steel components	0.1117
		E5	Road network density	0.0255
Social	0.0684	S1	Year-end residential completions	0.0389
		S2	Number of rural population	0.0142
		S3	Disposable income per resident in rural areas	0.0152
Technological	0.1648	T1	Number of prefabrication industrial bases	0.0965
		T2	Number of people working in the construction industry	0.0253
		T3	Number of construction general contract enterprises	0.0178
		T4	Number of construction enterprises	0.0252

Table 3. Distance between the potential of rural prefabrication implementation and the positive and negative ideal solutions in 32 counties of Chongqing.

County	D^+	D^-	County	D^+	D^-
Beibei	0.2959	0.0713	Kaiju	0.3106	0.0313
Yubei	0.2152	0.1718	Liangping	0.3120	0.0196
Banan	0.2489	0.1494	Chengkou	0.3149	0.0002
Fuling	0.2767	0.0890	Fengdu	0.3071	0.0245
Qijiang	0.2514	0.1589	Dianjiang	0.1862	0.2253
Dazu	0.3044	0.0401	Zhongxian	0.3050	0.0239
Changshou	0.2818	0.0827	Yunyang	0.3073	0.0242
Jiangjin	0.2031	0.1540	Fengjie	0.3068	0.0331
Hechuan	0.2887	0.0682	Wushan	0.3135	0.0086
Yongchuan	0.2479	0.1583	Wushi	0.3141	0.0059
Nanchuan	0.2771	0.0890	Qianjiang	0.3087	0.0301
Bishan	0.2953	0.0640	Wulong	0.3137	0.0103
Tongliang	0.2675	0.0944	Shizhu	0.3137	0.0090
Tongnan	0.2295	0.1114	Xiushan	0.3135	0.0105
Rongchang	0.2287	0.1591	Youyang	0.3142	0.0071
Wanzhou	0.3034	0.0440	Pengshui	0.3137	0.0082

Table 4. Results of the rural prefabrication implementation potential evaluation.

County	Political	Economic	Social	Technological	Comprehensive	Ranking
Dianjiang	0.6365	0.4133	0.5427	0.3758	0.5475	1
Yubei	0.3902	0.6232	0.5691	0.7757	0.4439	2
Jiangjin	0.2985	0.5277	0.8174	0.5893	0.4312	3
Rongchang	0.4248	0.4464	0.4503	0.2807	0.4103	4
Yongchuan	0.2568	0.5143	0.6168	0.6452	0.3897	5
Qijiang	0.1952	0.4968	0.4287	0.5972	0.3872	6
Banan	0.4309	0.3445	0.6199	0.3558	0.3751	7
Tongnan	0.3823	0.2403	0.5093	0.3896	0.3267	8
Tongliang	0.2411	0.3852	0.5311	0.4465	0.2610	9
Fuling	0.2103	0.3837	0.4659	0.6552	0.2434	10
Nanchuan	0.2243	0.2646	0.3280	0.2703	0.2431	11
Changshou	0.1023	0.4933	0.6307	0.1766	0.2268	12
Beibei	0.1866	0.4257	0.4451	0.2098	0.1942	13
Hechuan	0.1232	0.4035	0.6336	0.4032	0.1912	14
Bishan	0.1023	0.4125	0.4697	0.3067	0.1782	15
Wanzhou	0.1036	0.2070	0.5469	0.5070	0.1265	16
Dazu	0.1023	0.3242	0.4929	0.2895	0.1163	17
Fengjie	0.0000	0.1258	0.4715	0.3647	0.0975	18
Kaiju	0.0000	0.1072	0.6302	0.3239	0.0916	19
Qianjiang	0.1036	0.1311	0.2945	0.1566	0.0888	20
Fengdu	0.0708	0.1680	0.3409	0.2090	0.0739	21
Yunyang	0.0000	0.1582	0.4272	0.3234	0.0731	22
Zhong	0.0363	0.1926	0.4189	0.2023	0.0726	23
Liangping	0.0000	0.2404	0.4284	0.1875	0.0592	24
Xiushan	0.0000	0.1666	0.2145	0.1117	0.0323	25
Wulong	0.0000	0.1871	0.2259	0.0618	0.0318	26
Shizhu	0.0000	0.0939	0.2516	0.0979	0.0280	27
Wushan	0.0000	0.0884	0.1940	0.1266	0.0266	28
Pengshui	0.0000	0.0973	0.2346	0.0810	0.0253	29
Youyang	0.0000	0.0289	0.2432	0.0374	0.0222	30
Wushi	0.0000	0.0177	0.1089	0.1218	0.0184	31
Chengkou	0.0000	0.0000	0.0076	0.0000	0.0006	32

Table 5. Grading of evaluation results.

Value Range	Status	Level	Feature Description
0 < H < 0.2	worse	I	The level of implementation potential of rural prefabrication is extremely low and not suited to the promotion of prefabricated rural housing.
0.2 ≤ H < 0.4	bad	II	The level of implementation potential of rural prefabrication is low and barely suited to the promotion of prefabricated rural housing.
0.4 ≤ H < 0.6	normal	III	The level of implementation potential of rural prefabrication is general and basically suited to the promotion of prefabricated rural housing.
0.6 ≤ H < 0.8	good	IV	The level of implementation potential of rural prefabrication is good and more suited to the promotion of prefabricated rural housing.
0.8 ≤ H < 1.0	excellent	V	The level of implementation potential of rural prefabrication is very high and well suited to the promotion of prefabricated rural housing.

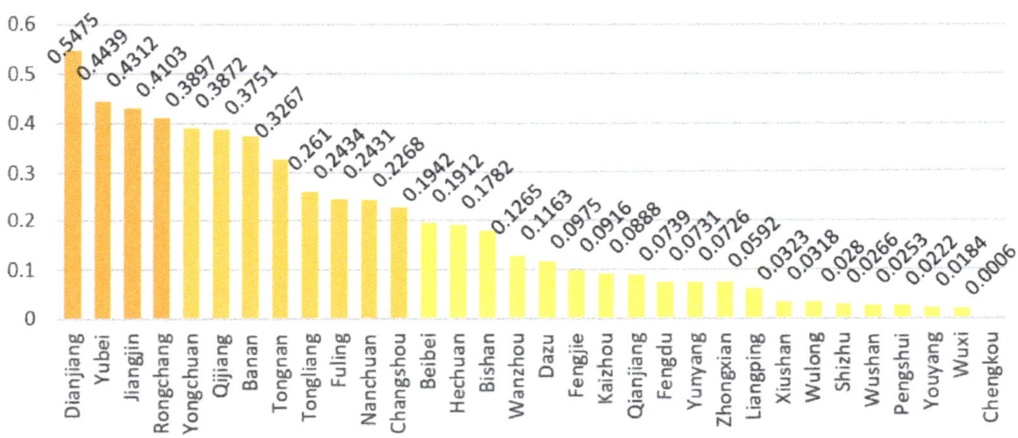

Figure 2. Ranking the comprehensive implementation potential of prefabricated rural housing in Chongqing by county.

ArcGIS was used to visualize the evaluation results to visually reflect the implementation potential of prefabricated rural housing in Chongqing, resulting in Figures 3 and 4.

3.2. Analysis of Comprehensive Implementation Potential Evaluation Results

Figure 3 shows the comprehensive evaluation results. There are 28 counties in Chongqing with comprehensive nearness degrees concentrated in (0.0, 0.4). This indicates great difficulties in implementing prefabricated rural housing in Chongqing as a whole. The potential of rural prefabricated implementation in Dianjiang, Yubei, Jiangjin, and Rongchang is at Level III, which meets the basic conditions for implementing prefabricated rural housing. Yongchuan, Qijiang, Banan, Tongnan, Tongliang, Fuling Nanchuan, and Changshou show and implementation potential at Level II and have the relative possibility of implementing prefabricated rural housing in Chongqing. The other 20 counties have the potential of implementing prefabricated rural housing below 0.2, and the development foundation is weak and not suitable for implementing prefabricated rural housing. In summary, some counties in western Chongqing have the possibility of implementing prefabricated rural housing due to their good economic and construction industry foundation.

Only Dianjiang in northeastern and southeastern Chongqing can implement prefabricated rural housing.

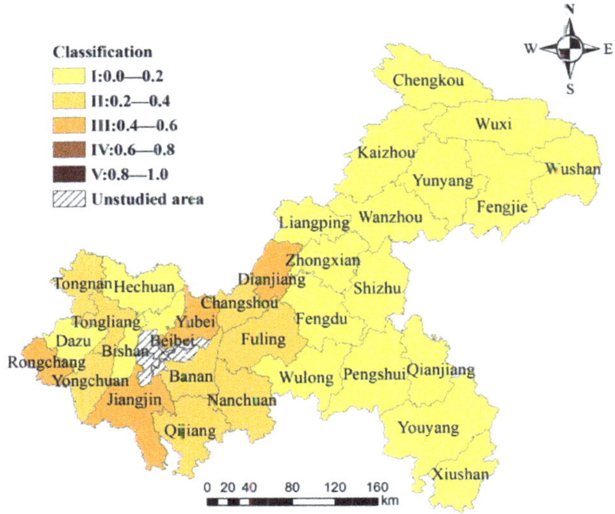

Figure 3. Results of the comprehensive evaluation of the implementation potential of prefabricated rural housing in Chongqing.

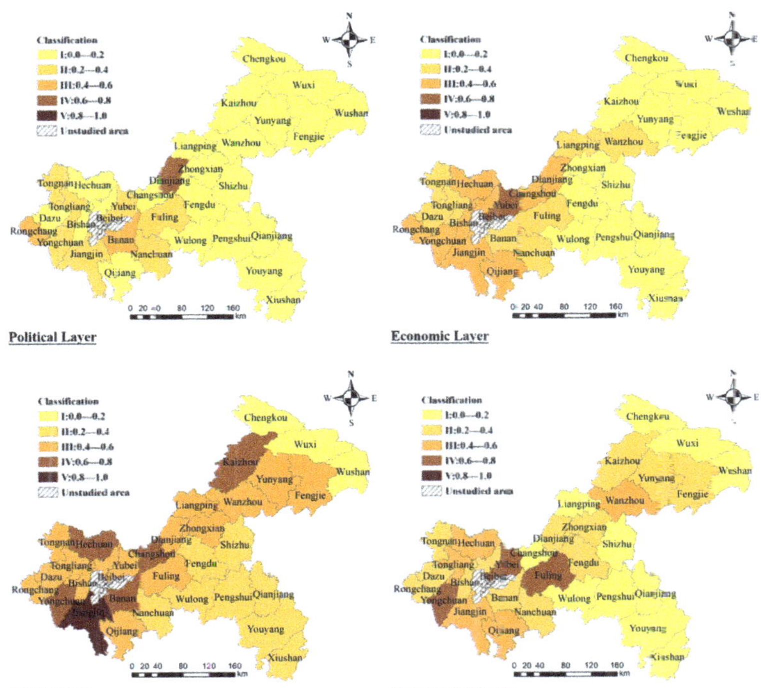

Figure 4. Results of the subsystem evaluation of the implementation potential of prefabricated rural housing in Chongqing.

3.3. Analysis of Subsystem Evaluation Results

Figure 4 shows the evaluation results of each subsystem. The evaluation results of the political level subsystem show that Dianjiang is located at Level IV, and Banan and Rongchang are located at Level III with relative policy advantages. In 2022, Chongqing subsidized RMB 500/m^2 and RMB 200/m^2 for new prefabricated rural houses for self-occupation or business, respectively. Under the policy guidance, Dianjiang, Tongnan, Rongchang, Jiangjin, and Banan have taken the lead in the construction of prefabricated rural housing demonstration projects. At the same time, Chongqing Municipality clarifies the proportion of prefabricated buildings to new buildings in each county: 20–25% in western Chongqing, 15–20% in northeastern Chongqing, and 15% in southeastern Chongqing. Banan, Rongchang, and Tongnan issued policies requiring the active promotion of green building materials applications and PC methods in rural areas. The policies of each county regarding the implementation of rural prefabrication are reflected in the evaluation results of the political layer subsystem.

The evaluation results of the economic layer subsystem show that the overall potential of western Chongqing is high. There are 61 prefabricated component production factories in Chongqing, involving the production of prefabricated concrete components, steel components, and new wall panel components, which are concentrated in western Chongqing and the counties adjacent to western Chongqing. This shows that areas with a high level of regional economic development and dense road networks attract more prefabricated enterprises to invest and set up factories, which will galvanize the development of local prefabricated rural houses.

The evaluation results of the social layer subsystem show that there is a significant rural housing demand in western and northeastern Chongqing (except in Chengkou, Wuxi, and Wushan counties). In contrast, Chengkou, Wuxi, and Wushan counties in southeast and northeastern Chongqing have weaker economic bases, the villagers have lower disposable income, and there is less new housing demand. The high cost of prefabricated rural houses creates difficulties in implementation for Chengkou, Wuxi, and Wushan counties.

The evaluation results of the technological layer subsystem show that the technological conditions for the promotion of prefabricated rural housing in western Chongqing are significantly better than those in northeastern and southeastern Chongqing. The 30 prefabricated industry bases in Chongqing are all located in western Chongqing, which could provide technological and workforce support for the implementation of rural prefabrication. Wanzhou, as the central county of northeastern Chongqing, has a relative advantage in the technical dimension of implementing prefabricated rural housing due to a large number of construction general contractors and construction enterprises.

4. Discussion

4.1. Empirical Findings

The weighting of the indicators in the evaluation system for the implementation potential of prefabricated rural housing is 0.4516, 0.3152, 0.0684, and 0.1648 for the political, economic, social, and technological subsystems, respectively, and the sum of the weighting of the political and economic subsystems is over 0.76. The potential for implementing prefabricated rural housing in Chongqing is closely related to policy orientation and the foundation of industrialization. The sum of the weights of the P1 and P3 indicators in the political subsystem is 0.3, which corresponds to the number of policies to incentivize the construction of prefabricated rural housing and the area of demonstration projects for prefabricated rural housing; the sum of the weights of the E2, E3, and E4 indicators in the economic subsystem is 0.28, which corresponds to the capacity of various prefabricated components. In summary, the announcement of policies related to prefabricated rural housing and the construction of prefabricated rural housing demonstration projects could promote the implementation of prefabricated rural housing in the region. At the same time, good prefabricated component plant capacity is the basis for promoting prefabricated rural housing in the region.

It can be seen from the results of the implementation potential of prefabricated rural housing in Chongqing that the potential for implementing prefabricated rural housing in western Chongqing is significantly higher than that of other regions, due to their relatively good economic and construction industrialization base and with the support of relevant policies. In contrast, northeastern and southeastern Chongqing (except Dianjiang) are disadvantaged in implementing prefabricated rural housing due to their insufficient policy efforts, weak economic base, and lack of technological support. The implementation potential of each region in Chongqing is ranked as follows: Western Chongqing >> Northeastern Chongqing > Southeastern Chongqing. The counties that should be prioritized for implementation are Dianjiang, Yubei, Jiangjin, and Rongchang.

4.2. Countermeasures and Suggestions

Previous studies on prefabricated building promotion strategies coincide in recognizing the necessity of government macro policies, which is also in line with the logic of this study to construct an evaluation index system for the potential of rural prefabrication implementation using PEST analysis. For example, Correia et al. [70] argued that government policy is the main driver of promoting prefabrication technology. Du et al. [71] argued that incentive policies and uniform industry standards are necessary to promote housing industrialization. Xue et al. [72] suggested that the government should increase the proportion of prefabrication in public projects and extend the pilot experience to other projects. The government should also promote communication among relevant stakeholders to achieve cost reduction.

Scholars have also provided insight into the economic, social, and technological means of implementing prefabricated buildings. From an economic perspective, some scholars suggested that introducing a competition mechanism into the PC market could effectively improve the quality of prefabricated buildings and reduce construction costs. For example, Lou et al. [73] suggested that expanding the number of component factories and improving the PC chain are beneficial to achieve fine management of PC. Chiang et al. [58] argued that the technology of prefabricated construction contracting firms must be valuable, rare, not completely limited, and irreplaceable to be a source of sustainable competitive advantage. From a social perspective, Moradibistouni et al. [74] argued that key to the promotion of prefabricated buildings is the attitude of potential users, which requires government investment and the construction of a large number of high-quality prefabricated demonstration projects. From a technical perspective, Lou et al. [73] proposed to improve technology in the prefabricated building industry by combining BIM technology. Mao et al. [75] argued that the improvement of component production capacity and technology level is an effective way to reduce costs in prefabricated buildings. Jiang et al. [54] argued that the government needs to develop unified technical standards, train skilled workers, and foster technology pioneer companies in government-led projects. Chang et al. [65] argued that there are few sunk cost barriers in China's emerging prefabrication industry and that new prefabrication factories should adopt high-performance green technologies and equipment, as well as cultivate a stable and skilled workforce. Furthermore, the contracting model of projects should be optimized to enhance collaboration in all phases of prefabricated construction.

Combining the results of previous research and empirical analysis, considering the differences between prefabricated rural and prefabricated urban buildings, the following countermeasures and suggestions are proposed for the implementation of prefabricated rural housing in Chongqing.

Political Dimension: Adopt mandatory policies to stipulate the proportion of prefabrication in rural demonstration projects and develop uniform and feasible technical and industry standards to guide residential construction. Adopt incentive policies to reduce the construction burden of farmers and attract corporate investment. Counties with a good industrial base located in western Chongqing are given priority to implement prefabricated rural housing construction. Other counties can adopt prefabricated technology in the renovation of rural houses to improve the life cycle of houses and gradually realize the industrialization of rural housing.

Economic Dimension: At present, there are 61 prefabricated building component manufacturing enterprises in Chongqing, of which 52 are located in western Chongqing, 9 in northeastern Chongqing, and 0 in southeastern Chongqing. It could take the lead in building prefabricated rural housing industry chains in western Chongqing, cultivate general contracting enterprises, and form good market competition. Dianjiang and Wanzhou in northeast Chongqing could arrange prefabricated rural housing industry bases to attract the construction of prefabricated component factories and optimize the construction cost of prefabricated rural housing. Southeastern Chongqing does not have the conditions for the development of rural residential industrialization for the time being.

Social Dimension: In the context of China's rural revitalization, villagers' living standards are increasingly improving and they are actively seeking a better quality environment. Chongqing's huge rural population base creates a massive rural construction market. The government can use prefabricated technology in rural public projects and housing construction to enhance villagers' understanding and recognition of prefabricated housing.

Technological Dimension: The government could set design selection standards for prefabricated rural housing, organize a special technical demonstration, promote the use of BIM technology, and cultivate prefabricated industry bases to promote technical exchanges; at the same time, strengthen technical training for rural grassroots builders, blending regional construction experience with prefabricated technology, and achieve increased rural employment opportunities while building prefabricated livable rural houses with regional characteristics.

4.3. Suggestions for Future Research

This study focuses on an approach based on objective means to solve evaluation problems and validates its effectiveness with case studies. Although the entropy weighted TOPSIS model has been successfully applied to a large number of cases, it also has limitations. For example, the evaluation system needs to be established by selecting quantitative indicators; the results of indicator weights are territorial; and the weights can be affected significantly when the dispersion of indicator data is too high. The development of MCDM methods is getting faster and faster at present. The emergence of new methods is expected to provide more reliable analysis results. The ordinal priority approach (OPA), for example, is a novel and potential alternative. The OPA is an emerging MCDM method proposed by Ataei et al. in 2020 [76]. This method is based on linear programming and ordinal relations to solve MCDM problems and is currently considered an effective, objective, and flexible method. The significant advantage of this method is that it does not require a standardization process, pairwise comparison, and data integrity. For instance, the OPA does not make use of a pairwise comparison matrix, decision-making matrix (no need for numerical input), normalization methods, and averaging methods for aggregating the opinions of experts (in group decision making) [77]. The current related research extends OPA, for example, Mahmoudi et al. [77] proposed OPA for Fuzzy Linguistic Information (OPA-F), which has extended its applicability to problems containing linguistic information. Mahmoudi et al. [78] also proposed the Grey Ordinal Priority Approach (OPA-G), showing that it can work without any linguistic variable or pairwise comparison-based data and has a high capability of dealing with greyness/uncertainty. Mahmoudi et al. [79] also proposed the Robust Ordinal Priority Approach (OPA-R), which can detect and decrease subjectivity in experts' opinions; calculate the weights of the experts, criteria, and projects associated with the most robust scenario.

On the other hand, previous research results provide a macro-strategic reference for the initial implementation of prefabricated rural housing in the region, and the shortcomings exposed by its subsystem evaluation (PEST) deserve further study. The entropy weighted TOPSIS model can be combined with grey relation analysis in the methodology. Grey relation analysis is a method for quantitative description and comparison of the development of a system and, using this method, the degree of influence of each sub-series on the parent series can be analyzed [80]. Therefore, using grey relation analysis can find the degree of relation between each subsystem and the parent system, analyze the potential influence of internal subsystems on the implementation of prefabricated rural housing by the magnitude of the relation, and conduct an internal horizontal evaluation of the evaluation system.

5. Conclusions

Rural prefabrication implementation in China has just started, and there is a lack of research to evaluate the potential of rural prefabrication implementation in counties. To address this research gap, this study proposes an evaluation method for the potential of rural prefabrication implementation in counties. Through a systematic review of previous studies on the evaluation of regional construction industrialization development, this study identifies 16 evaluation indicators in four dimensions: political, economic, social, and technological, and uses the entropy weighted TOPSIS model for indicator assignment and case empirical evidence to provide strategies and suggestions for rural prefabrication implementation in Chongqing.

The entropy weighted TOPSIS model overcomes the bias caused by personal factors in previous subjective assignment methods and reflects the difference between the potential of the evaluation object and the ideal level. The entropy weighted TOPSIS model also shows that the indicator weights are not static and will change not only with the evaluation region but also with the development of the industry. However, this paper is not fully developed in terms of indicator selection and evaluation system construction due to the influence of data accessibility, so the evaluation study on the implementation potential of county rural prefabrication still has some limitations. Chongqing, as the city with the largest number of counties under its jurisdiction in China, has uneven development across regions and is therefore typical. Choosing Chongqing as a research case allows us to judge the rationality and validity of the evaluation system. However, at the same time, because different cities are in different development stages and social environments, some indicators may differ to some extent between cities and countries. Therefore, it is valuable to continue to deepen the universality of the evaluation system and evaluation methods.

Author Contributions: Conceptualization, J.S. (Jingyuan Shi) and J.S. (Jiaqing Sun); methodology, J.S. (Jingyuan Shi); software, J.S. (Jiaqing Sun); validation, J.S. (Jingyuan Shi); formal analysis, J.S. (Jingyuan Shi) and J.S. (Jiaqing Sun); investigation, J.S. (Jiaqing Sun); resources, J.S. (Jingyuan Shi); data curation, J.S. (Jiaqing Sun); writing—original draft preparation, J.S. (Jingyuan Shi); writing—review and editing, J.S. (Jingyuan Shi) and J.S. (Jiaqing Sun); supervision J.S. (Jingyuan Shi) and J.S. (Jiaqing Sun); project administration, J.S. (Jingyuan Shi) and J.S. (Jiaqing Sun). All authors have read and agreed to the published version of the manuscript.

Funding: National College Students' Innovation and Entrepreneurship Training Program: Research on the Design Strategy of Prefabricated Rural Housing in Western Chongqing in the Context of Cultural, Agricultural, and Tourism Integration (202210618009).

Institutional Review Board Statement: Not applicable.

Informed Consent Statement: Not applicable.

Data Availability Statement: Not applicable.

Conflicts of Interest: The authors declare no conflict of interest.

References

1. GlobalABC 2022 Global Status Report for Buildings and Constructions 2022. Available online: https://globalabc.org/our-work/tracking-progress-global-status-report (accessed on 27 January 2023).
2. Jaillon, L.; Poon, C.S. Sustainable construction aspects of using prefabrication in dense urban environment: A Hong Kong case study. *Constr. Manag. Econ.* **2008**, *26*, 953–966. [CrossRef]
3. Shen, L.Y.; Tam, V.W.; Tam, L.; Ji, Y.B. Project feasibility study: The key to successful implementation of sustainable and socially responsible construction management practice. *J. Clean. Prod.* **2010**, *18*, 254–259. [CrossRef]
4. Zhang, W.; Lee, M.W.; Jaillon, L.; Poon, C.S. The hindrance to using prefabrication in Hong Kong's building industry. *J. Clean. Prod.* **2018**, *204*, 70–81. [CrossRef]
5. Dixit, M.K.; Fernández-Solís, J.L.; Lavy, S.; Culp, C.H. Identification of parameters for embodied energy measurement: A literature review. *Energy Build.* **2010**, *42*, 1238–1247. [CrossRef]
6. Wu, Z.; Luo, L.; Li, H.; Wang, Y.; Bi, G.; Antwi-Afari, M.F. An analysis on promoting prefabrication implementation in construction industry towards sustainability. *Int. J. Environ. Res. Public Health* **2021**, *18*, 11493. [CrossRef]
7. Steinhardt, D.A.; Manley, K. Adoption of prefabricated housing–the role of country context. *Sustain. Cities Soc.* **2016**, *22*, 126–135. [CrossRef]
8. Hu, X.; Chong, H.Y. Environmental sustainability of off-site manufacturing: A literature review. *Eng. Constr. Archit. Manag.* **2019**, *28*, 332–350. [CrossRef]
9. Cao, X.; Li, X.; Zhu, Y.; Zhang, Z. A comparative study of environmental performance between prefabricated and traditional residential buildings in China. *J. Clean. Prod.* **2015**, *109*, 131–143. [CrossRef]
10. Chen, X.; Shuai, C.; Wu, Y.; Zhang, Y. Analysis on the carbon emission peaks of China's industrial, building, transport, and agricultural sectors. *Sci. Total Environ.* **2020**, *709*, 135768. [CrossRef]
11. Tumminia, G.; Guarino, F.; Longo, S.; Ferraro, M.; Cellura, M.; Antonucci, V. Life cycle energy performances and environmental impacts of a prefabricated building module. *Renew. Sustain. Energy Rev.* **2018**, *92*, 272–283. [CrossRef]
12. Li, X.J.; Xie, W.J.; Jim, C.Y.; Feng, F. Holistic LCA evaluation of the carbon footprint of prefabricated concrete stairs. *J. Clean. Prod.* **2021**, *329*, 129621. [CrossRef]
13. Xinhua. Xinhua Headlines: China Moves to Explore Path toward Carbon Neutrality Goal. 2020. Available online: http://www.xinhuanet.com/english/2020-10/18/c_139449759.htm (accessed on 27 January 2023).
14. Shi, Q.; Wang, Z.; Li, B.; Hertogh, M.; Wang, S. Evolutionary Analysis of Prefabrication Implementation in Construction Projects under Low-Carbon Policies. *Int. J. Environ. Res. Public Health* **2022**, *19*, 12511. [CrossRef] [PubMed]
15. Hosseini, M.R.; Martek, I.; Zavadskas, E.K.; Aibinu, A.A.; Arashpour, M.; Chileshe, N. Critical evaluation of off-site construction research: A Scientometric analysis. *Autom. Constr.* **2018**, *87*, 235–247. [CrossRef]
16. Zhou, J.; He, P.; Qin, Y.; Ren, D. A selection model based on SWOT analysis for determining a suitable strategy of prefabrication implementation in rural areas. *Sustain. Cities Soc.* **2019**, *50*, 101715. [CrossRef]
17. Yu, Z. The Achievements, Problems and Improvements in the Development of the Rule of Law for Rural Environmental Protection. *IOP Conf. Ser. Earth Environ. Sci.* **2020**, *545*, 012001. [CrossRef]
18. Tan, Y. Vitalization of rural areas and environmental management. *IOP Conf. Ser. Mater. Sci. Eng.* **2020**, *782*, 052017. [CrossRef]
19. Evans, M.; Yu, S.; Song, B.; Deng, Q.; Liu, J.; Delgado, A. Building energy efficiency in rural China. *Energy Policy* **2014**, *64*, 243–251. [CrossRef]
20. Dong, J.; Jin, H. The design strategy of green rural housing of Tibetan areas in Yunnan, China. *Renew. Energy* **2013**, *49*, 63–67. [CrossRef]
21. Qi, F.; Musonda, B.M.; Shen, H.; Wang, Y. Geometric classification method of rural residences at regional scale. *Energy Build.* **2018**, *172*, 170–180. [CrossRef]
22. Wu, Y.; Li, H.F. Attention to rural green-building design in China. *Appl. Mech. Mater.* **2013**, *357*, 327–331. [CrossRef]
23. Li, X.P. Applied Research on Straw Bale in Northwest Rural Residential Building. *Appl. Mech. Mater.* **2012**, *204*, 3815–3818. [CrossRef]
24. He, B.J.; Yang, L.; Ye, M.; Mou, B.; Zhou, Y. Overview of rural building energy efficiency in China. *Energy Policy* **2014**, *69*, 385–396. [CrossRef]
25. Liu, W.; Spaargaren, G.; Mol, A.P.; Heerink, N.; Wang, C. Low carbon rural housing provision in China: Participation and decision making. *J. Rural Stud.* **2014**, *35*, 80–90. [CrossRef]
26. Rocchi, L.; Kadziński, M.; Menconi, M.E.; Grohmann, D.; Miebs, G.; Paolotti, L.; Boggia, A. Sustainability evaluation of retrofitting solutions for rural buildings through life cycle approach and multi-criteria analysis. *Energy Build.* **2018**, *173*, 281–290. [CrossRef]
27. Liu, Y. Introduction to land use and rural sustainability in China. *Land Use Policy* **2018**, *74*, 1–4. [CrossRef]
28. General Office of the State Council, PRC. Implementation Plan for Rural Construction Action. 2022. Available online: http://www.gov.cn/gongbao/content/2022/content_5695035.htm (accessed on 27 January 2023).
29. Liu, P.; Li, Q.; Song, L.; Jia, R. The index system for the development level evaluation of regional construction industrialization: A case study in Jiangsu, China. *Appl. Sci.* **2017**, *7*, 492. [CrossRef]
30. Dou, Y.; Xue, X.; Wang, Y.; Luo, X.; Shang, S. New media data-driven measurement for the development level of prefabricated construction in China. *J. Clean. Prod.* **2019**, *241*, 118353. [CrossRef]

31. Wang, T.; Wang, X.; Wang, L.; Au-Yong, C.P.; Ali, A.S. Assessment of the development level of regional industrialized building based on cloud model: A case study in Guangzhou, China. *J. Build. Eng.* **2021**, *44*, 102547. [CrossRef]
32. Jin, Z.; Xia, S.; Cao, H.; Geng, X.; Cheng, Z.; Sun, H.; Jia, M.; Liu, Q.; Sun, J. Evaluation and Optimization of Sustainable Development Level of Construction Industrialization: Case Beijing-Tianjin-Hebei Region. *Sustainability* **2022**, *14*, 8245. [CrossRef]
33. Kumar, A.; Sah, B.; Singh, A.R.; Deng, Y.; He, X.; Kumar, P.; Bansal, R.C. A review of multi criteria decision making (MCDM) towards sustainable renewable energy development. *Renew. Sustain. Energy Rev.* **2017**, *69*, 596–609. [CrossRef]
34. Zhao, D.Y.; Ma, Y.Y.; Lin, H.L. Using the entropy and TOPSIS models to evaluate sustainable development of islands: A case in China. *Sustainability* **2022**, *14*, 3707. [CrossRef]
35. Yalcin, A.S.; Kilic, H.S.; Delen, D. The use of multi-criteria decision-making methods in business analytics: A comprehensive literature review. *Technol. Forecast. Soc. Change* **2022**, *174*, 121193. [CrossRef]
36. Govindan, K.; Jepsen, M.B. ELECTRE: A comprehensive literature review on methodologies and applications. *Eur. J. Oper. Res.* **2016**, *250*, 1–29. [CrossRef]
37. Jahan, A.; Ismail, M.Y.; Sapuan, S.M.; Mustapha, F. Material screening and choosing methods–a review. *Mater. Des* **2010**, *31*, 696–705. [CrossRef]
38. Behzadian, M.; Kazemzadeh, R.B.; Albadvi, A.; Aghdasi, M. PROMETHEE: A comprehensive literature review on methodologies and applications. *Eur. J. Oper. Res.* **2010**, *200*, 198–215. [CrossRef]
39. Hatami-Marbini, A.; Tavana, M.; Moradi, M.; Kangi, F. A fuzzy group Electre method for safety and health assessment in hazardous waste recycling facilities. *Saf. Sci.* **2013**, *51*, 414–426. [CrossRef]
40. Seker, S.; Kahraman, C. Socio-economic evaluation model for sustainable solar PV panels using a novel integrated MCDM methodology: A case in Turkey. *Socio-Econ. Plan. Sci.* **2021**, *77*, 100998. [CrossRef]
41. Abdel-Basset, M.; Gamal, A.; Chakrabortty, R.K.; Ryan, M.J. Evaluation of sustainable hydrogen production options using an advanced hybrid MCDM approach: A case study. *Int. J. Hydrog. Energy* **2021**, *46*, 4567–4591. [CrossRef]
42. Opricovic, S.; Tzeng, G.H. Compromise solution by MCDM methods: A comparative analysis of VIKOR and TOPSIS. *Eur. J. Oper. Res.* **2004**, *156*, 445–455. [CrossRef]
43. Zavadskas, E.K.; Podvezko, V. Integrated determination of objective criteria weights in MCDM. *Int. J. Inf. Technol. Decis. Mak.* **2016**, *15*, 267–283. [CrossRef]
44. Shannon, C.E. A mathematical theory of communication. *Bell Syst. Tech. J.* **1948**, *27*, 379–423. [CrossRef]
45. Nijkamp, P. Stochastic quantitative and qualitative multicriteria analysis for environmental design. *Pap. Reg. Sci.* **1977**, *39*, 175–199. [CrossRef]
46. Huang, W.; Shuai, B.; Sun, Y.; Wang, Y.; Antwi, E. Using entropy-TOPSIS method to evaluate urban rail transit system operation performance: The China case. *Transp. Res. Part A Policy Pract.* **2018**, *111*, 292–303. [CrossRef]
47. Yu, X.; Suntrayuth, S.; Su, J. A comprehensive evaluation method for industrial sewage treatment projects based on the improved entropy-topsis. *Sustainability* **2020**, *12*, 6734. [CrossRef]
48. Bhowmik, C.; Kaviani, M.A.; Ray, A.; Ocampo, L. An integrated entropy-TOPSIS methodology for evaluating green energy sources. *Int. J. Bus. Anal. (IJBAN)* **2020**, *7*, 44–70. [CrossRef]
49. Kaynak, S.; Altuntas, S.; Dereli, T. Comparing the innovation performance of EU candidate countries: An entropy-based TOPSIS approach. *Econ. Res. Ekon. Istraživanja* **2017**, *30*, 31–54. [CrossRef]
50. Igliński, B.; Iglińska, A.; Cichosz, M.; Kujawski, W.; Buczkowski, R. Renewable energy production in the Łódzkie Voivodeship. The PEST analysis of the RES in the voivodeship and in Poland. *Renew. Sustain. Energy Rev.* **2016**, *58*, 737–750. [CrossRef]
51. Dave, M.; Watson, B.; Prasad, D. Performance and perception in prefab housing: An exploratory industry survey on sustainability and affordability. *Procedia Eng.* **2017**, *180*, 676–686. [CrossRef]
52. Gan, X.; Chang, R.; Zuo, J.; Wen, T.; Zillante, G. Barriers to the transition towards off-site construction in China: An Interpretive structural modeling approach. *J. Clean. Prod.* **2018**, *197*, 8–18. [CrossRef]
53. Mao, C.; Liu, G.; Shen, L.; Wang, X.; Wang, J. Structural equation modeling to analyze the critical driving factors and paths for off-site construction in China. *KSCE J. Civ. Eng.* **2018**, *22*, 2678–2690. [CrossRef]
54. Jiang, R.; Mao, C.; Hou, L.; Wu, C.; Tan, J. A SWOT analysis for promoting off-site construction under the backdrop of China's new urbanisation. *J. Clean. Prod.* **2018**, *173*, 225–234. [CrossRef]
55. O'Neill, D.; Organ, S. A literature review of the evolution of British prefabricated low-rise housing. *Struct. Surv.* **2016**, *34*, 191–214. [CrossRef]
56. Zhai, X.; Reed, R.; Mills, A. Addressing sustainable challenges in China: The contribution of off-site industrialisation. *Smart Sustain. Built Environ.* **2014**, *3*, 261–274. [CrossRef]
57. Kubečková, D.; Vrbová, M. Historical development of thermal protection of prefab residential housing and its future, an example of the Czech Republic. *Energies* **2021**, *14*, 2623. [CrossRef]
58. Chiang, Y.H.; Chan, E.H.W.; Lok, L.K.L. Prefabrication and barriers to entry—A case study of public housing and institutional buildings in Hong Kong. *Habitat Int.* **2006**, *30*, 482–499. [CrossRef]
59. Nanyam, V.N.; Basu, R.; Sawhney, A.; Vikram, H.; Lodha, G. Implementation of precast technology in India–opportunities and challenges. *Procedia Eng.* **2017**, *196*, 144–151. [CrossRef]
60. Arif, M.; Goulding, J.; Rahimian, F.P. Promoting off-site construction: Future challenges and opportunities. *J. Archit. Eng.* **2012**, *18*, 75–78. [CrossRef]

61. Zhao, L.; Mbachu, J.; Domingo, N. Exploratory factors influencing building development costs in New Zealand. *Buildings* **2017**, *7*, 57. [CrossRef]
62. Boyd, N.; Khalfan, M.M.; Maqsood, T. Off-site construction of apartment buildings. *J. Archit. Eng.* **2013**, *19*, 51–57. [CrossRef]
63. Li, Z.; Shen, G.Q.; Xue, X. Critical review of the research on the management of prefabricated construction. *Habitat Int.* **2014**, *43*, 240–249. [CrossRef]
64. Hou, L.; Tan, Y.; Luo, W.; Xu, S.; Mao, C.; Moon, S. Towards a more extensive application of off-site construction: A technological review. *Int. J. Constr. Manag.* **2022**, *22*, 2154–2165. [CrossRef]
65. Chang, Y.; Li, X.; Masanet, E.; Zhang, L.; Huang, Z.; Ries, R. Unlocking the green opportunity for prefabricated buildings and construction in China. *Resour. Conserv. Recycl.* **2018**, *139*, 259–261. [CrossRef]
66. Arditi, D.; Ergin, U.; Günhan, S. Factors affecting the use of precast concrete systems. *J. Archit. Eng.* **2000**, *6*, 79–86. [CrossRef]
67. Hwang, B.G.; Shan, M.; Looi, K.Y. Key constraints and mitigation strategies for prefabricated prefinished volumetric construction. *J. Clean. Prod.* **2018**, *183*, 183–193. [CrossRef]
68. Blismas, N.; Pasquire, C.; Gibb, A. Benefit evaluation for off-site production in construction. *Constr. Manag. Econ.* **2006**, *24*, 121–130. [CrossRef]
69. Du, T.; Xie, X.; Liang, H.; Huang, A.; Han, Q. County economy comprehensive evaluation and spatial analysis in Chongqing city based on entropy-weight TOPSIS and GIS. *Econ. Geogr.* **2014**, *34*, 40–47. (In Chinese)
70. Correia, J.M.; Sutrisna, M.; Zaman, A.U. Factors influencing the implementation of off-site manufacturing in commercial projects in Western Australia: A proposed research agenda. *J. Eng. Des. Technol.* **2020**, *18*, 1449–1468. [CrossRef]
71. Du, Q.; Bao, T.; Li, Y.; Huang, Y.; Shao, L. Impact of prefabrication technology on the cradle-to-site CO 2 emissions of residential buildings. *Clean Technol. Environ. Policy* **2019**, *21*, 1499–1514. [CrossRef]
72. Xue, H.; Zhang, S.; Su, Y.; Wu, Z. Factors affecting the capital cost of prefabrication—A case study of China. *Sustainability* **2017**, *9*, 1512. [CrossRef]
73. Lou, N.; Guo, J. Study on Key cost drivers of prefabricated buildings based on system dynamics. *Adv. Civ. Eng.* **2020**, *2020*, 8896435. [CrossRef]
74. Moradibistouni, M.; Vale, B.; Isaacs, N. Evaluating sustainability of prefabrication methods in comparison with traditional methods. In *Sustainability in Energy and Buildings 2018, Proceedings of the 10th International Conference in Sustainability on Energy and Buildings (SEB'18), Gold Coast, Australia, 24–26 June 2018*; Springer International Publishing: Berlin/Heidelberg, Germany, 2019; pp. 228–237.
75. Mao, C.; Xie, F.; Hou, L.; Wu, P.; Wang, J.; Wang, X. Cost analysis for sustainable off-site construction based on a multiple-case study in China. *Habitat Int.* **2016**, *57*, 215–222. [CrossRef]
76. Ataei, Y.; Mahmoudi, A.; Feylizadeh, M.R.; Li, D.F. Ordinal priority approach (OPA) in multiple attribute decision-making. *Appl. Soft Comput.* **2020**, *86*, 105893. [CrossRef]
77. Mahmoudi, A.; Javed, S.A.; Mardani, A. Gresilient supplier selection through fuzzy ordinal priority approach: Decision-making in post-COVID era. *Oper. Manag. Res.* **2022**, *15*, 208–232. [CrossRef]
78. Mahmoudi, A.; Deng, X.; Javed, S.A.; Zhang, N. Sustainable supplier selection in megaprojects: Grey ordinal priority approach. *Bus. Strategy Environ.* **2021**, *30*, 318–339. [CrossRef]
79. Mahmoudi, A.; Abbasi, M.; Deng, X. A novel project portfolio selection framework towards organizational resilience: Robust ordinal priority approach. *Expert Syst. Appl.* **2022**, *188*, 116067. [CrossRef] [PubMed]
80. Kuo, Y.; Yang, T.; Huang, G.W. The use of grey relational analysis in solving multiple attribute decision-making problems. *Comput. Ind. Eng.* **2008**, *55*, 80–93. [CrossRef]

sustainability

MDPI

Article

Evaluation of the Interventions to Built Heritage: Analysis of Selected Façades of Kaunas by Space Syntax and Sociological Methods

Kęstutis Zaleckis [1], Huriye Armağan Doğan [2,*] and Natanael Lopez Arce [1]

1 Faculty of Civil Engineering and Architecture, Kaunas University of Technology, 51367 Kaunas, Lithuania; kestutis.zaleckis@ktu.lt (K.Z.); arq.natanael.lopez@gmail.com (N.L.A.)

2 Institute of Architecture and Construction, Kaunas University of Technology, 44405 Kaunas, Lithuania

* Correspondence: huriye.dogan@ktu.lt

Abstract: This paper is an attempt to analyse the correlation between the perception of people and their evaluation regarding contemporary interventions and changes on the façades of cultural heritage buildings, which might affect cultural sustainability. The paper uses two different experimental methods for the analysis of the building façades from various eras in the city centre of Kaunas, which experienced interventions that affected the appearance of the structures. The first experiment performed is a sociological survey, and the second one is a façade analysis conducted by the space syntax method. The paper follows the theory of Nikos Salingaros for measuring the properties of the size distribution on the façades and implements Bill Hillier's methodology for symmetry index analysis. The research demonstrates some significant correlations between the results of Space Syntax modelling and the sociological survey answers, thus demonstrating the possibility of modelling and predicting changes in the perception of architectural transformations of the façades with potential usability in the monitoring of the transformation of cultural heritage objects, preservation of the cultural identity of a cityscape, etc.

Keywords: space syntax; façade configuration; visual perception; built heritage; Kaunas; cultural sustainability

Citation: Zaleckis, K.; Doğan, H.A.; Arce, N.L. Evaluation of the Interventions to Built Heritage: Analysis of Selected Façades of Kaunas by Space Syntax and Sociological Methods. *Sustainability* **2022**, *14*, 4784. https://doi.org/10.3390/su14084784

Academic Editors: Oleg Kapliński, Lili Dong, Agata Bonenberg and Wojciech Bonenberg

Received: 15 March 2022
Accepted: 14 April 2022
Published: 16 April 2022

1. Introduction

The expression of genius loci in architecture implies the reflection of memory and symbols, which serve to root the society that can be traced on the cultural heritage and the language of architecture. This property of architecture is essential for human beings to associate themselves with the place since it contributes to culture and cultural sustainability. As stated by Abusafieh, there is a significant link between culture and sustainability, and the rules, values, beliefs, and norms of the culture transfer the sustainability of vitality of the communities [1]. If people see the reflection of themselves in the environment they inhabit, they feel more comfortable in these environments. However, the interventions implemented with respect to cultural heritage due to the need for new spaces in developed cities can establish unavoidable transformations of heritage buildings in the city centres. Furthermore, they can negatively affect the requirement of the cultural heritage buildings to remain recognisable as historical marks, which might affect the cultural sustainability of these artefacts. Moreover, they can impact the language of architecture, which might give rise to inadmissible results and disrupt the existing architectural language permanently. Such a significant alteration or even demolition of cultural heritage is not a sustainable way of urban development in opposition to a way for cultural heritage to continue to resonate in future developments [2]. Therefore, it is essential to identify the impact of these changes on the perceptions of people.

In this regard, this research tries to analyse the correlation between the changes in the façades of cultural heritage artefacts with the perception of the people before and after the adaptive reuse applications, which involve contemporary interventions, and the impact of adaptive reuse on the volume of the façades. The analysis for identifying all these different aspects can be achieved by various methods which are used in architectural research. One of the most common methods which has been used in recent years is analysis by eye-tracking technology [3,4]. With this technology, it is possible to identify the parts of the façades which catch the attention of observers the most. However, identifying the reasoning behind this selection cannot be achieved by eye-tracking glasses. Therefore, more detailed research is required to determine and understand the perception process. In the course of perception, especially if the observer does not have enough knowledge or has uncertain evidence, it is likely that the observer will use prior knowledge to achieve an optimal solution [5]. Therefore, it would be beneficial to use the other possible approaches such as fractal analysis of the façades or pattern approach and apply them as the sources of prior knowledge for analysing, classifying, and predicting responses [6,7].

This paper follows the theory of Salingaros to measure the properties of the size distribution on the façades to analyse the changes, and the aesthetic synergy of the adjacent buildings is determined by implementing Hillier's methodology for symmetry index analysis [8,9]. The paper begins by examining the definition of a façade in architecture. This is followed by an explanation of the adaptive reuse of cultural heritage and how adaptive reuse strategies can affect the façades of buildings and sustainable development. The paper then gives information about the methodology and the method of the two experiments (the first one with a sociological survey and the second one is a façade analysis) and how they were implemented. Furthermore, it discusses the results of both experiments regarding the analysis of the building façades in Kaunas and determines the changes established by the interventions.

2. Literature Review

2.1. The Function of Façades in Architecture

According to the Cambridge Dictionary, a façade is the front of a building [10]. However, structures can contain different façades, which are not necessarily the front but also the side or the back of the building. Most of the time, people have their first connection with the design by the front façade, and as Pallasmaa describes, the door handle is the first handshake that people share with a building [11]. Especially when the front façade of the building is analysed, it is possible to state that it is the display of a design that provides information to observers or users in a similar way as one visual representation of an object with a visually complex design gives valuable information about the more visually complex object, as Dotson states [12]. Furthermore, according to various studies, the changes that appear on the façades can affect the preference, complexity, and impressiveness of the buildings for observers. They can change the familiarity and the liking of the structures [13,14]. As Mao et al. state, the façades of heritage buildings can also establish an impact on the behaviour of people and human activities in public spaces [15]. Therefore, even though the façades might seem as if they are one of the physical elements of the structures, they do have an important role both in the environment and in the perception of the structures for people.

It could be argued that the same façade might look differently from different observation points while changing proportions, making some parts invisible, etc. In this case, two aspects of façades as building faces should be mentioned:

- If the description of the façade is focused on the architectural pattern concept (Salingaros) instead of a detailed description of architectural form, then it opens a way to look for some general, fundamental features of architectural composition which are not sensitive to small changes in form because of changing observation points.
- Based on the logical analogy with the human face recognition process, it could be assumed that at least known façades, e.g., cultural heritage objects that are perceived

in situ and potentially through various media channels, could be recognised from various positions of observation.

In this regard, the message or the information that the front façade provides with the language it uses is essential because it establishes the initial impression that people need to understand or perceive the whole structure.

According to Gehl, there are two types of façades. The first type is the active façade which gives an adequate impression to observers or users since it contains the material whereby people can communicate with the language of the architecture [16]. The other one is the passive façade which does not cause any feelings or emotions. As Ellard states, the buildings which have passive façades are structures where people feel as if they are on the wrong side of the façade [17]. However, it might not just be a matter of being active or passive, but it might also be about the reflection it creates on people's perceptions. In that regard, it is essential to understand the languages that architecture uses for communicating with people on buildings' façades.

The urbanist and controversial theorist Salingaros states that architecture is established by two distinct, complementary languages: a pattern language and a form language [18]. The pattern language involves the interaction of human beings with their environment, and it is appropriate for local customs, society, and the climate where the building is. It is a set of repeatedly tried and true solutions inherited from the previous generations, which developed optimisations that create a sense of well-being for the people. However, Alexander was the first to propose the definition of a pattern language for architecture [19]. As he points out, while many, if not most, of the patterns in pattern language, are universal, there are an infinite number of existing individual patterns that can be included. According to him, each pattern language tends to reflect a different mode of life or customs or behaviours. Additionally, it is appropriate to specific climates, geographies, cultures, and traditions. Therefore, the pattern language of a building establishes the interpretation of the architecture and how architecture was formed in different regions by the effects of the local architecture and experience. Furthermore, it reflects the culture.

On the other hand, form language is defined by the elements of a building that establish the whole. The elements which determine the form are the floors, the walls, the windows, the doors, the ceilings, the partitions, and all the architectural components which together represent the style. In the accumulation of all the different elements in the form language, the building expresses its architectural style. Furthermore, every traditional architecture has its own form language as well. It has been established from various influences of daily life, traditions, and practical concerns, which act together to define the structures that take the most natural visual expressions of a specific culture. Architecture becomes an accumulation of the circumstances of culture and a signifier of the collective when establishing a system of relations between the differentiating elements. Therefore, the form language which was used in architecture is also affected and influenced by the culture, like the pattern language. Pattern language makes buildings more readable and understandable since it is possible to have a universal form language, but it is not that easy to have a pattern language valid in every culture since they have their own characteristics related to the region. When architecture utilises both of these languages in its design, it establishes a valid architecture, and furthermore, primarily with the effect of the pattern language, it establishes an environment for the people where they feel familiar with their surroundings. According to Alexander, it is possible to improve the patterns by testing them against experience by recognising how the patterns make people feel regarding the existence of the patterns in their surroundings [19]. However, in his research, Alexander did not describe patterns and how to test them quantitatively.

The quantitative approach regarding the description of the patterns has been designed by Salingaros. In this book called *A Theory of Architecture*, he explains the scientific basis of creating architectural forms, hierarchical cooperation, modularity, and the number of design choices in the formation of architecture [18]. As he states, architecture is an expression and, at the same time, the application of geometrical order; therefore, if the order can be understood, it will give the knowledge to understand the language of architecture and what it is trying to explain to users/observers as well. When the information quality which passes to the user/observer is rich, it is more likely that there will be an emotional bond established. However, the method Salingaros established has a high degree of interpretation regarding the selection of symmetries and asymmetries or which elements of the façade should be analysed.

Furthermore, the selection of the elements of the patterns for the identification of the symmetries is indefinite as well. When the elements of the patterns are identified by different people, their interpretations might be dissimilar. Therefore, it is possible to state that the same façade might give different results depending on the perception or the evaluation of the person who identifies the elements. As a result, his model establishes a level of subjectivity to some extent. In his approach, he mainly focuses on the structural order and the scaling rules, which are independent of architectural styles or architectural shapes as well.

According to Salingaros, there is an ideal scaling factor, which is approximately equal to the logarithmic constant 2.7, and it leads to the scaling coherence of the objects [18]. His proposal of the scaling rule derives from Alexander's scaling rule; however, he adds another dimension to it, which is the designation of the ideal number for the scale. On the one hand, his approach helps the architects or the people who want to understand what architecture is communicating by establishing a set of rules and order in the process. On the other hand, it does not consider other aspects of the perception of architecture, such as the colour, material, and texture. All these aspects have an impact on perception as well. Therefore, the method that should be followed is still ambiguous. However, according to Hillier, the meaning reflected in the façades of the buildings by the language that architecture is adopting can be identified [20]. Furthermore, limits can be set towards meaning by distinguishing the idea of meaning from aesthetics in architecture. Therefore, it might be possible to achieve quantitative data and measure the meaning or how the meaning is changing by analysing the architecture, specifically the façades affected by the interventions of adaptive reuse.

In the last decades, specifically in recent years, interventions regarding cultural heritage buildings in the city centre started to be seen more in Kaunas, Lithuania. One of the reasons for this can be explained by the nomination for Inscription on the UNESCO World Heritage List. In the course of the preparation of the protection and management plans for the city centre, the demolition of existing buildings or construction of new structures in this area started to be controlled and regulated by different institutions [21]. Furthermore, additions and interventions to the heritage buildings are administered and authorised as long as they are coordinated with the relevant governmental organisations responsible for cultural heritage protection. Therefore, with the requirement for more space in this valuable area of the city, heritage revitalisation and all the new additions to these buildings started to be seen more often [22]. Moreover, adaptive reuse projects in the city accelerated and monitoring façade changes became more essential [23,24]. As a result, the structures which were used in this research were selected from the cultural heritage buildings from various eras in the city centre of Kaunas, which had different interventions regarding their façades.

2.2. *Adaptive Reuse and Built Heritage*

Over the centuries, the concept and treatment of heritage and approaches to the conservation of it have changed as values have changed as well. As Vecco states, the monument is no longer considered alone but is now taken in its context in heritage studies [25]. Therefore, the adoption of an integrated approach towards heritage, its evaluation and its preservation does not merely affect the structure itself but also has an impact on the environment. Adaptive reuse is a commonly used method that provides a new function to an existing structure; therefore, the structure adjusts to the current needs. Even though most of the time, the reason for this action is due to the requirements of the market and financial gain, in contemporary conditions, adaptive reuse is implemented with respect to the cultural heritage buildings for their protection. According to Haldrup and Bærenholdt, heritage has traditionally been bound with the conservation of the imagined past [26]. However, heritage is not only about the past, but it also has a reflection in the present. By the method of adaptive reuse, it is possible to keep the progression of the artefacts and, at the same time, the environment, which can help people to associate with them better and more easily.

In the book called *Uses of Heritage*, Smith states that there is no such thing as 'heritage', and heritage has to be experienced for it to be heritage [27]. Therefore, heritage must be a part of daily life, and it should contribute to the genius loci of the environment. According to Vecco, genius loci is the intangible quality of a material place, which can be perceived both physically and spiritually through visible tangible and perceivable non-material features [28]. Therefore, while adaptive reuse can provide the continuity of the material characteristics of the heritage, its outcome, which is cultural sustainability, can contribute to its non-material aspects. It is crucial to indicate that one of the leading characteristics of heritage is that it is a carrier rather than a solid concept. It only endures when used on a daily basis and perceived by society itself. In that regard, adaptive reuse provides both aspects for the cultural heritage buildings to be experienced.

However, adaptive reuse has another constraint which is its collaboration with the sustainable development of the environment. Most of the time, the meaning of sustainability is merely associated with the sustainability of nature, recycling, and self-sufficiency; however, sustainability has many different aspects which have a direct impact on the built environment as well. As stated by various authors, the main assets of sustainable development are society, the environment, and the economy. However, according to Hawkes, sustainable development contains a fourth pillar, which is culture [29]. The main concern of sustainability and sustainable development is the protection and continuity of resources that are irreplaceable. In that regard, it is possible to state that cultural heritage is irreplaceable when it vanishes as well. It is crucial to maintain the continuity of cultural heritage to maintain the culture and, at the same time, maintain the built environment. According to Hristova, a city remembers through its buildings; thus, the preservation of the old urban fabric is analogous to the preservation of memories in the human mind [30]. Therefore, a city is a collective memory of its people, and it is a way of remembering which is associated with objects and places. Associations that people obtain through the built environment help establish potential stimuli for people to remember, which is one of the crucial impacts of architecture on people. Therefore, when the built environment carries its own characteristics, it stays recognisable to the people, which supports sustainability. Sustainability derives from the ability of continuum. Therefore, adaptive reuse of buildings also has the same impact on the environment since it helps the structures to continue their lives and helps to keep the environment the way it is. However, some of the interventions with respect to the structures can change the perception of the building as well as the proportion and the symmetry, which has an impact on the intelligibility of the structure. As Rabun and Kelso state, a building to which the adaptive reuse will be applied with a change of use must be evaluated from both the exterior and the interior, and the assessment of it must be done in a comprehensive manner [31]. Furthermore, it is essential to pursue the acceptance of the artefact in its environment. Therefore, modern additions, which might affect the recognition of the original elements, need to be omitted. As a result, there are various factors that have a direct and indirect effect on the process of adaptive reuse and the built environment and its perception.

To understand these effects, a pilot study was conducted in Kaunas, Lithuania, by two experiments. The first experiment is based on a sociological survey, which checks how well people perceive changes of the building façades. In contrast, the second experiment attempts to model those changes of the façades in a mathematical way. Statistical analysis of relations between the results of both experiments is employed to identify if and how human perception could be reflected and potentially predicted before any modifications of the façades based on mathematical analysis. Changes of the exterior of buildings are chosen as the most subjectively perceived and the least function-based aspect. The methodology of both experiments is described in more detail in the third section.

3. Methodology and Experiments

As Hillier states, building façades are physical shapes that are capable of being understood as communicators of information. However, to understand the shapes, the shapes need to be identified and recognised by the observer [9]. According to him, the recognition of the shape of an object occurs in two stages. The first stage of recognition is the syntactic stage, and the second stage of recognition is the semantic stage. In the first stage, people tend to determine the object by the identification of the elements that they perceive in its configuration; however, in the second stage, people attach meaning to the object or they interpret what they see. In that regard, it might not be possible to measure the attached meanings or interpretations; however, the syntactic stage of recognition can be measured by analysing the configuration as the symmetry index.

Symmetry is a concept acquired from mathematics, specifically from the group theory by Miller and Carter [32,33]. However, it is widely used in other disciplines, such as physics, chemistry, biology, psychology, art, and architecture. According to the APA Dictionary of Psychology, symmetry is "the mirrorlike correspondence of parts on opposite sides of a centre, providing balance and harmony in the proportions of objects, and it is considered an aesthetically pleasing quality" [34]. Furthermore, as Hodgson states, symmetry can be a persisting feature for the perception of the visual world since it provides valuable means which can be encoded for the purpose of efficient recognition of the objects [35]. Therefore, symmetry is a distinctive peculiarity for the perception of objects, and furthermore, it is essential for analysing and understanding nature, art, and architecture.

According to Mitra and Pauly, symmetry and structural regularity in architectural design are not coincidental [36]. Most of the time, it is the consequence of economical, manufactural, functional, or aesthetic considerations that make the structure universally appealing. Therefore, symmetry can affect the visual perception of the architecture in our environment by its peculiarity of reflecting the nature and natural orders that can influence the recognition of the objects and the sense of beauty.

The symmetry of architectural objects can be measured by the symmetry index, which Hillier has suggested. According to him, the symmetry index is a considerable ratio in which low and high values can demonstrate similarities and differences in how the parts relate to the whole [20]. Therefore, by measuring the symmetry indexes of the buildings before and after the cultural heritage intervention and comparing them to the results of surveys on human perception of the selected façades, it might be possible to identify how the perception of the building has changed and if it could be reflected by the offered space syntax indexes. In that regard, two experiments are conducted in this research.

The first experiment involved a sociological survey performed by online communication tools, and the second experiment involved checking the façades by Space Syntax based modelling. The experiments contained the analysis of the façades of eight buildings in Kaunas, Lithuania, which underwent an adaptive reuse process. All eight buildings are located in the new town section of Kaunas city centre, and they have undergone small- or large-scale interventions such as additions to the roof or additions to the building itself, which have affected their appearance. The experiments aim to understand the effects of the interventions, which are later added to the buildings. Therefore, all eight buildings are analysed by the depthmap program before and after the adaptive reuse to detect the changes in the symmetry indexes of the structures. At the same time, a sociological survey is performed to understand how people's perception is affected by the interventions. At the end of both analyses, the results are compared. At the same time, correlations between the survey answers (attractiveness and perceived symmetry) and space syntax indexes (symmetry index plus some other offered indicators) were calculated. There was no separate analysis of correlations before and after changes of the façades. At the moment, the research aims to check if space syntax indexes can reflect certain objective features of the façades that correspond to the subjective perception.

3.1. Experiment I: Sociological Survey

3.1.1. Design

The social survey prepared for this research is a questionnaire which contains 16 photographs of 8 different buildings which underwent a change from their original appearances due to different forms and scales of interventions. The buildings selected for this research are cultural heritage buildings from various eras in the city centre of Kaunas. While it was possible to find the prior status photographs of some of the buildings, with other ones, it was not possible to find any documentation regarding the earlier situation due to there either being no photographs or the width of the street not allowing us to see the whole building. Therefore, in 6 of the buildings, the additions were deleted with the help of various software. The survey aims to examine people's perception of the buildings before and after the adaptive reuse.

In the design of the survey, a qualitative approach and non-probability sampling were adopted. The goal of adopting a non-probability sampling was not to achieve objectivity in selecting samples or attempting to make generalizations (i.e., statistical inferences) from the sample studied by the broader population of interest. Therefore, generalizations from the sample to the population under study are secondary considerations. With a purposive and convenience sampling technique, 66 participants in total took part through an online survey platform. The selection of the participants adopted snowball sampling, which is a non-probability sampling method commonly used in the process of collecting data from participants. It is one of the most common methods of sampling in qualitative research, central to which are the characteristics of networking and referral. The researchers usually start with a small number of initial contacts (seeds) who fit the research criteria and are invited to become participants in the research. The agreeable participants are then asked to recommend other contacts who fit the research criteria and who might also be willing participants, who then, in turn, recommend other potential participants, and so on. Researchers, therefore, use their social networks to establish initial links, with sampling momentum developing from these, capturing an increasing chain of participants. Sampling usually finishes once either a target sample size or saturation point has been reached" [27]. The potential participants who were willing to participate in the research with an average knowledge of cultural heritage were asked a set of questions. The participants were asked to transfer the experiment to their circles that fulfilled the same criteria. Therefore, the focus group involved people who are already interested in architecture and the changes happening in their cities. Participants were heterogeneous regarding age, ranging between 12–75 years old, and heterogeneous regarding place of origin. In the process of data collection, the data regarding gender, age, and education level were collected. However, we decided to use these criteria in broader research which would focus on the analysis of these specific variables. Therefore, it was found to be irrelevant to the experiment as the network of people created by the snow sampling method did not demonstrate enough differences between age groups. Furthermore, the main aim was not to measure different people's reactions but the reactions to the buildings.

The survey was structured in three main sections and four subsections for the second and third sections. The main difference between the second and third sections was that while the second section reflected the status of the buildings before the interventions, the third section reflected the current status of the buildings.

In the implementation of the survey, 16 photographs were presented to the participants (Figure 1).

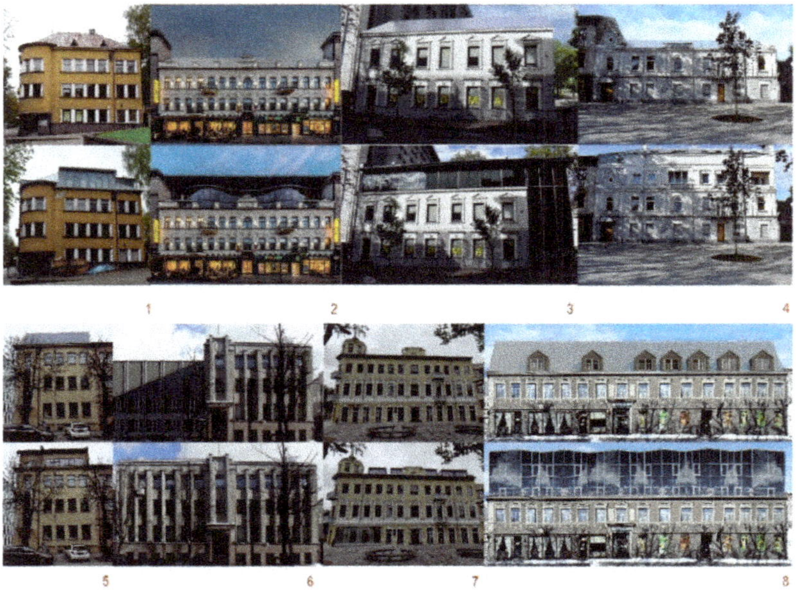

Figure 1. Presented photographs with the sample numbers below. Each column demonstrates before and after version of a building with its sample number.

The buildings for the experiment were selected according to the following criteria:

- cultural heritage objects;
- the recent renovation of the buildings, including façades with the old part still visible in architectural composition;
- possibility of describing the façades mathematically as a set of polygons with clear boundaries created by architectural elements.

As it was not possible to obtain the photographs of the same buildings in frontal view before reconstruction, in order to avoid additional influence on the survey results because of the technique differences, it was decided to make visualizations for both situations before and after the reconstructions. Fifteen of the photographs were modified with the objective to model in 3D software the previous appearance before the modifications were completed; the 3D modelling was combined with rendering and postproduction in photo editing software to recreate the previous stage of the building as close as possible in appearance to the photographs, literature information, and online sources (Table 1).

In order to analyze the perception of the changes in the buildings and not make it difficult to see them or perceive them from a different angle, the digital recreation was kept in the same position and angle from the actual renovation state of the buildings (Figure 1). The modification of the façades contains the following steps:

Evaluation of the buildings, which requires 3D modelling or just image modification.

1. For 3D modelling: a. Measure in Google Earth distance, b. Import photograph in Google Sketchup, c. Usage of Google Sketchup tool [Match new photo] to set position, angle and draw distances to compare to the measured distances from Google Earth, d. Model the roof as a surface, e. Export the model to Blender, f. In Blender, set tool for textures and sun position, g. Rendering by cycles, h. Save image file as PNG, i. Import to GIMP, j. Masking tool in the original photograph, k. Modify the tool for the PNG photographs to position in the original photograph and l. Save picture.

2. For the photographs which do not require modelling in 3D, the GIMP program is used as follows: a. The masking tool, b. Layer tool, c. Match colour, d. Position, e. Copy, f. Modify layer and g. Save file.

After the modification of the photographs, the first section of the survey was organised with general questions about gender, age, and education level to obtain demographic characteristics of the survey participants. However, the data of this section were not used in the analysis of this article.

The second and third sections were organised based on each photograph (named sample), followed by four sub-sections. These sections aimed to measure the capacity to recognise the building's level of attractiveness and the composition features of the façade, such as symmetry, rhythm (known as repetition in our survey), and the elements that provide a level of hierarchy in the façade.

In both sections, the first subsections were based on a scale from 1 to 10 regarding the building's attractiveness level. This was set using the Likert scale, where the lowest value corresponded to 1 and was equal to not attractive, and the highest value, equal to 10, corresponded to very attractive (Figure 2).

Table 1. Description of the 3D modelling process of the photographs.

Photograph	Style	Construction Date	Addition/ Renovation Date	Addition/Renovation Description	3D Model Object + Rendering	Photo Edition Modification
Sample 1	Modernism	1938	2020	Roof shape modification, windows addition	-	-
Sample 2	Historicism	1870	2002	Roof shape modification, building height, fourth floor, curtain panel, terrace, balcony, handrail	Roof	Sky, nearest building
Sample 3	Historicism	1901	2006	Roof shape modification, building height, curtain panel	Roof	Sky, nearest building, walls, surroundings
Sample 4	Historicism	1902	1999	Roof height, third floor, windows, doors, balcony, drains	Roof	Sky, nearest building, walls, surroundings
Sample 5	Modernism	1937	2013	Roof height, fourth floor, windows, doors, balcony, drains, handrails	Roof	Sky, nearest building, walls, surroundings
Sample 6	Modernism	1938	1982	Roof height, third floor, windows, doors, balcony, drains, handrails	-	Sky, nearest building, walls, surroundings
Sample 7	Historicism	1896	2021	Roof, windows, doors, balcony, curtain panel	-	Sky, nearest building, walls, windows, doors, surroundings
Sample 8	Historicism	1897	2011	Roof height, third floor, fourth floor, windows, doors, balcony, handrails	-	Sky, nearest building, walls, windows, doors, surroundings

Figure 2. Sample sheet of the social survey regarding attractiveness.

The second subsection contained the same scale from 1 to 10, and it aimed to understand the perception of symmetry of the façade. Again, the Likert scale was applied where the lower value corresponded to 1, which is equal to completely asymmetrical, and the value of 10 corresponded to highly symmetrical. In this subsection, a red segmented line in the center of the building was added using photo editing software to guide the participants over the main façade of the buildings (Figure 3).

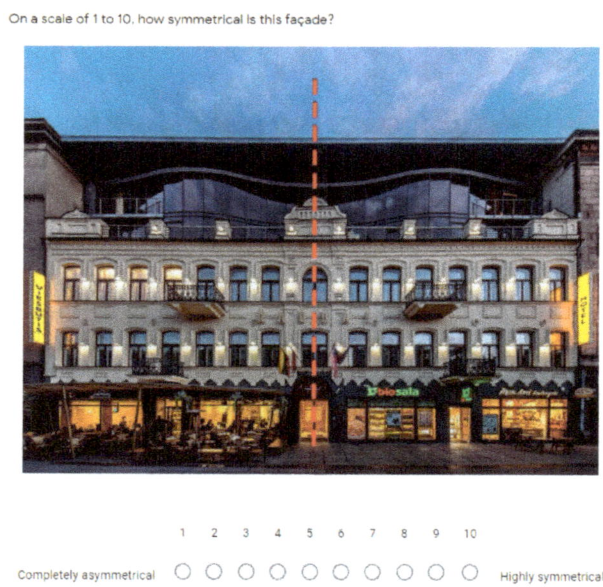

Figure 3. Sample sheet of the social survey regarding symmetry.

The third subsections were open questions regarding the repetitive elements in the buildings, and the fourth subsections were open questions related to the most striking features. All the elements mentioned above belong to the second section, summarised in two-scale questions and two open questions.

3.1.2. Analysis of the Data

According to the sociological survey, the additions or the changes on the façades impacted the attractiveness levels of the buildings. While attractiveness decreased in the 1st, 3rd, 4th, 5th, 7th, and 8th buildings, it increased for the 6th building. However, attractiveness levels stayed the same for the 2nd building. The common characteristics of the buildings where the attractiveness levels decreased can be identified as the height of these structures. The additions can be easily detected from the street level in all of these structures. Furthermore, the additions are significant when they are analyzed in the sense of proportion. Therefore, it might be possible to state that the human eye level and proportion have a particular impact on the perception of these additions.

On the other hand, decreased symmetry of the 3rd and the 4th buildings affected the attractiveness levels of these structures in a negative way. However, even though the symmetry perception increased for the 1st and 8th buildings, their attractiveness levels still decreased. The increased symmetry affected the perception and the attractiveness of the 6th building in a positive way. However, it did not have a specific outcome for the 2nd building since both symmetry and attractiveness stayed the same for this structure. The perception of symmetry levels decreased for the 4th and the 5th buildings, which also diminished the attractiveness levels (Table 2).

Table 2. Results of the sociological survey: Decrease ⇩ Increase ⇧ Stable ⇔.

Building	Attractiveness Level before the Addition	Attractiveness Level after the Addition	Result	Perceived Symmetry Level Before the Addition	Perceived Symmetry Level after the Addition	Result
Sample 1	7	3	⇩	4	5	⇧
Sample 2	9	9	⇔	10	10	⇔
Sample 3	5	3	⇩	9	3	⇩
Sample 4	8	5	⇩	6	4	⇩
Sample 5	7	5	⇩	4	3	⇩
Sample 6	7	8	⇧	1	10	⇧
Sample 7	8	7	⇩	7	7	⇔
Sample 8	8	1	⇩	4	8	⇧

For the 1st, 2nd, and 3rd buildings, the roof's addition was found to be one of the most striking elements of the buildings after the additions. However, the participants also stated that the contrast between the old and new established a striking effect. On the other hand, the diverse nature of the 4th building was found to be striking, and according to one participant, the building was found to be irregular and asymmetrical. According to the participants at the 7th building, the tower was described as significant and striking; however, after the interventions, the tower was found to lose its significance. Furthermore, the glass windows which were added to the ground floor caught participants' attention. For the 8th building, participants described the roof windows as striking because they do not follow any axis, and they look random. However, after the intervention, the roof is mainly described as aggressive, although still striking due to its establishing a contrast.

3.2. Experiment II: Façade Analysis

3.2.1. Design

The aim of the façade analysis was to check if Space Syntax based modelling could reflect subjectively perceived features of the buildings, such as perceived symmetry and attractiveness.

In the process of the preparation of the data, the façades of the buildings are measured by a laser meter, and the data are transferred to the computer as a drawing by the Autodesk AutoCAD program. After the required drawings are assembled, every façade is divided into plains by drawing polygons to create convex spaces formed by the elements of the façades, such as windows, doors, decorations, etc. Based on the mathematical graph model, each polygon of the façade is treated as a node. Links between the adjacent nodes are formed if they have common boundaries. Distance from one node to its neighbor while crossing the common boundary is considered one topological step in further calculation. The various graph centralities are calculated within the above described network of nodes and links—for example, the mead depth of a precise node is calculated as a sum of its distances to all the other nodes within the network measured in topological steps; betweenness of a node is calculated as a sum of the shortest routes in terms of a number of topological steps between all the pairs of the nodes within the network which cross the calculated node; connectivity of a precise node is calculated as a sum of the neighboring nodes with common links, etc.

The methodology was grounded in the façade modelling approach based on the convex graph. According to Hillier, a convex graph could be constructed based either on the tessellation of the façade into uniramous parts or on its divisions according to architectural details [9,20]. The tessellation approach is the most developed in Hillier's research, but it is oriented towards the analysis of general form instead of its details. The second approach was chosen as more appropriate in a selected urban setting where all buildings have relatively similar volumes and numerous different architectural details. In this case, a little generalized, each part of the architectural façade defined either by architectural details (e.g., decorative elements, lines) or formed by the borders of the other elements (e.g., openings of the windows or doors) was marked as a convex space defined by a polygon. Each convex space was converted into a mathematical graph node. Edges of the graph or links between the nodes were formed according to the following rule: convex spaces/nodes should have a common boundary in the form of a line in order to be connected by a common edge (Figure 4). The procedure of modelling was conducted while using depthmap software [38].

Eight recently renovated historical buildings (Figure 5) in the Kaunas New Town heritage area were selected as representatives of two architectural styles: historicism and interwar modernism. The choice was grounded by the wish to increase the sensitivity of modelling by focusing on changes to the same façades before and after renovations. Therefore, space syntax modelling was conducted for both situations before and after the renovations.

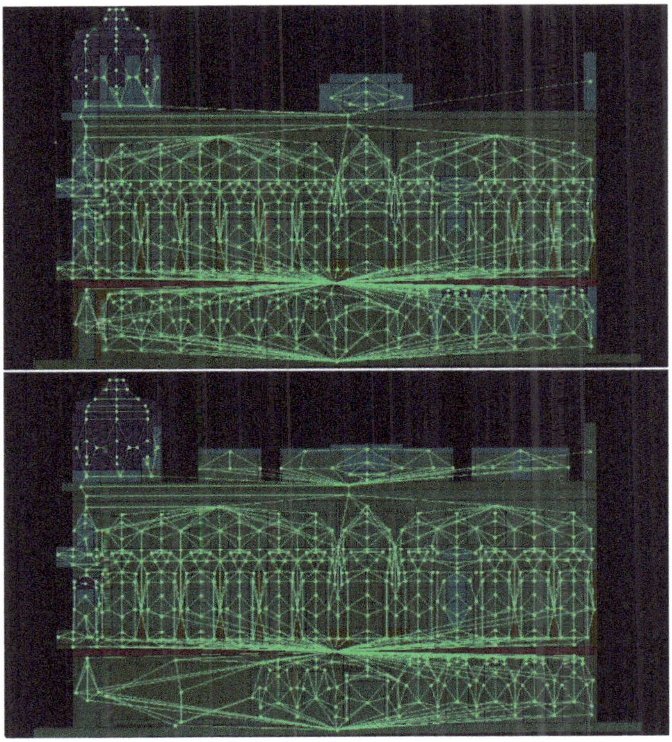

Figure 4. The mathematical graph is shown on the façade of the analysed building.

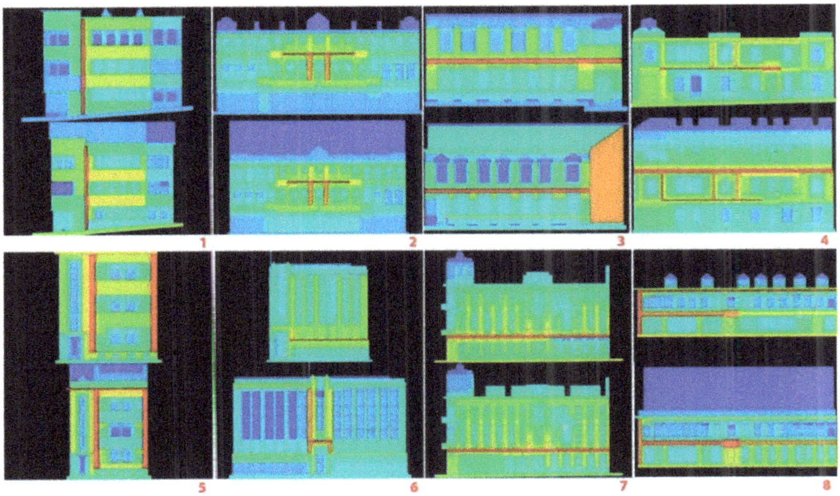

Figure 5. Analysed building façades with Integration values marked by colours—red colour means higher numerical values, blue—the lowest, yellow—the interim (for integration meaning look below). The façade after renovation is shown beneath the ones before renovation. Each column demonstrates before and after version of a building with its sample number below

Based on the mathematical graph, various analysis types of centralities of the nodes that are traditional for Space Syntax were calculated, including the following:

- Node count (NC)—total number of the nodes in the graph within radius n;
- Connectivity—number of neighbouring nodes which have a common edge with the calculated node; the mean values in order to represent general properties of a whole façade were used (Con mean);
- Choice or betweenness centrality as a sum of shortest hypothetical journeys between all pairs of nodes that cross the calculated node; the shortest distance was found on the basis of the smallest number of topological steps—one change of node is equal to one step; the normalised mean Ch values (Ch mean norm) were chosen in order to have better possibilities for comparison between different façades;
- Mean Depth or sum of the topological distances from the calculated node to the rest of the nodes on the network. The mean value of this index is used in order to represent an analysed façade in general (MD);
- Integration (Mean Integration as a generalising index for a whole façade was used, marked as MI) as the normalised version of closeness centrality is inversely proportional to the sum of distances from the calculated node to all the other nodes in the network. The distance was calculated using topological but not metrical steps, where movement of the visual focus from one node-architectural element to its neighbouring node is considered one step. Accordingly, the distance between two nodes equals the number of nodes on the possibly straightest line between them.

A number of secondary, additionally calculated indexes were used, starting with symmetry index (SI). "The ratio of the total number of elements to the number of elements that . . . (have identical integration or mean depth (MD) values and) . . . index the degree of balanced asymmetry in shape" (Hillier, 2007:89). In other words, the calculation is conducted while dividing the number of different MD values by a number of elements of a façade that are represented by mathematical graph nodes—for example, if we have 10 nodes with such values as 1,1,1,2,3,3,5,5,6,6, then SI = 4/10.

One additional index was proposed in order to address the architectural composition of the investigated façades more precisely:

Symmetry index 2 (SI2) is equal to the ratio of elements with at least one node with the same MD value. SI2, more precisely, can represent part of the elements, which could be seen as symmetric, and its max value is equal to 1. If the above-mentioned example with 10 elements is used, then SI2 = 9/10.

Finally, in order to have a more complex view, some more results of calculations were used in the analysis as follows:

- Standard Deviation of Integration (Dev Int)—more minor deviation means that in topological terms or distances measured in nodes or elements of the façades, they differ less and could be seen as more compact compositions;
- There are various synergies between the other indicators, such as SI* (multiplied) SI2; SI*MI (Mean Integration); SI2*MI; SI*SI2*MI; SI/Dev; SI2/Dev; SI*Dev; SI2*Dev; NC*SI; NC/Con; Con*MI; NC*SI2, etc.

Statistical analysis of relations between space syntax variables and the survey results was conducted while calculating Pearson correlations.

3.2.2. Analysis of the Data

Many statistical methods, including Pearson correlation, require that dependent and independent variables are approximately normally distributed. In statistical analysis, " . . . independent variables are variables that are manipulated or are changed by researchers and whose effects are measured and compared . . . The other variable(s) are also considered the dependent variable(s). The dependent variables refer to that type of variable that measures the effect of the independent variable(s) on the test units" [39]. The space syntax indexes were seen as independent and survey results as dependent variables.

Both the Kolmogorov–Smirnov Test of Normality and the Shapiro–Wilk test on all collected data were performed. According to the results of the Kolmogorov–Smirnov Test, p-values of all variables exceed 0.05 while varying from 0.21327 (attractiveness) to 0.97384 (Mean Connectivity) and do not differ significantly from that which is normally distributed. The mean *p*-value of all variables is 0.696392. According to the Shapiro–Wilk normality test, the majority of variables do not differ significantly from that which is normally distributed, with *p*-values exceeding 0.05 and varying from 0.0529034 (NC*SI2) to 0.491186 (Connectivity mean). The mean *p*-value for those variables is equal to 0.327459. Five variables have *p*-values lower than 0.05: SD—0.0334651; SI2/Dev Int—0.00050362; NC—0.0296501; NC/SI—0.040611; NC*SI—0.0345589. The bigger part of the data of these five variables still demonstrated the feature of normal distribution, so they were used in calculation with some extra concern towards the results.

The calculated Pearson correlation between the syntactic indexes and both perceived symmetry and attractiveness of the façades is presented in Table 1. Despite a pretty significant number of both positive and negative correlations, only four of 42 results are significant at the level of 0.05 (marked in red in Table 3). On the other hand, another four results could be seen as being near significant (marked in yellow in Table 3), and even if, speaking statistically, there is not enough evidence to say that these correlations appeared not by accident, they could be used for some insights for future research.

Table 3. Pearson correlations between syntactic indicators and both—perceived symmetry and attractiveness in all façades (before and after intervention). Correlation marked by red colour is significant at the 0.05 level. Nearly significant correlations are marked in yellow.

	Perceived Symmetry		Attractiveness	
	Pearson Correlation	**Sig. (2-Tailed)**	**Pearson Correlation**	**Sig. (2-Tailed)**
SI	−0.159	0.556	−0.371	0.157
MI	−0.380	0.146	−0.364	0.166
SI2	−0.046	0.866	−0.242	0.367
MD	−0.280	0.294	−0.203	0.451
Dev Int	−0.502	0.048	−0.343	0.194
SI*SI2	−0.141	0.602	−0.347	0.188
SI*MI	−0.286	0.284	−0.428	0.098
SI2*MI	−0.199	0.460	−0.325	0.220
SI*SI2*MI	−0.231	0.388	−0.386	0.140
SI/Dev Int	0.379	0.148	0.033	0.903
SI2/Dev Int	0.385	0.141	0.116	0.668
SI*Dev Int	−0.417	0.108	−0.420	0.105
SI2*Dev Int	−0.392	0.134	−0.395	0.130
NC	0.474	0.064	0.377	0.150
Con mean	−0.318	0.230	−0.232	0.338
Ch norm mean	−0.484	0.058	−0.311	0.241
NC/SI	0.401	0.124	0.405	0.119
NC*SI	0.543	0.030	0.319	0.229
NC/Con	0.495	0.051	0.411	0.114
Con mean*MI	−0.516	0.041	−0.452	0.079
NC*SI2	0.614	0.011	0.371	0.157

The significant strong or moderate, depending on scale, correlations are following:

- −0.502 between perceived symmetry and Dev Int of MI. This means façades that are more "compact" in terms of connectivity and topological distances could be perceived as symmetrical by people and this is not necessarily related to strict architectural symmetries.
- 0.543 between perceived symmetry and NC*SI. This means that architectural composition consisting of a more significant number of elements and having a more extensive index of "flexible" symmetry expressed by SI could be perceived as more symmetrical.

However, additional consideration should be given to this index in the future as NC*SI values demonstrate normal distribution only partially.

- −0.516 between perceived symmetry and Con*MI. This means more topological connections together with more prominent syntactic integration (more compact composition in topological terms and contacts between the forms) in façades are perceived as less symmetrical. For example, this might be explained by the statement that too big a number of compositional connections and contacts between various forms makes a façade less perceivable as symmetrical.
- 0.614 between perceived symmetry and NC*SI2. This means that a more significant percentage of elements in the composition, which corresponds to Hillier's idea of "flexible" symmetry, together with a more significant number of elements, is related to the perceived symmetry in a façade.
- Nearly significant correlations with the perceived symmetry do not add anything essentially new to the ones described above except correlation with normalised choice (−0.484; *p*-value 0.058). The possible explanation for this depends on the answer to the question, "what syntactic choice means in architectural composition?" and requires further investigation.
- In the case of attractiveness, no significant correlations were found, thus possibly reflecting the more subjective nature of this feature of architectural compositions. On the other hand, nearly significant negative correlations of 0.452 (*p*-value 0.079) with Con*MI might speculatively identify a potential relation between attractiveness and more topologically scattered composition.

4. Discussion and Conclusions

Adaptive reuse of built heritage is a challenging topic since it does not only contain the implementation of the physical changes to the structures, but it also affects social and cultural sustainability. When an adaptive reuse project of a heritage building is prepared, the architects need to consider the resulting changes to the urban fabric and the perception of both the experts and the non-experts. While the opinion of the experts can be more related to the design itself, it might be possible to state that the non-experts' opinions are more subjective due to their perception of the environment at a personal level. Every intervention and alteration can add a different layer to the structure and can influence the proportion of the design. In that regard, it is essential to understand the potential impact of these changes.

According to the sociological survey performed in this study, symmetry and attractiveness of a structured look is related even if, according to the collected data, no significant correlations between them were found. Therefore, when the structure becomes more asymmetrical, it is perceived as less attractive by the participants. However, if the intervention followed the same axis and the same symmetry of the main design, as long as the addition was not dominant, it did not affect the attractiveness of the building. Therefore, it is possible to state that, in most cases, the symmetry was perceived in a more classical way by the survey participants regarding architectural compositions based on central vertical axial symmetry. However, the space syntax models used in the second experiment can be more beneficial for understanding more complex situations—not only the axial symmetry but also different types of symmetries.

In most cases, windows had an impact on the whole of the participants while they were making their decisions regarding repetitiveness. Furthermore, the decorative elements tended to give the impression of repetitiveness to the participants.

According to the façade analysis of the second experiment, the perceived features of architectural composition could differ depending on architectural style. However, the presented research results demonstrate a few essential things:

- Hillier's proposed façade analysis method, at least in the tested sample and two presented architectural styles, could be related to human perception of architecture, thus proving the possibility to model and predict human reactions to architectural changes.

- The original methodology based on the symmetry index alone could be productively expanded by adding more indexes.
- In all cases, the syntactic indexes were sensitive even to formally small changes of architectural composition introduced by the renovations, so they could be potentially used for monitoring or evaluation of modification of the objects of cultural heritage, for example, while identifying acceptable limits of changes of indexes, predicting acceptability of changes by observers, etc.
- Potential connections between attractiveness and visually presented patterns of integration values of the façades could be noted in some cases. However, they were not caught by statistical analysis while using integration values but are worth investigating in the future.

The results support the idea of the continuation of the presented research and open new perspectives for predictive modelling, control, and monitoring of evolutional changes of immovable cultural heritage. Therefore, this research is preliminary, and it does not provide final answers. Rather, it is merely confirming the new directions for research and raising more targeted discussion.

Author Contributions: Conceptualisation, K.Z., H.A.D., and N.L.A.; methodology, K.Z., H.A.D and N.L.A.; software, K.Z., H.A.D.; validation, K.Z., H.A.D. and N.L.A.; formal analysis, K.Z., investigation, K.Z, H.A.D. and N.L.A.; resources, K.Z., H.A.D.; data curation, K.Z., H.A.D. and N.L.A.; writing—original draft preparation, K.Z, H.A.D. and N.L.A.; writing—review and editing, K.Z., H.A.D.; visualisation, N.L.A. All authors have read and agreed to the published version of the manuscript.

Funding: This research received no external funding.

Institutional Review Board Statement: Not applicable.

Informed Consent Statement: Not applicable.

Data Availability Statement: Not applicable.

Conflicts of Interest: The authors declare no conflict of interest.

References

1. Abusafieh, S. From genius loci to sustainability: Conciliating between the spirit of place and the spirit of time A case study on the old city of Al-Salt. In *Innovative Renewable Energy*; Springer: New York, NY, USA, 2019; pp. 141–163.
2. Tweed, C.; Sutherland, M. Built cultural heritage and sustainable urban development. *Landsc. Urban Plan.* **2007**, *83*, 62–69. [CrossRef]
3. Doğan, H.A. Implementation of eye tracking technology on cultural heritage research and practice. *J. Creativity Games* **2019**, *7*, 16–21. [CrossRef]
4. De la Fuente Suárez, L.A. Subjective experience and visual attention to a historic building: A real-world eye-tracking study. *Front. Archit. Res.* **2020**, *9*, 774–804. [CrossRef]
5. Joyce, J. Bayes' Theorem. In *The Stanford Encyclopedia of Philosophy*, Spring 2019 ed.; Zalta, E.N., Ed.; Metaphysics Research Lab, Stanford University: Stanford, CA, USA, 2003.
6. Yannick, J. A review of the presence and use of fractal geometry in architectural design. *Environ. Plan. B Plan. Des.* **2011**, *38*, 814–828.
7. Okuyucu, Ş.E.; Baştaş, M.S. Analysis based on fractal geometry of traditional housing facades: Afyonkarahisar traditional housing facade examples, Turkey. *J. Appl. Nanosci.* **2022**. Available online: https://link.springer.com/content/pdf/10.1007/s13204-021-02226-3.pdf (accessed on 1 March 2022). [CrossRef]
8. Salingaros, N.A.; Klinger, A. A pattern measure. *Environ. Plan. B Plan. Des.* **2000**, *27*, 537–547. Available online: https://arxiv.org/html/1108.5508 (accessed on 17 January 2022).
9. Hillier, B. *Space is the Machine: A Configurational Theory of Architecture*; UCL Publishing: London, UK, 1996.
10. Cambridge Dictionary. Available online: https://dictionary.cambridge.org/dictionary/english/facade (accessed on 5 May 2020).
11. Pallasmaa, J. *The Eyes of the Skin*; John Wiley & Sons Hoboken, NJ, USA, 1996.
12. Dotson, J.P.; Beltramo, M.; McDonnell, F.E.; Smith, R.C. Modelling the Effect of Images on Product Choices. Available online: https://ssrn.com/abstract=2282570 (accessed on 12 April 2019).
13. Akalın, A.; Yıldırım, K.; Wilson, C.; Kilicoglu, O. Architecture and engineering students' evaluations of house façades: Preference, complexity and impressiveness. *J. Environ. Psychol.* **2009**, *29*, 124–132. [CrossRef]

14. Imamoglu, C. Complexity, liking and familiarity: Architecture and non-architecture Turkish students' assessments of traditional and modern house facades. *J. Environ. Psychol.* **2000**, *20*, 5–16. [CrossRef]
15. Mao, Y.; Qi, J.; He, B. Impact of the heritage building façade in small-scale public spaces on human activity: Based on spatial analysis. *Environ. Impact Assess. Rev.* **2020**, *85*, 1–13. [CrossRef]
16. Gehl, J.; Kaefer, L.J.; Reigstad, S. *Close Encounters with Buildings*; Centre for Public Space Research/Realdania Research, Institute of Planning, School of Architecture, The Royal Danish Academy of Fine Arts: Copenhagen, Denmark, 2005.
17. Ellard, C. *Places of the Heart: The Psychogeography of Everyday Life*; Bellevue Literary Press: New York, NY, USA, 2015.
18. Salingaros, N.A. *A Theory of Architecture*; Umbau-Verlag Harald Püschel: Berlin, Germany, 2006.
19. Alexander, C. *A Pattern Language: Towns, Buildings, Construction*; Oxford University Press: Oxford, UK, 1977.
20. Hiller, B. Is architectural form meaningless? *J. Space Syntax.* **2011**, *2*, 125–153.
21. Modernist Kaunas: Architecture of Optimism, 1919–1939, Nomination for Inscription, on the UNESCO World Heritage List Nomination Dossier. Available online: https://modernizmasateiciai.lt/wp-content/uploads/2018/11/Modernist-Kaunas.-Nomination-Dossier-2021.pdf (accessed on 1 April 2022).
22. Structum. Available online: https://structum.lt/straipsnis/tikslas-naujas-senu-statiniu-gyvenimas/?fbclid=IwAR3Mkhgs66IAznyAb4F-AFtLukXmEwLCCQp8YjAQo_0OiLvEijfI9fg1Sns (accessed on 8 April 2022).
23. Laužikas, R.; Žižiūnas, T.; Kuncevičius, A.; Šmigels kas, R.; Amilevičius, D. Nekilnojamojo kultūros paveldo monitoringas taikant 3D ir dirbtinio intelekto technologijas. *Archaeol. Litu.* **2019**, *20*, 151–166. [CrossRef]
24. Lrytas: Kodėl Kaune ant Vertingų Pastatų Dygsta Parazitai? Atsakymas—Nuliūdins. Available online: https://www.lrytas.lt/bustas/architektura/2020/05/13/news/kodel-kaune-ant-vertingu-pastatu-dygsta-parazitai-atsakymas-nuliudins-14820962 (accessed on 9 April 2022).
25. Vecco, M. A definition of cultural heritage: From the tangible to the intangible. *J. Cult. Herit.* **2010**, *11*, 321–324. [CrossRef]
26. Haldrup, M.; Bærenholdt, J. Heritage as Performance. In *The Palgrave Handbook of Contemporary Heritage Research*; Waterton, E., Watson, S., Eds.; Macmillan Publishers: Hampshire, UK, 2015; pp. 52–68.
27. Smith, L. *Uses of Heritage*; Routledge Press: Oxford, UK, 2006.
28. Vecco, M. Genius loci as a meta-concept. *J. Cult. Herit.* **2020**, *11*, 225–231. [CrossRef]
29. Hawkes, J. *The Fourth Pillar of Sustainability: Culture's Essential Role in Public Planning*; Common Ground Publishing: Melbourne, Australia, 2001.
30. Hristova, Z. *The Collective Memory of Space: The Architecture of Remembering and Forgetting*; Ryerson University Press: Toronto, ON, Canada, 2006.
31. Rabun, J.; Kelso, R. *Building Evaluation for Adaptive Reuse and Preservation*; John Wiley & Sons Publishers: Hoboken, NJ, USA, 2009.
32. Miller, W. *Symmetry Groups and Their Applications*; University of Minnesota Academic Press: Minneapolis, MN, USA, 1972.
33. Carter, N.C. *Visual Group Theory*; Mathematical Association of America Press: Cambridge, MA, USA, 2009.
34. Van den Bos, G. Symmetry. In *APA Dictionary of Psychology*; American Psychological Association Press: Waghington, DC, USA, 2007.
35. Hodgson, D. The first appearance of symmetry in the human lineage: Where perception meets art. *J. Symmetry* **2011**, *3*, 37–53. [CrossRef]
36. Mitra, N.J.; Pauly, M. Symmetry for architectural design. *Adv. Archit. Geom.* **2008**, 13–16. Available online: http://www.architecturalgeometry.org/aag08/aag08proceedings-papers_and_poster_abstracts.pdf (accessed on 17 September 2019).
37. Parker, C.; Scott, S.; Geddes, A. Snowball Sampling. In *Research Methods*; Atkinson, P., Delamont, S., Cernat, A., Sakshaug, J.W., Williams, R.A., Eds.; SAGE Publishing: Thousand Oaks, CA, USA, 2019.
38. The Bartlet School of Architecture: DepthmapX: Visual and Spatial Network Analysis Software. Available online: https://www.ucl.ac.uk/bartlett/architecture/research/space-syntax/depthmapx (accessed on 9 April 2022).
39. Independent and Dependent Variables at Complete Dissertation by Statistic Solution: Expert Guidance every Step of the Way. Available online: https://www.statisticssolutions.com/independent-and-dependent-variables/ (accessed on 6 April 2022).

sustainability MDPI

Article

An Energy-Saving-Oriented Approach to Urban Design—Application in the Local Conditions of Poznań Metropolitan Area (Poland)

Wojciech Bonenberg [1,*], Wojciech Skórzewski [1,*], Ling Qi [2,*], Yuhong Han [3], Wojciech Czekała [4] and Mo Zhou [1]

1 Faculty of Architecture, Poznań University of Technology, 60-965 Poznań, Poland
2 Faculty of Architecture, Civil and Transporatation Engineering, Beijing University of Technology, Beijing 100124, China
3 College of Art and Design, Beijing University of Technology, Beijing 100124, China
4 Department of Biosystems Engineering, Poznań University of Life Sciences, 60-624 Poznań, Poland
* Correspondence: wojciech.bonenberg@put.poznan.pl (W.B.); wojciech.skorzewski@put.poznan.pl (W.S.); 63651106@bjut.edu.cn (L.Q.); Tel.: +48-61-665-3260 (W.S.); +86-137-1766-1348 (L.Q.)

Abstract: This article discusses the impact of urban layout on the energy performance of residential buildings. A comparative analysis of multiple variants of land development differing in building layout only, with all other features being the same, including the building envelope properties and technical equipment, was carried out. The research was conducted in two selected locations in Poznań metropolitan area (Poland), which is located in the Dfb climate zone (humid continental—warm summer subtype). For each location, the following variants of building layout were considered: parallel buildings (12 variants with an orientation towards the sides of the world rotated in steps of 15 degrees), perimeter frontage buildings, and comb-shaped buildings with semi-open courtyards (4 variants with courtyards open to each side of the world). The calculation of annual end uses for heating and cooling was conducted as well as the peak values. All calculations were performed using OpenStudio Application Release v. 1.2.1 software with the SketchUp plugin. The results showed that the proper arrangement of buildings on the urban plot may result in significant energy savings. The considered variants differed in terms of annual end uses for heating and cooling even by approx. 15%, and the peak values on the hottest days were 4–10 times lower in comparison with the least advantageous variants. The results show the slight advantage of compact development over free-standing development in terms of total end uses as well as the south and north orientation of facades over the east and west in terms of peak solar heat gain values.

Keywords: urban planning; building layout; energy demand; solar energy; climate

Citation: Bonenberg, W.; Skórzewski, W.; Qi, L.; Han, Y.; Czekała, W.; Zhou, M. An Energy-Saving-Oriented Approach to Urban Design—Application in the Local Conditions of Poznań Metropolitan Area (Poland). *Sustainability* **2023**, *15*, 10994. https://doi.org/10.3390/su151410994

Academic Editor: Ali Bahadori-Jahromi

Received: 30 March 2023
Revised: 4 July 2023
Accepted: 11 July 2023
Published: 13 July 2023

1. Introduction

1.1. Problem Highlighting

Energy shortages are a significant technical, social, and cultural challenge for modern cities. A contemporary society dependent on energy supply cannot dispense with energy-oriented urbanism. Popular interest in energy is encapsulated by reflections in rising energy prices and temporary shortages of energy availability for different groups of consumers. Energy has become essential to modern urban life. One of the goals of this study is to illustrate the possibility of saving energy through targeted action in creating appropriate design solutions in urban planning. Our research focuses on the Dfb climate zone according to the Köppen–Geiger climate classification [1]. Energy conservation in urban planning is associated with a reduction in the operating costs of urban structures and makes an important contribution to reducing CO_2 emissions. These elements have implications for environmental sustainability and climate change mitigation [2–4].

The issue of energy efficiency in urban planning is clearly related to advances in building material technology, access to renewable energy sources, smart grid technology, and energy-efficient urban transportation. Energy efficiency in urban planning is also related to social and cultural determinants. It is closely related to architecture, defined as "the art of creating space for humans" [5–7].

Until recently, energy efficiency in particular has not been the focus of urban planners. Traditionally, the issues of land use came to the fore with the separation of basic urban functions such as places of residence, places of work, and transport links between the functional zones of the city. Against this backdrop, the following basic research streams emerged: the sociological stream, studying the city from the side of social conditions, the natural–ecological stream, emphasizing the problems of natural balance within urban structures, the cultural trend, considering the development of urban space against the background of cultural processes, and the economic stream, related to the analysis of economic and spatial conditions determining the competitiveness of urban space.

A review of recent scientific studies seems to confirm the observation that we are witnessing the emergence of a new trend in urban studies, which is energy-saving oriented urban planning. Within the framework of this direction, not only are the energetic determinants of the development of urban structures analyzed but also the problems of energy conservation and efficiency with the use of appropriate urban planning tools, such as composition, the placement of functional zones, transportation facilities, and changes in socio-cultural habits. An important task is to ensure the energy self-sufficiency of cities and efforts to become independent of external energy sources. It analyzes elements such as fuel chains in the urban structure (energy sources, the location of energy production, transmission networks, impacts during operation, consequences for the environment and the health of residents, etc.), emissions, and other environmental stresses. In this context, the basic strategies of a sustainable urban approach can be distinguished [8–10]:

(a) Economic:
  - Investment risk;
  - Investment efficiency;
  - Consumer preferences (residents, tenants, property owners, and users);
  - Public support for renewable energy sources.

(b) Social and cultural:
  - Spatial behavior of residents and its impact on energy consumption;
  - Social acceptance of energy-saving solutions;
  - Cultural conditions;
  - Social potential of the area (ability to create innovative pro-environmental solutions).

(c) Natural:
  - Environmental risks associated with energy production and transmission;
  - Natural stability (species diversity, spatial distribution, species dominants, stratification, etc.);
  - The location of tall trees in the immediate vicinity of buildings and their impact on climatic comfort in residential environments [11].

(d) Structural:
  - Technical standard of development (thermal insulation envelope of buildings);
  - Energy-efficient methods of transportation (public transportation and bicycles);
  - Optimization of the length of technical infrastructure routes in the urban structure;
  - Optimization of the spatial arrangement of buildings.

The strategies of a sustainable urban approach mentioned above are reflected in energy-saving urban design methods, with them being taken into consideration in certification standards or algorithms for calculating the energy performance of buildings. Among the issues listed above, an important structural issue in urban planning, the arrangement of buildings on urban plots, which our article is focused on, is strictly related to the concept of passive solar heat gains. They are included in the energy balance of buildings as passive heat sources. The level of passive solar heat gains, which is the amount of solar radiation passing through windows into the building interior, is dependent on geometric factors related to urban planning:

- The solar rays incidence angle, which depends on the orientation of buildings towards the directions of the world;
- The degree of shading on the glazed parts of facades, which depends on the locations of the buildings and other shading objects in the surroundings [12].

The methodology for calculating passive solar heat gains is based on the formula [13]:

$$Qs = r \cdot g \cdot Ag \cdot G \text{ [kWh/year]} \quad (1)$$

In this equation, Qs—total amount of passive solar gains; r—reduction coefficient, including the solar incidence angle, shading, and dirt; g—total solar energy transmittance of the glazing; Ag—glazing area; and G—total solar radiation during the heating season), and it has been implemented in several algorithms used in building energy certification standards and energy simulation software [13–17].

This makes urban planning one of the significant fields of energy efficiency research [18–20].

1.2. State of the Art—Review of the Literature and Methodologies

The integration of energy efficiency into the urban design process has been studied from many perspectives. Among the fundamental works for the development of the theory of solar urban planning, there are the studies by Amado and Poggi [21,22]. The research conducted in Copenhagen, Denmark by Strømann-Andersen and Sattrup [23] on the impact of the size of urban canyons on access to solar radiation shows the significant influence of distances between buildings on the total energy use for heating, cooling, and artificial lighting, reaching up to +30% for office buildings and +19% for residential buildings.

The research results by Deng et al. [24] indicate a quantitative correlation between the placement of buildings and the microclimate and energy performance of a building on the scale of an urban project. In addition, the research provides several recommendations for urban planners and designers such as strategies for reducing the UHI effect and decreasing energy use. Independently, the research results of Wang et al. [25] showed that the arrangement of urban blocks and the vegetation configuration can significantly reduce the concentration of air pollutants and improve the microclimate. The correlation of various parameters of urban morphology (the type of building area, the amount of space between buildings, etc.) with urban air temperature was the subject of a study by Tong et al. [26] in Tianjin, China. Similarly, multiple parameters influencing the energy performance of buildings in different climate zones in China were investigated in the work of Zhao et al. [27].

There have also been trials on the use of machine learning to find the most appropriate architectural form in terms of reducing carbon footprints [28]. One of the co-authors performed his own simulations of the distribution of solar radiation on the surfaces of building facades depending on the spatial arrangement of buildings and greenery including its role in protecting against overheating in summer, using SketchUp (version 19.1 174) software with the DL Light add-on [11,29].

Some authors have made attempts to identify the key indicators of access to solar energy in housing estates in different climatic zones [30–33]. The last year has seen particularly intense interest from research teams on solar access planning issues [34], solar energy systems with energy storage [35], the analysis of urban initiatives affecting energy consumption and the absorption of solar energy [36], and the location of urban blocks in neighborhood units in relation to solar radiation [37].

Many authors emphasize that the early phase of architectural and urban design is critical to energy efficiency. This draws attention to problems such as integrating the impacts and costs of energy-saving solutions in the early design phase [38], bridging the gap between research and the early design phases [39], the optimization of the urban features of housing developments in terms of energy efficiency [40], and the impact of energy conservation on the integration of social and technical conditions in the age of the knowledge society [41,42]. The presented review of the latest research allows us to confirm the timeliness of the research topic, which concerns not only the technical aspects of the energy efficiency of housing estates but also touches on a wide spectrum of architectural, urban, and compositional issues and inhabitants' communities.

Previous studies present a differentiated approach in terms of the following fields: research materials (location and spatial models), building typologies, climate zone and data, simulation software and algorithms, and output data. In terms of research materials, the studies can be divided into two main groups: the first are based on theoretical models [43–45], the second are based on real locations. In the second group, for real locations, there are also studies relating to the existing development [46] or considering hypothetical development models [47,48]. Among the proposed building layout typologies taken into consideration, the majority of authors take into consideration a few basic factors: tower (high-rise), courtyard, row (parallel), and some other additional factors, like point blocks.

M.M. Akrofi and M. Okitasari [49] prepared a systematic review of the state-of-the-art in the field of integrating solar energy considerations into urban planning, in which the authors indicated the main fields of interest of researchers, characterized the geographic distribution of the studies, and described the main conceptions of solar urban planning. In their conclusions, they also defined research gaps and future research directions, including geographical gaps, socio-technical gaps, and the need for theories. This work can be placed in the first group—geographical gaps, taking into account the small number of studies on this subject in Poland to date (only two). That means also adapting the research to the local climate, laws, and social conditions. In particular, the presented research takes into consideration the following local circumstances:

- The local climatic conditions;
- The requirements of Polish construction laws and technical conditions, referring to the insolation time, distances between buildings, building envelope parameters, etc.;
- Local spatial planning constraints;
- The shape of the plot;
- The existing built environment in the direct vicinity.

The original contribution of this study to the state of the art is to demonstrate how the local conditions mentioned above modify certain regularities shown by theoretical models. The scientific question posed is whether different local circumstances may lead to different or contradictory recommendations regarding preferred variants of the mutual arrangement of buildings or whether, in all circumstances, the same regularities are kept. For this reason, the same set of variants was tested in different locations in order to prepare the comparative analysis.

Additionally, unlike previous research, our models were created in the Sketchup software environment, the use of which is widespread among architects [50], along with freeware OpenStudio add-on.

2. Materials and Methods

2.1. Research Objective and Overview

The methodological scheme of the study is shown in Figure 1.

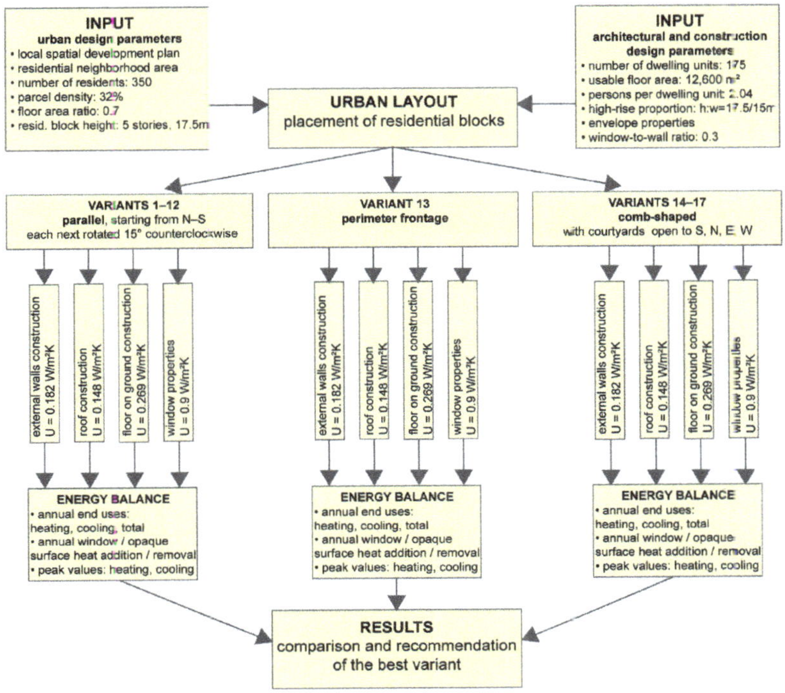

Figure 1. Methodological scheme of the study (data for research area no. 1).

The objective of the research is to identify the possibilities to improve the energy performance of buildings by changing their spatial arrangement only. For this, comparative analysis of the energy performance of a few variants of land development is performed. To achieve the research goal, the following principles are adopted:

- The only difference between the considered variants is the spatial arrangement of the buildings.
- All variants of land development and all adopted properties of buildings have to meet the minimum requirements according to the Polish law regulations and standards.
- The rest of the adopted features and parameters of the buildings are the same in all variants to ensure that the variants are comparable and to eliminate the impact of all other variables except the spatial arrangement of the buildings.

All of the adopted parameters of the buildings are described in detail below.

As far as function is concerned, multi-family residential buildings were chosen to easily determine standards for them and to compare them.

The dimensions of the building models were adjusted to the standards that allow one to fit the functional living units inside. The width of the building was set at 15 m, which corresponds to the two-bay layout of the building interior, assuming a solar penetration depth of about 6 m from both sides of the building. This is the area dedicated for rooms intended for human occupation. An additional 1.50 m was left for internal corridors (minimum 1.40 m required due to fire protection regulations for emergency escape roads). Approximately 0.50 m was added for each of the two external walls and 2 × approx. 0.25 m for the internal construction walls (which gives a total of 1.50 m for the walls' thickness). The adopted average height of the story (measured floor-to-floor) was 3.00–3.50 m to ensure clear height for a story larger than the minimum required 2.50 m for apartments (recommended approx. 2.80 m). The remaining height was intended for construction, acoustic insulation, and the finishing layers of floor slabs and for a possible space reserve for installations (e.g., mechanical ventilation ducts if needed) located above the suspended ceiling.

The height of the buildings was set at 5 stories, which is the maximum permissible according to the local spatial development plan. This gives an overall height of 17.5 m. However, there was an additional 10-story variant added, which exceeds the permissible value, in order to check if other solutions not provided for in the plan would not be better from the point of view of energy efficiency.

The window-to-wall area ratio was set at 0.3, resulting from an estimation of the minimum required window-to-floor area ratio, which is 1:8 according to the Regulation of the Minister of Infrastructure on technical specifications for buildings and their location [51]. While the room width is 6 m, for each running meter of the facade length, there is 6 m^2 of floor area, which gives a minimum of 0.75 m^2 for the window area. This means that the window-to-wall area ratio should be at least 0.214, so 0.3 meets this requirement.

The other assumptions for the newly designed plot are as follows:

- A built-up area of approx. 3600 m^2 (except for the additional 10-story variant, with a built-up area of approx. 2400 m^2), which means approx. 18,000 m^2 of gross floor area and approx. 12,600 m^2 of usable floor area;
- For each apartment, meeting sunlight requirements of at least 3 h on equinox days (21 March and 23 September), according to the Polish Regulation of the Minister of Infrastructure on technical specifications for buildings and their location [51].

2.2. Research Material

The theoretical models of the designed development were placed in two exemplary locations in Poznań metropolitan area (Figures 2 and 3):

- Research area no. 1—A proposal for the design of a local spatial development plan in Poznań, located in the neighborhood of Grunwaldzka and Ułańska Street in Poznań. One of the quarters of the designed development in this plan is taken as the material for research due to its dimensions, which allows for the placement of various layouts of buildings, and due to its location in the neighborhood of the typical development representative of central districts in Poznań. The direct surroundings of the given plot are designed as built-up with compact frontage buildings with a height of approx. 17 m and 5 stories. The selected plot of land has a shape similar to a rectangle, and the area is approx. 11,250 m^2. The adopted parameters of the designed development variants: 5 stories, a floor height of 3.50 m, and a parcel density of 32%.
- Research area no. 2—A plot near Wschodnia Street in Luboń. This location was chosen as a representative of the location in the neighborhood of the typical land development of Poznań suburban area, containing mixed single-family and multi-family housing. The selected area is approx. 27,250 m^2. The adopted parameters of the designed development variants: 5 stories, a floor height of 300 m, and a parcel density of 27%.

Taking into account all of the assumptions described in the previous chapter, for each research area, 17 variant land development concepts differing in the layout of buildings were prepared. The concepts are the following (Figures 4 and 5):

- Variants 1–12: Parallel buildings along the north–south axis;
- Variant 13: Perimeter frontage buildings along the streets, around the quarter, with one large courtyard inside;
- Variants 14–17: Comb-shaped layouts of buildings with the courtyards open to different sides of the plot: south, north, east, and west.

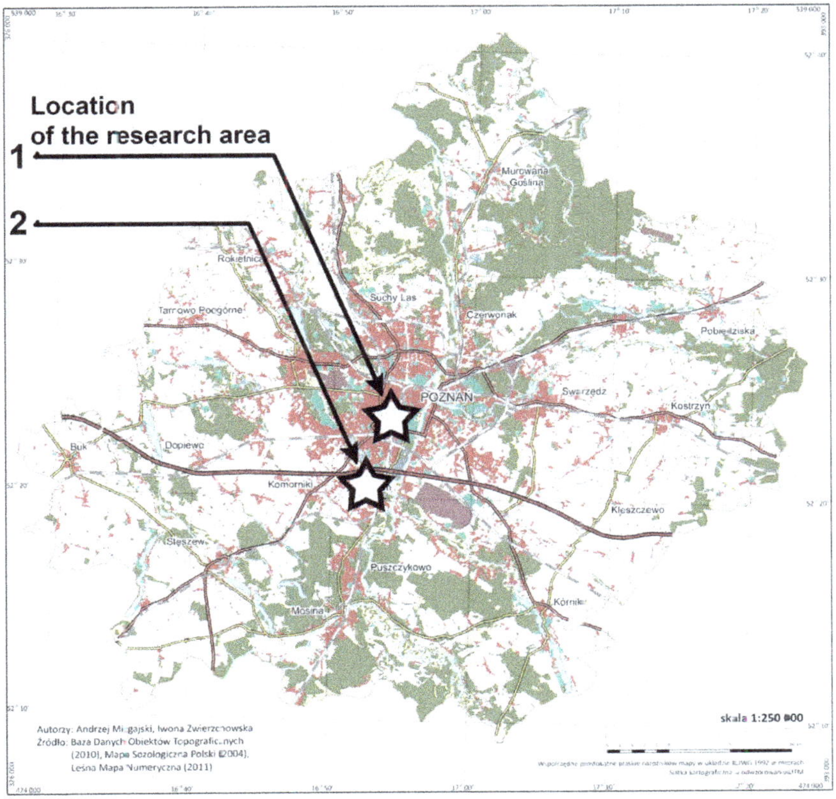

Figure 2. Locations of the research areas (no. 1 and 2) on the map of Poznań metropolitan area.

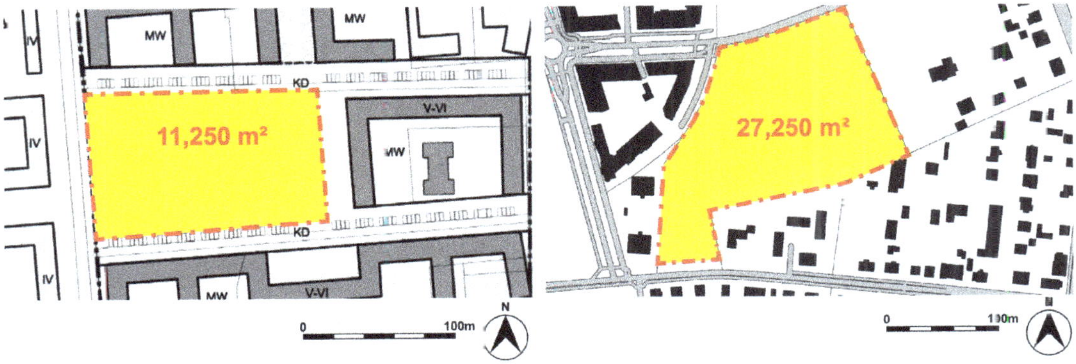

(a) (b)

Figure 3. Research areas (research plots marked in yellow): (a) Research area no. 1: Proposal for the design of a local spatial development plan near Grunwaldzka and Ułańska Street in Poznań and (b) Research area no. 2: Plot near Wschodnia Street in Luboń.

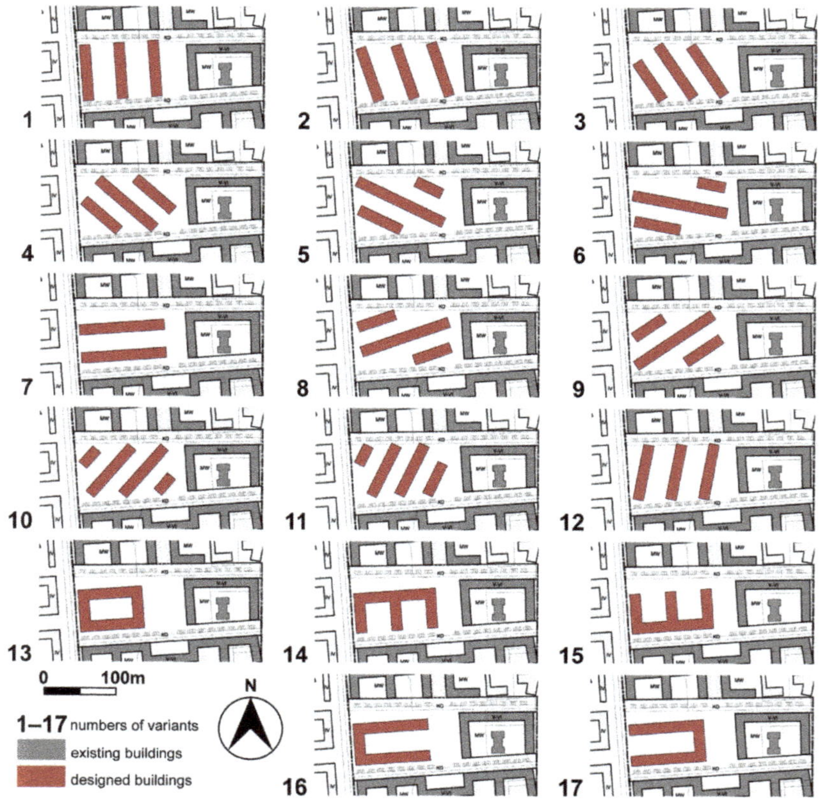

Figure 4. Variants of land development for comparative analysis in research area no. 1.

2.3. Research Methods and Tools

The following software was used for the research:

- SketchUp 2020 for 3D modeling;
- OpenStudio Application Release v. 1.2.1/OpenStudio SDK (core) Version 3.2.1 with SketchUp Plugin Version v. 1.4.0 for energy performance calculation.

All variants were modeled using the OpenStudio add-on within SketchUp, defining all the needed properties of building elements necessary to run the energy performance calculations.

The following properties of the 3D models were defined:

- Thermal zones: type, 189.1–2009, midrise apartment, Apartment CZ4-8
- Loads: people, 0.03 people/m^2; interior lights, 10.65 W/m^2; electric equipment, 3.88 W/m^2

The structure of the external walls, windows, roofs, and floors was specified to meet the requirements of the regulation on technical specifications for buildings as far as the heat transfer coefficient (U-value) of the building envelope is concerned, with the use of materials commonly used in housing construction in Poland. The construction sets of all the external partitions are specified in Tables 1–3.

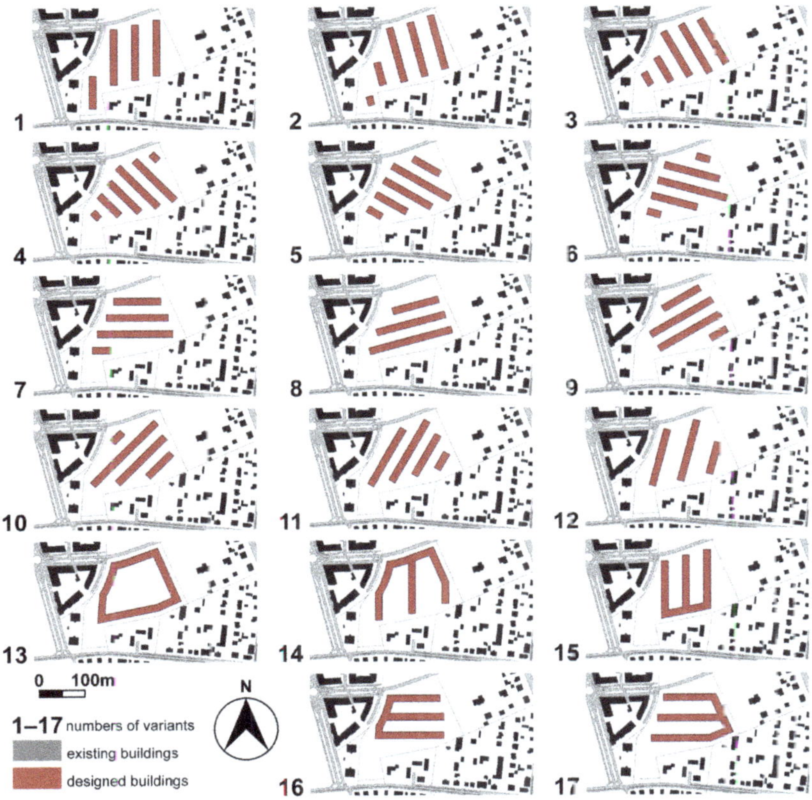

Figure 5. Variants of land development for comparative analysis in research area no. 2.

Table 1. Construction set of external walls.

External Wall		
Material	**Λ (W/mK)**	**d (cm)**
(Exterior, air)		
Gypsum	0.160	1.0
Graphite Styrofoam EPS	0.031	15.0
Silicate blocks	0.510	18.0
Gypsum	0.160	1.0
(Interior)		
Heat transfer coefficient U (W/m^{2}K)	0.182	

Table 2. Construction set of roofs.

External Roof		
Material	**λ (W/mK)**	**d (cm)**
(Exterior, air)		
Metal roofing	45.006	0.2
Graphite Styrofoam EPS	0.031	20.0

Table 2. *Cont.*

External Roof		
Material	**λ (W/mK)**	**d (cm)**
Concrete slab	1.700	20.0
Gypsum	0.160	1.0
(Interior)		
Heat transfer coefficient U (W/m^{2}K)	0.148	

Table 3. Construction set of floors on the ground.

Floor on Ground		
Material	**λ (W/mK)**	**d (cm)**
(Interior, finishing layers)		
Extruded polystyrene XPS	0.035	12.0
Concrete slab	1.700	20.0
(Exterior, ground contact)		
Heat transfer coefficient U (W/m^{2}K)	0.269	

The adopted properties of windows are the following:

- Heat transfer coefficient U_w (W/m^{2}K) = 0.9
- Solar heat gain coefficient g = 0.55
- Visible transmittance L_t = 0.75

Using the properties and settings listed above, the following values are calculated:

1. Annual end uses:
    - For the purposes of heating
    - For the purposes of cooling
    - Total
2. Annual building sensible heat gain components:
    - Window heat addition
    - Window heat removal
    - Opaque surface conduction and other heat removal
3. Peak cooling sensible heat gain components:
    - Window heat addition
4. Peak heating sensible heat gain components:
    - Window heat removal
    - Opaque surface conduction and other heat removal

3. Results

For each considered variant, the energy balance of the buildings was calculated, including the following:

- The overall energy demand for heating, cooling, interior lighting, interior equipment, and fans, calculated annually and monthly;
- The peak energy demand for heating, cooling, interior lighting, interior equipment, and fans, calculated for the extreme values in each month.

The monthly values of both overall and peak energy demand for the first variant of the building layout are presented in Figures 6 and 7. Analogous simulations were carried out for the other variants.

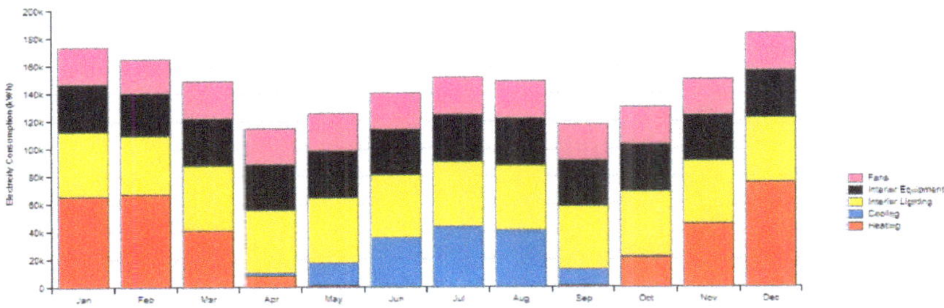

Figure 6. Results for research area no. 1, variant 1: overall energy demand.

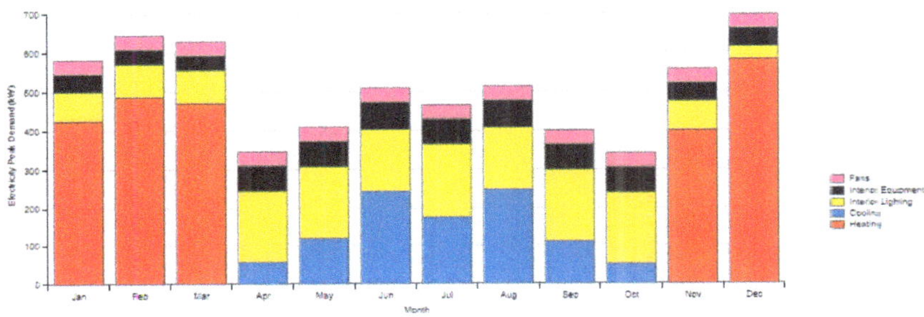

Figure 7. Results for research area no. 1, variant 1: peak energy demand.

The results for all of the variants are in Tables 4–7. The tables display the most important values in terms of energy demand, heat transfer, and the use of solar passive heat gains, including annual and peak values. The same results are shown in the charts (Figures 8–11) as well.

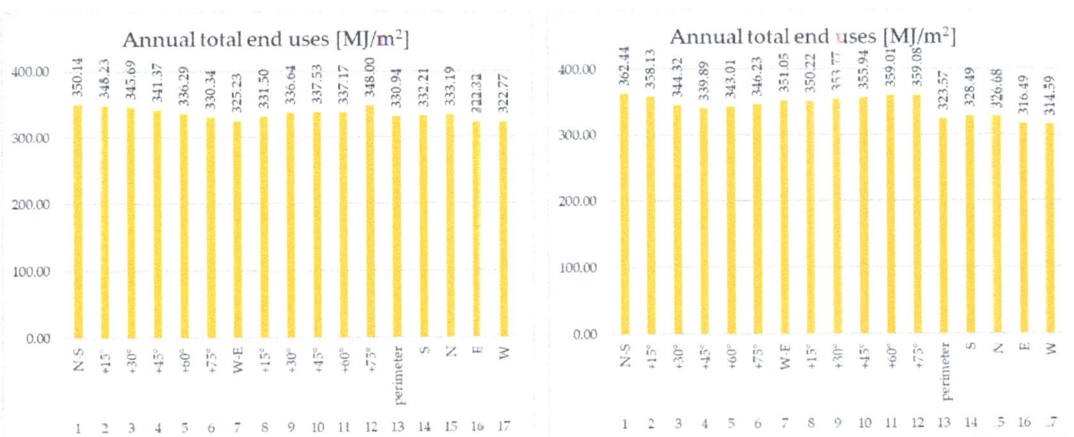

(a) (b)

Figure 8. Comparison of variants 1–17—annual total end uses: (**a**) research area no. 1 and (**b**) research area no. 2.

Table 4. Results for research area no. 1, variants 1–17: annual values.

Research Area No. 1				Total End Uses			Sensible Heat Gain Components		
Variant no.	**Layout Orientation**		**Usable Floor Area [m²]**	**Heating [MJ/m²]**	**Cooling [MJ/m²]**	**Total [MJ/m²]**	**Window Heat Addition [MJ/m²]**	**Window Heat Removal [MJ/m²]**	**Opaque Surface Conduction and Other Heat Removal [MJ/m²]**
1	parallel	N-S	18,000	65.71	30.35	350.14	112.10	−42.05	−129.65
2		+15°	18,000	65.79	29.82	348.23	110.26	−42.02	−129.23
3		+30°	18,000	65.67	28.98	345.69	108.32	−42.06	−128.78
4		+45°	18,000	65.41	28.45	341.37	108.86	−41.99	−128.50
5		+60°	18,000	64.72	28.82	336.29	115.76	−42.03	−129.33
6		+75°	18,000	65.09	26.85	330.34	108.09	−41.77	−127.65
7		W-E	18,000	62.97	24.57	325.23	108.67	−39.83	−127.06
8		+15°	18,000	63.40	26.03	331.50	111.85	−41.85	−128.35
9		+30°	18,000	63.08	27.68	336.64	112.12	−41.70	−127.68
10		+45°	18,000	57.19	30.74	337.53	116.40	−44.11	−131.86
11		+60°	18,000	55.01	31.50	337.17	113.37	−44.14	−131.79
12		+75°	18,000	65.48	29.91	348.00	111.97	−42.01	−129.47
13	perimeter		18,000	66.56	24.73	330.94	102.45	-40.13	−129.39
14	comb-shaped	S	18,000	65.18	25.26	332.21	97.91	−37.96	−127.50
15		N	18,000	66.15	25.14	333.19	95.16	−38.03	−127.69
16		E	18,000	61.70	24.00	322.32	98.01	−37.19	−123.05
17		W	18,000	61.81	24.08	322.77	97.67	−37.21	−123.01

Table 5. Results for research area no. 1, variants 1–17: peak values.

Research Area No. 1				Peak Cooling		Peak Heating		
Variant no.	**Layout Orientation**		**Usable Floor Area [m^2]**	**Date**	**Window Heat Addition [W/m^2]**	**Date**	**Window Heat Removal [W/m^2]**	**Opaque Surface Conduction and Other Heat Removal [W/m^2]**
1	parallel	N-S	18,000	05-AUG	31.11	12-DEC	−3.29	−6.75
2		+15°	18,000	05-AUG	31.04	12-DEC	−3.33	−6.71
3		+30°	18,000	05-AUG	28.70	12-DEC	−3.37	−6.65
4		+45°	18,000	05-AUG	27.05	12-DEC	−3.35	−6.71
5		+60°	18,000	05-AUG	23.11	12-DEC	−3.28	−6.73
6		+75°	18,000	05-AUG	16.54	12-DEC	−3.33	−6.66
7		W-E	18,000	29-JUN	7.46	12-DEC	−3.09	−6.44
8		+15°	18,000	29-JUN	15.08	12-DEC	−3.31	−6.52
9		+30°	18,000	29-JUN	18.08	12-DEC	−3.29	−6.40
10		+45°	18,000	05-AUG	24.55	12-DEC	−3.49	−6.52
11		+60°	18,000	05-AUG	26.74	12-DEC	−3.50	−6.46
12		+75°	18,000	05-AUG	30.29	12-DEC	−3.30	−6.70
13	perimeter		18,000	05-AUG	15.60	12-DEC	-3.21	−7.04
14	comb-shaped	S	18,000	05-AUG	20.33	12-DEC	−3.04	−7.14
15		N	18,000	05-AUG	20.10	12-DEC	−2.83	−7.42
16		E	18,000	05-AUG	8.19	12-DEC	−2.94	−6.09
17		W	18,000	05-AUG	8.31	12-DEC	−2.95	−6.13

Table 6. Results for research area no. 2, variants 1–17: annual values.

Research Area No. 1				**Total End Uses**			**Sensible Heat Gain Components**		
Variant No.	**Layout Orientation**		**Usable Floor Area [m²]**	**Heating [MJ/m²]**	**Cooling [MJ/m²]**	**Total [MJ/m²]**	**Window Heat Addition [MJ/m²]**	**Window Heat Removal [MJ/m²]**	**Opaque Surface Conduction and Other Heat Removal [MJ/m²]**
1	parallel	N-S	32,250	85.39	29.96	362.44	98.48	−39.70	−147.36
2		+15°	32,250	79.17	31.17	358.13	100.27	−40.86	−148.30
3		+30°	32,250	65.63	32.88	344.32	96.61	−40.83	−148.11
4		+45°	32,250	64.07	32.61	339.89	96.48	−42.10	−149.27
5		+60°	32,250	75.33	29.68	343.01	95.96	−40.45	−144.83
6		+75°	32,250	84.83	26.62	346.23	94.84	−40.84	−147.63
7		W-E	32,250	91.39	25.29	351.05	92.66	−39.57	−145.66
8		+15°	32,250	89.34	25.41	350.22	92.28	−38.55	−145.57
9		+30°	32,250	87.13	27.09	353.77	95.34	−39.78	−146.71
10		+45°	32,250	86.07	27.55	355.94	96.97	−39.96	−147.88
11		+60°	32,250	84.60	29.96	359.01	98.62	−39.67	−145.66
12		+75°	29,250	81.89	30.90	359.08	103.81	−38.62	−144.97
13	perimeter		36,316	58.43	25.52	323.57	100.33	−32.46	−123.03
14	comb-shaped	S	33,257	59.23	26.51	328.49	94.95	−31.51	−122.11
15		N	32,610	58.28	25.67	326.68	91.05	−31.28	−120.44
16		E	33,389	60.99	22.19	316.49	87.93	−31.41	−121.12
17		W	36,104	58.41	22.38	314.59	89.74	−31.12	−119.77

Table 7. Results for research area no. 2, variants 1–17: peak values.

Research Area No. 1				Peak Cooling		Peak Heating		
Variant No.	Layout Orientation		Usable Floor Area [m^2]	Date	Window Heat Addition [W/m^2]	Date	Window Heat Removal [W/m^2]	Opaque Surface Conduction and Other Heat Removal [W/m^2]
1	parallel	N-S	18,000	05-AUG	25.40	12-DEC	−4.34	−7.39
2		+15°	18,000	05-AUG	26.38	12-DEC	−4.47	−7.36
3		+30°	18,000	05-AUG	25.64	12-DEC	−4.46	−7.36
4		+45°	18,000	05-AUG	24.22	12-DEC	−4.59	−7.44
5		+60°	18,000	05-AUG	20.18	12-DEC	−4.43	−7.01
6		+75°	18,000	05-AUG	15.24	12-DEC	−4.44	−7.08
7		W-E	18,000	29-JUN	5.94	12-DEC	−4.30	−6.86
8		+15°	18,000	29-JUN	7.11	12-DEC	−4.19	−6.90
9		+30°	18,000	29-JUN	14.44	12-DEC	−4.33	−7.06
10		+45°	18,000	29-JUN	15.90	12-DEC	−4.33	−6.86
11		+60°	18,000	05-AUG	21.21	12-DEC	−4.35	−7.19
12		+75°	18,000	05-AUG	23.69	12-DEC	−4.24	−6.85
13	comb-shaped	perimeter	18,000	29-JUN	13.73	12-DEC	−2.44	−6.51
14		S	18,000	05-AUG	21.21	12-DEC	−2.41	−5.65
15		N	18,000	05-AUG	21.76	12-DEC	−2.40	−5.08
16		E	18,000	05-AUG	6.54	12-DEC	−2.45	−5.48
17		W	18,000	05-AUG	2.64	12-DEC	−2.39	−5.08

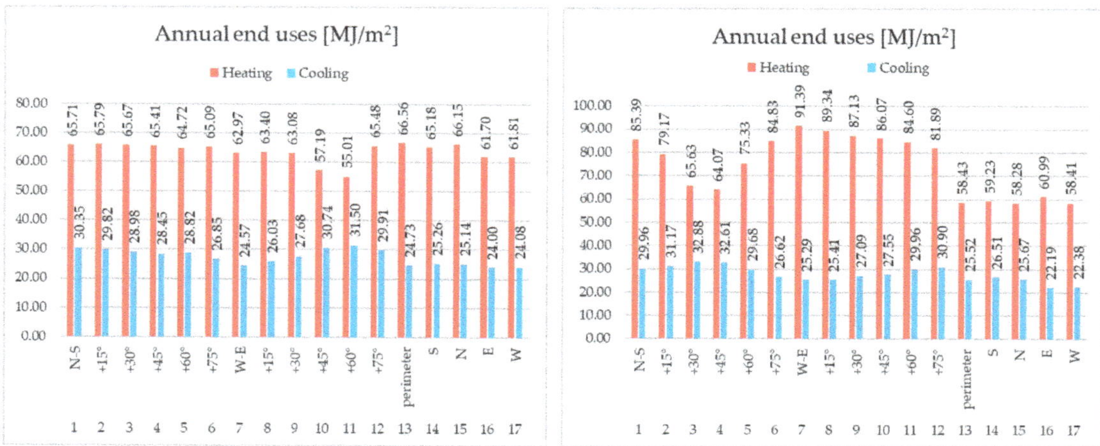

(a) (b)

Figure 9. Comparison of variants 1–17—annual end uses for heating and cooling: (**a**) research area no. 1 and (**b**) research area no. 2.

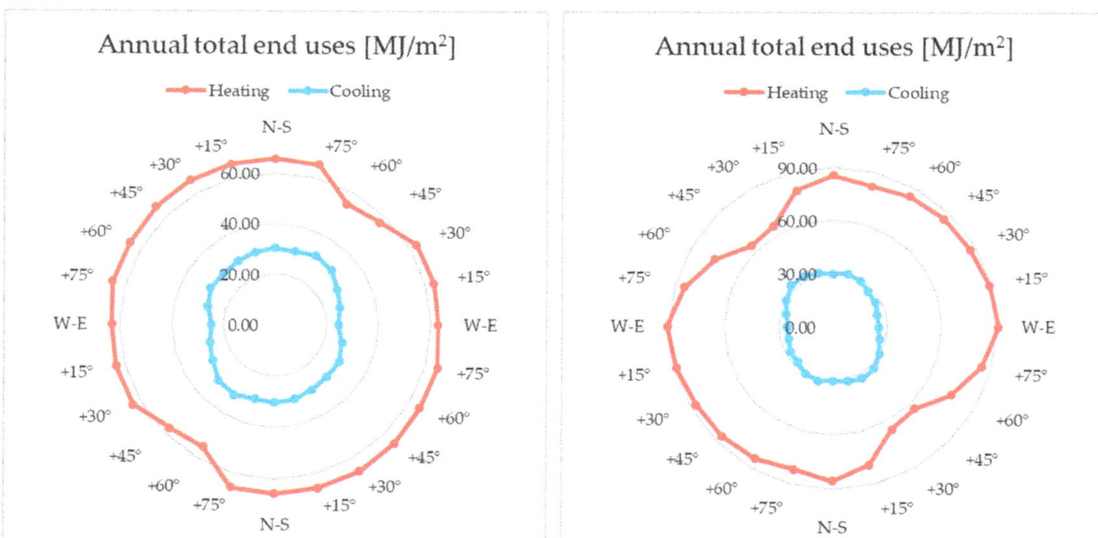

(a) (b)

Figure 10. Comparison of variants 1–12 (parallel buildings)—annual end uses for heating and cooling: (**a**) research area no. 1 and (**b**) research area no. 2.

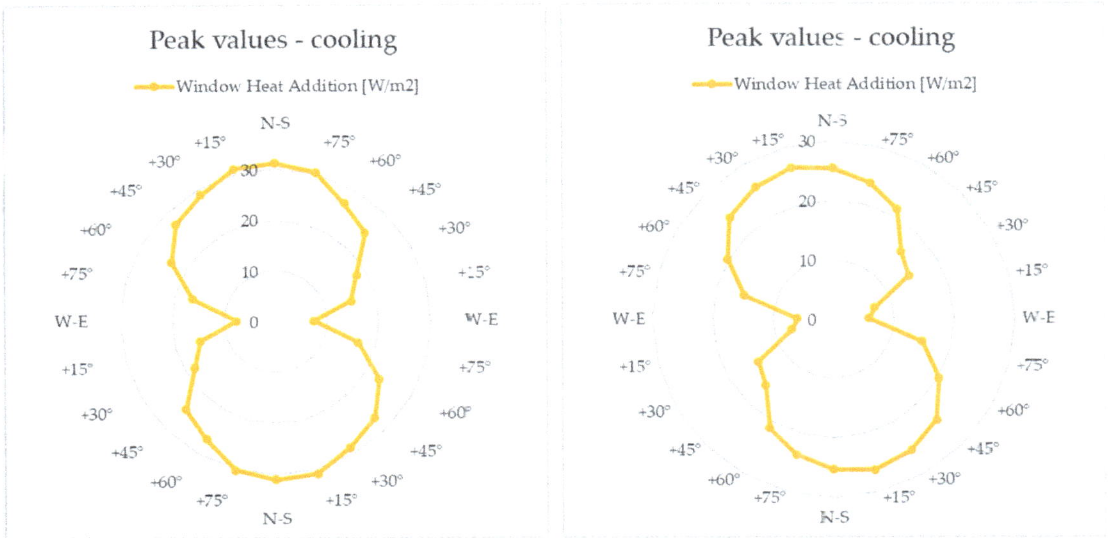

(**a**) (**b**)

Figure 11. Comparison of variants 1–12 (parallel buildings)—peak values of energy demand for cooling: (**a**) research area no. 1 and (**b**) research area no. 2.

In terms of annual total end uses (Figure 8), the results vary from 322.77 MJ/m^2 to 350.14 MJ/m^2 for research area no. 1 and 314.59 MJ/m^2 to 362.44 MJ/m^2 for research area no. 2. The difference between the highest and lowest value is between 8.5% and 15.2%. In both cases, we can see the advantage of the frontage building layouts (perimeter and comb-shaped) over the parallel layouts. Also, in both cases, the most profitable variants are comb-shaped with semi-open courtyards oriented to the east or west direction. The least advantageous are parallel buildings along the north–south axis.

If we look at the components of energy consumption, the demand for heating and cooling purposes (Figures 9 and 10), we can make the following observations.

We can see significant differences between the variants in terms of energy demand for heating; depending on the case (location), they range from 21.0% (area no. 1) to 56.8% (area no. 2). In both cases, there is a certain orientation, which brings significant savings in energy use for heating purposes.

As far as the energy demand for cooling is concerned, the differences between the most and least favorable spatial layout is between 31.2% to 48.2%. In all cases, energy demand for cooling was lower in the perimeter and comb-shaped developments, especially with semi-open courtyards oriented to the east or west directions, while the highest values appear in variants with parallel buildings along the north–south axis. Similar observations are even more visible in the results of peak demand for cooling; as is shown in the charts (Figure 11). the peak values in variants with dominant facades exposed to the east and west are 4 to 10 times higher than in the case of those with mostly south- and north-facing facades.

These results correspond with the results of previous research on the amount of solar energy in building facades, which showed that the eastern and western facades are the most vulnerable to overheating. South-facing facades are more advantageous in this respect, providing the right amount of solar gain in winter while contributing less to interior overheating in summer due to the high incidence angle of sunrays.

Ensuring that the facades of the building are shaded by other buildings located in its direct neighbourhood (or other parts of the same building) reduces the amount of solar energy delivered to the facades to varying degrees depending on the distance of the shading object from the facade, the orientation of the facade in relation to the directions, and the seasons. The compactness of building forms is an important factor in terms of the amount of heat transfer losses. In particular, we observed the following:

- Frontage or comb-shaped layouts result in lower energy demand than free-standing linear layouts.
- In the case of linear layouts, the orientation towards the sides of the world is crucial; orientation along the north–south axis is not recommended due to the highest risk of overheating as well as the overall energy demand. The changes in energy consumption for heating and cooling occurring with the change in direction are opposite to each other.
- East and west facades are exposed to the greatest risk of overheating.
- The best possibility to benefit from passive solar gains in winter is on facades exposed to the south; layouts with larger distances between facades located in that direction are preferred.

4. Discussion

The results of the study show the answer to the research question, the purpose of which was to assess how the urban arrangement (composition) of multifamily residential buildings affects energy consumption. In the comparative studies, only differences in the urban layout were taken into account. The other urban, architectural, and technical parameters remained identical for all variants. There are a number of studies analyzing urban development in terms of thermal effects dependent on the surface materials, variation in the building facades, the ratio of the window area to the exterior walls [52], green ground cover [53], and the morphology of vegetated and built surfaces [54]. In our model, all of these factors were unified in order to objectively assess only the impact of the blocks' arrangement itself on the energy efficiency of an urban plot. Studies have shown that the arrangement of blocks on an urban plot alone has a significant impact on energy consumption.

Many previous studies have confirmed this thesis but for different climate zones or other parameters of buildings. Some of them were theoretical models abstracted from a specific location, while our study is about adapting these computational models to specific local conditions.

In order to make the comparison of the results possible and reliable, the placement of the study among the existing research on solar urban planning should be indicated. The study is a multi-variant comparative analysis of different building layouts. A similar approach is used in a few other studies on energy efficiency in urban planning [43,45,46,48,55,56]. However, there are some limitations, which cause difficulties in comparing the results, which are the different assumptions, the scope of considered variants, surrounding developments, climate zones, etc.

In terms of the research material, the study refers to real locations and existing neighborhoods, while the proposed variants of the building layout are based on theoretical models adjusted to local conditions, including the plot shape and the requirements of planning and construction regulations that affect, among others, the distance between buildings which must ensure the minimum insolation time. When adapting the theoretical assumptions of solar-energy-saving urban planning and incorporating it into planning practice, taking into account its previous experience is very important for the implementation of this idea [57–59].

From this point of view, the results of this work can be compared with Loeffler and Geier's [46] study, which contains four variants of the development of an exemplary plot in Vienna; however, only two of them match with our research: the perimeter and row structured buildings. Their results confirm the same observations: lower energy demand is achieved in the case of perimeter buildings in comparison to row layouts. Another analogy is an influence of orientation towards the sides of the world, but the difference (up to 2.3%) is smaller in comparison to our study (about 6.6–7.7%). On the other hand, a study on theoretical models conducted by Giostra et al. [44] showed this difference on the level of approx. 10%.

If we take into consideration heating only, it is even 19.4–42.6% depending on the case, but it is compensated for by the opposite results in cooling. The results for the cooling and heating components, which change depending on the orientation, are contrary to each other, which coincides with the similar observation by Strømann-Andersen and Sattrup [45], but they refer to the density of buildings.

Existing research for the European climate (Dfb) has included composition problems as a fragment of a more comprehensive effort related to optimization of urban forms, energy balance, and environmental quality [43]. Other studies confirm the difference we found between staggered forms (variants 1–12) and compact forms (variants 13–17) [44].

The urban layout of residential blocks also affects the air flow between buildings. This factor affects the energy balance and the concentration of pollutants in the air. In the case of tall buildings, the speed of air movement is different at the ground and top floor levels. This problem is pointed out by Negin et al. [60].

On the other hand, comparing our results with analogous results for warm climate zones, the sheer impact of the arrangement of blocks on a plot in Poznań is more significant than in warm climates (e.g., Mediterranean districts) [45]. As our research has shown, the shape factor is noticeably significant in the area of the city of Poznań (in the Central European climate zone Dfb). To a lower extent, this is related to the temperate oceanic climate (Cfb), as confirmed by the study of Yannas, and Rodríguez-Álvarez [61,62]. This comparison is for reference only, provided that, in fact, there are also other conditions that affect the energy performance of buildings in warmer climatic zones, e.g., the specificity of touristic cities [63].

Our additional comparative analyses also confirmed this assumption. As part of our extended research team, we conducted analogous studies for the same five variants of plot development but in other climatic zones. For comparison, we made the theoretical assumption that the plot is located in:

- Beijing (39.9243, 116.3881)—Köppen–Geiger climate zone (Dwa)
- Palermo (38.1156, 13.3556)—Köppen–Geiger climate zone (Bdf)

Figure 12 shows the comparison of the results of variants one–five calculated for Poznań (Dfb), Palermo (Bdf), and Beijing (Dwa). It can be seen that the urban arrangement factor for the five analyzed development variants is most significant for the Poznań location.

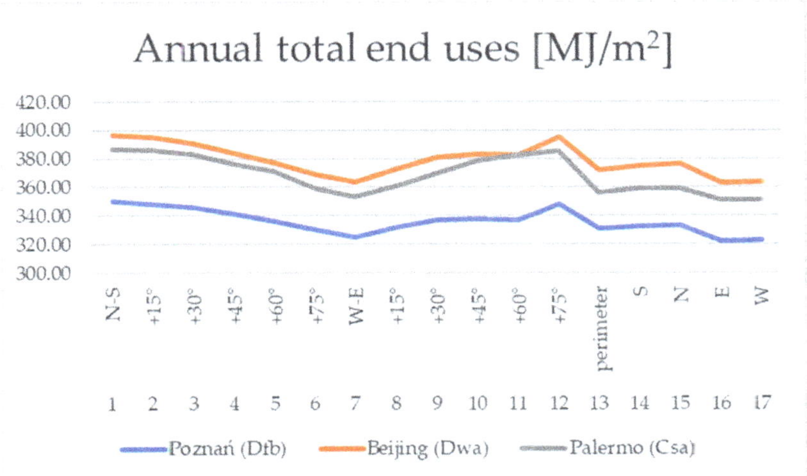

Figure 12. Comparison of the results calculated for the different climate zones.

5. Summary and Conclusions

The research presented here shows that in the initial (early) stages of a project, the arrangement of residential blocks on a plot is important for both urban composition and energy efficiency. At later stages of design work, more factors should be analyzed to seek energy savings.

The results demonstrate that the spatial arrangement can be an important factor influencing the energy performance of a building, especially due to its impact on the potential amount of passive solar heat gains. The proper building layout may result in significant energy savings. The considered variants differed in terms of annual end uses for heating and cooling even by up to 47.85 MJ/m^2, which is approx. 15.2%. The peak values on the hottest days were lower by 4 to even 10 times in comparison with the least advantageous variants, depending on the location.

Thus, urban layout is important for assessing thermal comfort in a residential environment, thus expanding the spectrum of environmental indicators considered in the urban planning process. This leads to the following conclusions:

- At the early stages of design, urban planning should analyze the arrangement of blocks in terms of energy efficiency on par with the aspect of composition, landscape, functionality, transportation, etc.
- Further stages of urban design should incorporate additional energy analyses of urban ventilation (wind speed), the shape and color of building facades, insulation materials, the layout of greenery, etc.
- Final approval of the urban project should come as a result of a multi-variant analysis of the space design concept, which consists of examining various layouts of the plot in terms of energy savings, in order to select the best variant.

The original contribution of this study to the state of the art is to demonstrate how local conditions can modify certain regularities shown by theoretical models. These specific variables are, in particular: the climatic conditions, the local planning constraints, the shape of the plot, and the existing built environment. The rules for ideal models of spatial arrangements must be adapted to the local conditions in each case, and the method shown may be helpful in selecting the most energy-efficient spatial arrangement.

Author Contributions: W.B.: validation, resources, writing—original draft, writing—review and editing, and project administration; W.S.: conceptualization, methodology, software, investigation, and writing—original draft; L.Q.: project administration and funding acquisition; Y.H.: project administration and funding acquisition; W.C.: resources and writing—review and editing; M.Z.: project administration and supervision. All authors have read and agreed to the published version of the manuscript.

Funding: Beijing University of Technology: International Research Cooperation Talent Introduction and Cultivation Project No. 2021C10 and Project No.3C035001202301 Fund Title: The Grand Canal National Cultural Park Interactive Landscape Living Cultural resource management Mining and Design Strategy Research. Poznan University of Technology: research subvention no. 0111/SBAD/2203.

Institutional Review Board Statement: Not applicable.

Informed Consent Statement: Not applicable.

Data Availability Statement: Documentation referring to the OpenStudio software are availabable at: https://www.openstudio.net (accessed on 2 February 2023), and the weather files used in the research are available at: https://energyplus.net/weather (accessed on 2 February 2023).

Conflicts of Interest: The authors declare no conflict of interest. The funders had no role in the design of the study; in the collection, analysis, or interpretation of the data; in the writing of the manuscript; or in the decision to publish the results.

References

1. Sismanidis, P.; Bechtel, B.; Perry, M.; Ghent, D. The Seasonality of Surface Urban Heat Islands across Climates. *Remote Sens.* **2022**, *14*, 2318. [CrossRef]
2. Abdollahzadeh, N.; Biloria, N. Urban microclimate and energy consumption: A multi-objective parametric urban design approach for dense subtropical cities. *Front. Archit. Res.* **2022**, *3*, 11. [CrossRef]
3. Zou, Y.; Deng, Y.; Xia, D.; Guo, J.; Zhong, Z. Comprehensive analysis on the energy resilience performance of urban residential sector in hot-humid area of China under climate change. *Sustain. Cities Soc.* **2023**, *88*, 104233. [CrossRef]
4. Abdollahzadeh, N.; Biloria, N. Outdoor thermal comfort: Analyzing the impact of urban configurations on the thermal performance of street canyons in the humid subtropical climate of Sydney. *Front. Archit. Res.* **2021**, *2*, 10. [CrossRef]
5. Abd Elsalam, M.F. Breaking through the classical determinants in the field of hyper urban planning. *Constr. Innov.* **2021**, *21*, 818–835. [CrossRef]
6. Ronchi, S.; Salata, S.; Arcidiacono, A. Which urban design parameters provide climate-proof cities? An application of the Urban Cooling InVEST Model in the city of Milan comparing historical planning morphologies. *Sustain. Cities Soc.* **2020**, *63*, 102459. [CrossRef]
7. Klemm, C.; Wiese, F. Indicators for the optimization of sustainable urban energy systems based on energy system modeling. *Energy Sustain. Soc.* **2022**, *12*, 3. [CrossRef]
8. Hillman, A.; Fisher, U.; Shapiro, M.A. Systematic Methodology for Design of Sustainable Urban Neighborhood Energy Infrastructure. *Sustainability* **2022**, *14*, 259. [CrossRef]
9. Hachem-Vermette, C.; Singh, K. Energy Systems and Energy Sharing in Traditional and Sustainable Archetypes of Urban Developments. *Sustainability* **2022**, *14*, 1356. [CrossRef]
10. Khan, K.; Szopik Depczyńska, K.; Dembińska, I.; Ioppolo, G. Most Relevant Sustainability Criteria for Urban Infrastructure Projects-AHP Analysis for the Gulf States. *Sustainability* **2022**, *14*, 14717. [CrossRef]
11. Skórzewski, W. Potential of using greenery to reduce overheating of buildings in Polish climate conditions. *Sci. Rev. Eng. Environ. Sci.* **2019**, *28*, 619–631. [CrossRef]
12. Danis, J.; Mishra, S.; Rempel, A.R. Direct heat flux sensing for window shading control in passive cooling systems. *Energy Build.* **2022**, *261*, 111950. [CrossRef]
13. Feist, W.; Munzenberg, U.; Thumulla, J. *Podstawy Budownictwa Pasywnego*; Polski Instytut Budownictwa Pasywnego: Gdańsk, Poland, 2009.
14. Dequaire, X. Passivhaus as a low-energy building standard: Contribution to a typology. *Energy Effic.* **2012**, *5*, 377–391. [CrossRef]
15. Schnieders, J.; Feist, W.; Rongen, L. Passive Houses for different climate zones. *Energy Build.* **2015**, *105*, 71–87. [CrossRef]
16. Huang, H.; Binti Wan Mohd Nazi, W.I.; Yu, Y.; Wang, Y. Energy performance of a high-rise residential building retrofitted to passive building standard—A case study. *Appl. Therm. Eng.* **2020**, *181*, 115902. [CrossRef]
17. Fedorczak-Cisak, M.; Furtak, M.; Surówka, M. Possibilities of achieving the nZEB building standard (nearly zero energy building) and the passive building standard for newly designed buildings in Poland. IOP conference series. *Mater. Sci. Eng.* **2020**, *960*, 32095. [CrossRef]
18. Schnieders, J.; Eian, T.D.; Filippi, M.; Florez, J.; Kaufmann, B.; Pallantzas, S.; Paulsen, M.; Reyes, E.; Wassouf, M.; Yeh, S. Design and realisation of the Passive House concept in different climate zones. *Energy Effic.* **2020**, *13*, 561–1604. [CrossRef]
19. Zou, Y.; Xiang, K.; Zhan, Q.; Li, Z. A simulation-based method to predict the life cycle energy performance of residential buildings in different climate zones of China. *Build. Environ.* **2021**, *193*, 07663. [CrossRef]
20. Ismail, K.A.R.; Lago, T.G.S.; Lino, F.A.M.; Mondlane, M.V.; Teles, M.P.R. Experimental investigation on ventilated window with reflective film and development of correlations. *Sol. Energy* **2021**, *230*, 421–434. [CrossRef]
21. Amado, M.; Poggi, F. Solar Urban Planning: A Parametric Approach. *Energy Procedia* **2014**, *48*, 1539–1548. [CrossRef]
22. Amado, M.; Poggi, F. Towards solar urban planning: A new step for better Energy performance. *Energy Procedia* **2012**, *30*, 1261–1273. [CrossRef]
23. Strømann-Andersen, J.; Sattrup, P.A. The urban canyon and building energy use: Urban density versus daylight and passive solar gains. *Energy Build.* **2011**, *43*, 2011–2020. [CrossRef]
24. Deng, J.Y.; Wong, N.H.; Zheng, X. The Study of the Effects of Building Arrangement on Microclimate and Energy Demand of CBD in Nanjing, China. *Procedia Eng.* **2016**, *169*, 44–54. [CrossRef]
25. Wang, F.; Sun, B.; Zheng, X.; Ji, X. Impact of Block Spatial Optimization and Vegetation Configuration on the Reduction of $PM_{2.5}$ Concentrations: A Roadmap towards Green Transformation and Sustainable Development. *Sustainability* **2022**, *14*, 11622. [CrossRef]
26. Tong, S.; Wong, N.H.; Jusuf, S.K.; Tan, C.L.; Wong, H.F.; Ignatius, M.; Tan, E. Study on correlation between air temperature and urban morphology parameters in built environment in northern China. *Build. Environ.* **2018**, *127*, 239–249. [CrossRef]
27. Zhao, M.; Künzel, H.M.; Antretter, F. Parameters influencing the energy performance of residential buildings in different Chinese climate zones. *Energy Build.* **2015**, *96*, 64–75. [CrossRef]
28. Płoszaj-Mazurek, M. Machine Learning-Aided Architectural Design for Carbon Footprint Reduction. *Builder* **2020**, *276*, 35–39. [CrossRef]
29. Skórzewski, W. Urban layout and energy savings. *Acta Sci. Pol. Archit.* **2020**, *19*, 3–10. [CrossRef]

30. Czachura, A.; Gentile, N.; Kanters, J.; Wall, M. Identifying Potential Indicators of Neighbourhood Solar Access in Urban Planning. *Buildings* **2022**, *12*, 1575. [CrossRef]
31. Guo, F.; Wang, Z.; Dong, J.; Zhang, H.; Lu, X.; Lau, S.S.Y.; Miao, Y. Spatial Differences in Outdoor Thermal Comfort during the Transition Season in Cold Regions of China. *Buildings* **2022**, *12*, 720. [CrossRef]
32. Speroni, A.; Mainini, A.G.; Zani, A.; Paolini, R.; Pagnacco, T.; Poli, T. Experimental Assessment of the Reflection of Solar Radiation from Façades of Tall Buildings to the Pedestrian Level. *Sustainability* **2022**, *14*, 5781. [CrossRef]
33. Wang, S.; Yi, Y.K.; Liu, N.X. Multi-objective optimization (MOO) for high-rise residential buildings' layout centered on daylight, visual, and outdoor thermal metrics in China. *Build. Environ.* **2021**, *205*, 108263. [CrossRef]
34. Kanters, J.; Gentile, N.; Bernardo, R. Planning for solar access in Sweden: Routines, metrics, and tools. *Urban Plan. Transp. Res.* **2021**, *9*, 347–367. [CrossRef]
35. Leyla, K.; Farkhondeh, J.; Mousa, M.; Behnam, M. Design, evaluation, and optimization of an efficient solar-based multi-generation system with an energy storage option for Iran's summer peak demand. *Energy Convers. Manag.* **2021**, *242*, 114324.
36. Dwijendra, N.K.A.; Rahardja, U.; Kumar, N.B.; Patra, I.; Zahra, M.M.A.; Finogenova, Y.; Guerrero, J.W.G.; Izzat, S.E.; Alawsi, T. An Analysis of Urban Block Initiatives Influencing Energy Consumption and Solar Energy Absorption. *Sustainability* **2022**, *14*, 14273. [CrossRef]
37. Veisi, O.; Shakibamanesh, A. Analysis of Solar Radiation towards Optimization and Location of the UrbanBlocks in the Neighborhood Units. *Environ. Sci. Sustain. Dev.* **2022**, *7*, 66–90. [CrossRef]
38. Zeng, R.; Chini, A.; Ries, R. Innovative design for sustainability: Integrating embodied impacts and costs during the early design phase. *Eng. Constr. Archit. Manag.* **2021**, *28*, 747–764. [CrossRef]
39. AboWardah, E.S. Bridging the gap between research and schematic design phases in teaching architectural graduation projects. *Front. Archit. Res.* **2020**, *9*, 82–105. [CrossRef]
40. Bahgat, R.; Reffat, R.M.; Elkady, S.L. Energy Efficiency Design Guide for Optimal Urban Features of Open Spaces in Residential Complexes. *ARCHive-SR* **2019**, *3*, 136–152. [CrossRef]
41. GUT LightLab; Faculty of Architecture; Gdansk University of Technology, Healthier and Environmentally Responsible Sustainable Cities and Communities. A New Design Framework and Planning Approach for Urban Illumination. *Sustainability* **2022**, *14*, 14525. [CrossRef]
42. Shan, X.; Deng, Q.; Tang, Z.; Wu, Z.; Wang, W. An integrated data mining-based approach to identify key building and urban features of different energy usage levels. *Sustain. Cities Soc.* **2022**, *77*, 103576. [CrossRef]
43. Natanian, J.; Aleksandrowicz, O.; Auer, T. A parametric approach to optimizing urban form, energy balance and environmental quality: The case of Mediterranean districts. *Appl. Energy* **2019**, *254*, 113637. [CrossRef]
44. Giostra, S.; Masera, G.; Monteiro, R. Solar Typologies: A Comparative Analysis of Urban Form and Solar Potential. *Sustainability* **2022**, *14*, 9023. [CrossRef]
45. Sattrup, P.A.; Strømann-Andersen, J. Building typologies in Northern European cities: Daylight, solar access, and building energy use. *J. Archit. Plan. Res.* **2013**, *30*, 56–76.
46. Loeffler, R.; Geier, S.; Oesterreicher, D. Identification and quantification of urban planning related framework conditions on energy and resource-efficiency of urban building developments. *Earth Environ. Sci.* **2020**, *588*, 052038. [CrossRef]
47. Amado, M.; Poggi, F. Solar energy integration in urban planning: GUUD model. *Energy Procedia* **2014**, *50*, 277–284. [CrossRef]
48. Lobaccaro, G.; Frontini, F. Solar energy in urban environment: How urban densification affects existing buildings. *Energy Procedia* **2014**, *48*, 1559–1569. [CrossRef]
49. Akrofi, M.M.; Okitasari, M. Integrating solar energy considerations into urban planning for low carbon cities: A systematic review of the state-of-the-art. *Urban Gov.* **2022**, *2*, 157–172. [CrossRef]
50. Grid Report for Architecture–Fall 2021–Architecture Software. Available online: https://read.uberflip.com/i/1417145-g2-grid-report-architecture-software/0? (accessed on 2 July 2023).
51. Regulation of the Minister of Infrastructure of 12, 04, 02 on technical specifications for buildings and their location. *J. Laws Repub. Pol.* **2002**, 690, Erratum in *J. Laws Repub. Pol.* **2022**, 1225.
52. Nazarian, N.; Kleissl, J. CFD simulation of an idealized urban environment: Thermal effects of geometrical characteristics and surface materials. *Urban Clim.* **2015**, *12*, 141–159. [CrossRef]
53. Naserikia, M.; Hart, M.A.; Nazarian, N.; Bechtel, B. Background climate modulates the impact of land cover on urban surface temperature. *Sci. Rep.* **2022**, *12*, 15433. [CrossRef]
54. Nice, K.A.; Nazarian, N.; Lipson, M.J.; Hart, M.A.; Seneviratne, S.; Thompson, J.; Naserikia, M.; Godic, B.; Stevenson, M. Isolating the impacts of urban form and fabric from geography on urban heat and human thermal comfort. *Build. Environ.* **2022**, *224*, 109502. [CrossRef]
55. Llaguno-Munitxa, M.; Shu, X.; Mistry, B. Multivariate analysis of the influence between building design and energy performance, sociodemographic metrics, and the intra-urban environment. *J. Phys.* **2021**, *2069*, 012056.
56. Yeo, I.A.; Lee, E. Quantitative study on environment and energy information for land use planning scenarios in eco-city planning stage. *Appl. Energy* **2018**, *230*, 889–911. [CrossRef]
57. Medeiros, E. Urban SUNstainability: A multi-dimensional policy evaluation framework proposal. *Cidades* **2020**, *40*, 117–133.
58. Kanters, J.; Wall, M. Experiences from the urban planning process of a solar neighbourhood in Malmö, Sweden. *Urban Plan. Transp. Res.* **2018**, *6*, 54–80. [CrossRef]

59. Formolli, M.; Croce, S.; Vettorato, D.; Paparella, R.; Scognamiglio, A.; Mainini, A.G.; Lobaccaro, G. Solar Energy in Urban Planning: Lesson Learned and Recommendations from Six Italian Case Studies. *Appl. Sci.* **2022**, *12*, 2950. [CrossRef]
60. Negin, N.; Martilli, A.; Norford, L.; Kleissl, J. Impacts of Realistic Urban Heating. Part II: Air Quality and City Breathability. *Bound.-Layer Meteorol.* **2018**, *168*, 321–341.
61. Yannas, S.; Rodríguez-Álvarez, J. Document details-Domestic overheating in a temperate climate: Feedback from London Residential Schemes. *Sustain. Cities Soc.* **2020**, *59*, 102189. [CrossRef]
62. Rodriguez-Alvarez, J. Urban Energy Index for Buildings (UEIB): A new method to evaluate the effect of urban form on buildings' energy demand. *Landsc. Urban Plan.* **2016**, *148*, 170–187. [CrossRef]
63. Lopes, H.S.; Remoaldo, P.C.; Ribeiro, V.; Martín-Vide, J. A comprehensive methodology for assessing outdoor thermal comfort in touristic city of Porto (Portugal). *Urban Clim.* **2022**, *45*, 101264. [CrossRef]

MDPI
St. Alban-Anlage 66
4052 Basel
Switzerland
www.mdpi.com

Sustainability Editorial Office
E-mail: sustainability@mdpi.com
www.mdpi.com/journal/sustainability

www.ingramcontent.com/pod-product-compliance
Lightning Source LLC
Chambersburg PA
CBHW061621220926
51850CB00005B/569
* 9 7 8 3 0 3 6 5 8 5 5 3 6 *